Nanoprobes for Tumor Theranostics

Nanoprobes for Tumor Theranostics

Editor

Zheyu Shen

MDPI • Basel • Beijing • Wuhan • Barcelona • Belgrade • Manchester • Tokyo • Cluj • Tianjin

Editor
Zheyu Shen
Southern Medical University
China

Editorial Office
MDPI
St. Alban-Anlage 66
4052 Basel, Switzerland

This is a reprint of articles from the Special Issue published online in the open access journal *Biosensors* (ISSN 2079-6374) (available at: https://www.mdpi.com/journal/biosensors/special_issues/nanoprob_tumor).

For citation purposes, cite each article independently as indicated on the article page online and as indicated below:

LastName, A.A.; LastName, B.B.; LastName, C.C. Article Title. *Journal Name* **Year**, *Volume Number*, Page Range.

ISBN 978-3-0365-6159-2 (Hbk)
ISBN 978-3-0365-6160-8 (PDF)

© 2022 by the authors. Articles in this book are Open Access and distributed under the Creative Commons Attribution (CC BY) license, which allows users to download, copy and build upon published articles, as long as the author and publisher are properly credited, which ensures maximum dissemination and a wider impact of our publications.

The book as a whole is distributed by MDPI under the terms and conditions of the Creative Commons license CC BY-NC-ND.

Contents

About the Editor vii

Jiaoyang Zhu and Zheyu Shen
Nanoprobes for Tumor Theranostics
Reprinted from: *Biosensors* 2022, 12, 1022, doi:10.3390/bios12111022 1

Yue Pan, Zhili Wang, Jialing Ma, Tongping Zhou, Zeen Wu, Pi Ding, Na Sun, Lifen Liu, Renjun Pei and Weipei Zhu
Folic Acid-Modified Fluorescent-Magnetic Nanoparticles for Efficient Isolation and Identification of Circulating Tumor Cells in Ovarian Cancer
Reprinted from: *Biosensors* 2022, 12, 184, doi:10.3390/bios12030184 5

Umar Azhar, Qazi Ahmed, Saira Ishaq, Zeyad T. Alwahabi and Sheng Dai
Exploring Sensitive Label-Free Multiplex Analysis with Raman-Coded Microbeads and SERS-Coded Reporters
Reprinted from: *Biosensors* 2022, 12, 121, doi:10.3390/bios12020121 19

Huan Chen, Yiming Zhang, Lulu Li, Rui Guo, Xiangyang Shi and Xueyan Cao
Effective CpG Delivery Using Zwitterion-Functionalized Dendrimer-Entrapped Gold Nanoparticles to Promote T Cell-Mediated Immunotherapy of Cancer Cells
Reprinted from: *Biosensors* 2022, 12, 71, doi:10.3390/bios12020071 33

Shuang Yin, Yongdong Liu, Sheng Dai, Bingyang Zhang, Yiran Qu, Yao Zhang, Woo-Seok Choe and Jingxiu Bi
Mechanism Study of Thermally Induced Anti-Tumor Drug Loading to Engineered Human Heavy-Chain Ferritin Nanocages Aided by Computational Analysis
Reprinted from: *Biosensors* 2021, 11, 444, doi:10.3390/bios11110444 43

Hamzah Al-madani, Hui Du, Junlie Yao, Hao Peng, Chenyang Yao, Bo Jiang, Aiguo Wu and Fang Yang
Living Sample Viability Measurement Methods from Traditional Assays to Nanomotion
Reprinted from: *Biosensors* 2022, 12, 453, doi:10.3390/bios12070453 61

Yongjie Chi, Peng Sun, Yuan Gao, Jing Zhang and Lianyan Wang
Ion Interference Therapy of Tumors Based on Inorganic Nanoparticles
Reprinted from: *Biosensors* 2022, 12, 100, doi:10.3390/bios12020100 97

Yonghong Tan, Peiying Liu, Danxia Li, Dong Wang and Ben Zhong Tang
NIR-II Aggregation-Induced Emission Luminogens for Tumor Phototheranostics
Reprinted from: *Biosensors* 2022, 12, 46, doi:10.3390/bios12010046 111

Wangbo Jiao, Tingbin Zhang, Mingli Peng, Jiabao Yi, Yuan He and Haiming Fan
Design of Magnetic Nanoplatforms for Cancer Theranostics
Reprinted from: *Biosensors* 2022, 12, 38, doi:10.3390/bios12010038 127

Dong Li, Jie Pan, Shuyu Xu, Shiying Fu, Chengchao Chu and Gang Liu
Activatable Second Near-Infrared Fluorescent Probes: A New Accurate Diagnosis Strategy for Diseases
Reprinted from: *Biosensors* 2021, 11, 436, doi:10.3390/bios11110436 149

Liangxi Zhu, Jingzhou Zhao, Zhukang Guo, Yuan Liu, Hui Chen, Zhu Chen and Nongyue He
Applications of Aptamer-Bound Nanomaterials in Cancer Therapy
Reprinted from: *Biosensors* **2021**, *11*, 344, doi:10.3390/bios11090344 . . . . **175**

About the Editor

Zheyu Shen

Zheyu Shen received his Ph.D. from the Institute of Process Engineering, Chinese Academy of Sciences, under the supervision of Prof. Guanghui Ma and Prof. Toshiaki Dobashi (Gunma University, Japan). After his postdoctoral studies with Prof. Kazuhiro Kohama (Gunma University, Japan) and Prof. Sheng Dai (The University of Adelaide, Australia), he was appointed an Associate Professor (2012) and promoted to a Full Professor (2015) at the Ningbo Institute of Materials Technology and Engineering, Chinese Academy of Sciences. After three years of visiting as a scholar at the National Institutes of Health (USA) under the supervision of Prof. Xiaoyuan Chen, he joined the Southern Medical University as a Full Professor in 2019. His lab focuses on MRI contrast agents, tumor ferroptosis therapy, and immunotherapy.

Editorial

Nanoprobes for Tumor Theranostics

Jiaoyang Zhu and Zheyu Shen *

School of Biomedical Engineering, Southern Medical University, 1023 Shatai South Road, Guangzhou 510515, China
* Correspondence: sz@smu.edu.cn

Citation: Zhu, J.; Shen, Z. Nanoprobes for Tumor Theranostics. *Biosensors* 2022, 12, 1022. https://doi.org/10.3390/bios12111022

Received: 9 November 2022
Accepted: 14 November 2022
Published: 16 November 2022

Publisher's Note: MDPI stays neutral with regard to jurisdictional claims in published maps and institutional affiliations.

Copyright: © 2022 by the authors. Licensee MDPI, Basel, Switzerland. This article is an open access article distributed under the terms and conditions of the Creative Commons Attribution (CC BY) license (https://creativecommons.org/licenses/by/4.0/).

This Special Issue of *Biosensors*, entitled "Nanoprobes for Tumor Theranostics", aims to report the research progress of using nanoprobes for the diagnosis and therapy of tumors, and promote their applications. We hope to attract the attention of researchers from various disciplines of chemistry, materials science, oncology, biology, and medicine. Currently, numerous scientific challenges in tumor theranostics remain, which must be surmounted to improve human health and quality of life. This Special Issue aims to contribute to furthering creative progress in the field via publishing ten papers reporting the synthesis, applications, and advances of several nanoprobes for tumor theranostics.

Circulating tumor cells (CTCs), released from tumor sites into the peripheral blood, have been recognized as promising biomarkers for cancer prognosis, treatment monitoring, and metastasis diagnosis. However, the number of CTCs in the peripheral blood is low, making it a technical challenge to isolate, enrich, and identify CTCs from patient blood samples. Yue Pan et al. [1] developed a simple, effective, and inexpensive strategy to capture and identify CTCs from the blood samples of patients with ovarian cancer (OC) using the folic acid (FA) and antifouling-hydrogel-modified fluorescent magnetic nanoparticles. The hydrogel showed a good antifouling property against peripheral blood mononuclear cells (PBMCs). The FA was coupled to the hydrogel surface as the targeting molecule for CTC isolation, maintained a good capture efficiency for SK-OV-3 cells (95.58%), and successfully isolated 2–12 CTCs from 10 OC patients' blood samples. The FA-modified fluorescent magnetic nanoparticles were thus successfully used for the capture and direct identification of CTCs from the blood samples of OC patients.

Accurate detection of multiple analytes from a single measurement, requiring complex encoding systems, is critical in modern bioanalysis. Umar Azhar et al. developed a novel bioassay with Raman-coded antibody supports (polymer microbeads with different Raman signatures) and surface-enhanced Raman-scattering (SERS)-coded nanotags (organic thiols on a gold nanoparticle surface with different SERS signatures) as a model fluorescent, label-free, bead-based multiplex immunoassay system [2]. The developed homogeneous immunoassays included two surface-functionalized monodisperse Raman-coded microbeads of polystyrene and poly(4-tert-butylstyrene) as the immune solid supports, and two epitope-modified nanotags (self-assembled 4-mercaptobenzoic acid or 3-mercaptopropionic acid on gold nanoparticles) as the SERS-coded reporters. Such multiplex Raman/SERS-based microsphere immunoassays could selectively identify specific paratope–epitope interactions from one sample solution under the illumination of a single laser, and thus hold great promise for future suspension multiplex analysis in a diverse range of biomedical applications.

Cell-based immunotherapy has become one of the most promising ways to completely eliminate cancer. The major challenge is to effectively promote a proper immune response, inducing T cells to kill the cancer cells. Huan Chen et al. [3] investigated the effect of T-cell-mediated immunotherapy trigged by Au DENPs-MPC (zwitterion 2-methacryloyloxyethyl phosphorylcholine (MPC)-functionalized dendrimer-entrapped gold nanoparticles) loading oli-godeoxynucleotides (ODNs) of an unmethylated cytosine guanine dinucleotide (CPG). They first synthesized Au DENPs MPC, evaluated its capability to compress and transfect

CpG-ODN to bone marrow dendritic cells (BMDCs), and investigated the potential of using T cells stimulated by matured BMDCs to inhibit tumor cell growth. The developed Au DENPs-MPC could apparently reduce the toxicity of Au DENPs, and enhanced transfer of CpG-ODN to the BMDCs to stimulate maturation, as demonstrated by the 44.41–48.53% increase in different surface maturation markers. The transwell experiments certificated that ex vivo-activated T cells display excellent antitumor properties, effectively inhibiting the growth of tumor cells. These results suggest that Au DENPs-MPC can deliver CpG-ODN efficiently to enhance the antigen presentation ability of BMDCs to activate T cells, indicating that T-cell-based immunotherapy mediated by Au DENPs-MPC loaded with CpG-ODN is an exceedingly promising treatment for cancer.

Diverse drug loading approaches for human heavy-chain ferritin (HFn), a promising drug nanocarrier, have been established. However, the antitumor drug loading ratio and protein carrier recovery yield are bottlenecks for future clinical application. The mechanisms behind drug loading have not been elaborated. Shuang Yin et al. [4] introduced a thermally induced approach to loading the antitumor drug doxorubicin hydrochloride (DOX) into HFn and two functionalized HFns, HFn-PAS-RGDK, and HFn-PAS. Optimal conditions were obtained through orthogonal tests. All three HFn-based proteins achieved a high protein recovery yield and drug loading ratio. Size exclusion chromatography (SEC) and transmission electron microscopy (TEM) results show the majority of DOX-loaded protein (protein/DOX) retained its nanocage conformation. Computational analysis, molecular docking followed by molecular dynamic (MD) simulation, revealed the mechanisms of DOX loading and the formation of by-products by investigating noncovalent interactions between DOX and the HFn subunit and possible binding modes of DOX and HFn after drug loading. In in vitro tests, DOX in protein/DOX entered the tumor cell nucleus and inhibited tumor cell growth.

Living-sample viability measurement is an extremely common process in medical, pharmaceutical, and biological fields, especially drug pharmacology and toxicology detection. Nowadays, there are a number of chemical, optical, and mechanical methods that have been developed in response to the growing demand for simple, rapid, accurate, and reliable real-time living-sample viability assessment. In parallel, the development trend of viability measurement methods (VMMs) has increasingly shifted from traditional assays towards the innovative atomic force microscope (AFM) oscillating sensor method (referred to as nanomotion), which takes advantage of the adhesion of living samples to an oscillating surface. A comprehensive review of the most common VMMs, laying emphasis on their benefits and drawbacks, as well as evaluating their potential utility, was provided by Hamzah Al-madani et al. [5] In addition, they discussed the nanomotion technique, focusing on its applications, sample attachment protocols, and result display methods. Furthermore, challenges and future perspectives with regard to nanomotion were commented on, mainly emphasizing scientific restrictions and development orientations.

The recent development of ion interference therapy (IIT) based on inorganic nanoparticles was introduced by Yongjie Chi et al. [6] They summarized the advantages and disadvantages of this treatment and the challenges of future development, hoping to provide a reference for future research. As an essential substance for cell life activities, ions play an important role in controlling the cell osmotic pressure balance, intracellular acid–base balance, signal transmission, biocatalysis, and so on. The imbalance of ion homeostasis in cells can seriously affect cells activities, cause irreversible damage to cells, and induce cell death. Therefore, artificially interfering with the ion homeostasis in tumor cells has become a new means by which to inhibit the proliferation of tumor cells. This treatment is called IIT. Although some molecular carriers of ions have been developed for intracellular ion delivery, inorganic nanoparticles are widely used in ion interference therapy because of their greater capacity for ion delivery and superior biocompatibility compared with molecular carriers.

As an emerging and powerful material, aggregation-induced emission luminogens (AIEgens), which can simultaneously provide precise diagnosis and efficient therapeutics, have exhibited significant superiorities in the field of phototheranostics. Of particular interest is phototheranostics based on AIEgens with emissions in the second near-infrared (NIR-II) range (1000–1700 nm), which has promoted the feasibility of their clinical applications by virtue of numerous preponderances benefiting from the extremely long wavelength. Yonghong Tan et al. [7] summarized the past 3 years of advances in the field of phototheranostics based on NIR-II AIEgens, including the strategies of constructing NIR-II AIEgens and their applications in different theranostic modalities (FLI-guided PTT, PAI-guided PTT, and multimodal-imaging-guided PDT–PTT synergistic therapy); in addition, a brief conclusion including perspectives and challenges in the field of phototheranostics is provided at the end.

Smart nanomedicines that are capable of diagnosis and therapy (theranostics) in one-nanoparticle systems are highly desirable for improving cancer treatment outcomes. Magnetic nanoplatforms are ideal for cancer theranostics, because of their diverse physiochemical properties and biological effects. In particular, a biocompatible iron oxide nanoparticle-based magnetic nanoplatform exhibits multiple magnetic-responsive behaviors under an external magnetic field and can realize the integration of diagnosis (magnetic resonance imaging, ultrasonic imaging, photoacoustic imaging, etc.) and therapy (magnetic hyperthermia, photothermal therapy, controlled drug delivery and release, etc.) in vivo. Furthermore, due to considerable variation among tumors and individual patients, iron oxide nanoplatforms, designed via the coordination of diverse functionalities for efficient and individualized theranostics, are urgently needed. Wangbo Jiao et al. [8] presented an up-to-date overview on iron oxide nanoplatforms, including both iron oxide nanomaterials and those that respond to an externally applied magnetic field, with an emphasis on their applications in cancer theranostics.

Second near-infrared (NIR-II) fluorescent imaging has been widely applied in biomedical diagnosis due to its high spatiotemporal resolution and deep tissue penetration. In contrast to the "always on" NIR-II fluorescent probes, activatable NIR-II fluorescent probes specifically target biological tissues, demonstrating a higher imaging signal-to-background ratio and a lower detection limit. Therefore, it is of great significance to utilize disease-associated endogenous stimuli (such as pH values, enzyme existence, hypoxia condition, and so on) to activate NIR-II probes and achieve switchable fluorescent signals for specific deep bioimaging. Dong Li et al. [9] introduced recent strategies and mechanisms for activatable NIR-II fluorescent probes and their applications in biosensing and bioimaging. Moreover, potential challenges and perspectives with regard to activatable NIR-II fluorescent probes were also discussed.

Aptamers, owing to their small size, low toxicity, good specificity, and excellent biocompatibility, have been widely applied in biomedical areas. Therefore, the combination of nanomaterials with aptamers offers a new method for cancer treatment. Liangxi Zhu et al. [10] briefly introduced this topic. They discussed the application of aptamers for the treatment of breast, lung, and other cancers. Finally, perspectives on challenges and future applications of aptamers in cancer therapy were discussed.

In summary, this Special Issue includes four research papers reporting the most recent research progress in CTC detection using fluorescent magnetic nanoparticles, the SERS-coded multiplex immunoassay system, T-cell-mediated immunotherapy, and human heavy-chain ferritin (HFn) drug nanocarriers, and six review papers reporting on living-sample viability measurement methods, IIT based on inorganic nanoparticles, NIR-II AIEgens, iron oxide nanoplatforms, activatable NIR-II fluorescent probes, and aptamers. Altogether, these papers present the most promising emerging nanoprobes in medicine and clinical research for tumor theranostics.

Author Contributions: Conceptualization, Z.S. and J.Z.; Validation, Z.S. and J.Z.; Writing—original draft preparation, J.Z.; Writing—review and editing, Z.S.; Supervision, Z.S.; Project administration, Z.S. All authors have read and agreed to the published version of the manuscript.

Funding: This work was financially supported by National Natural Science Foundation of China (32271374).

Acknowledgments: Z.S. is very pleased to have acted as a guest editor for this Special Issue. He would like to thank all the authors who submitted papers and the reviewers for their hard work. J.Z. and Z.S. are grateful to the *Biosensors* Editorial Office for the opportunity to publish this Special Issue and to all the staff of MDPI for their friendly and valuable support in the reviewing and organization of this Special Issue.

Conflicts of Interest: The authors declare no conflict of interest.

References

1. Pan, Y.; Wang, Z.; Ma, J.; Zhou, T.; Wu, Z.; Ding, P.; Sun, N.; Liu, L.; Pei, R.; Zhu, W. Folic Acid-Modified fluorescent-magnetic nanoparticles for efficient isolation and identification of circulating tumor cells in ovarian cancer. *Biosensors* **2022**, *12*, 184. [CrossRef] [PubMed]
2. Azhar, U.; Ahmed, Q.; Ishaq, S.; Alwahabi, Z.; Dai, S. Exploring sensitive label-free multiplex analysis with raman-coded microbeads and SERS-coded reporters. *Biosensors* **2022**, *12*, 121. [CrossRef] [PubMed]
3. Chen, H.; Zhang, Y.; Li, L.; Guo, R.; Shi, X.; Cao, X. Effective CpG delivery using zwitterion-functionalized dendrimer-entrapped gold nanoparticles to promote T cell-mediated immunotherapy of cancer cells. *Biosensors* **2022**, *12*, 71. [CrossRef] [PubMed]
4. Yin, S.; Liu, Y.; Dai, S.; Zhang, B.; Qu, Y.; Zhang, Y.; Choe, W.; Bi, J. Mechanism study of thermally induced anti-tumor drug loading to engineered human heavy-chain ferritin nanocages aided by computational analysis. *Biosensors* **2021**, *11*, 444. [CrossRef] [PubMed]
5. Hamzah, A.; Du, H.; Yao, J.; Peng, H.; Yao, C.; Jiang, B.; Wu, A.; Yang, F. Living sample viability measurement methods from traditional assays to nanomotion. *Biosensors* **2022**, *12*, 453. [PubMed]
6. Chi, Y.; Sun, P.; Gao, Y.; Zhang, J.; Wang, L. Ion interference therapy of tumors based on inorganic nanoparticles. *Biosensors* **2022**, *12*, 100. [CrossRef] [PubMed]
7. Tan, Y.; Liu, P.; Li, D.; Wang, D.; Tang, B. NIR-II aggregation-induced emission luminogens for tumor phototheranostics. *Biosensors* **2022**, *12*, 46. [CrossRef] [PubMed]
8. Jiao, W.; Zhang, T.; Peng, M.; Yi, J.; He, Y.; Fan, H. Design of magnetic nanoplatforms for cancer theranostics. *Biosensors* **2022**, *12*, 38. [CrossRef] [PubMed]
9. Li, D.; Pan, J.; Xu, S.; Fu, S.; Chu, C.; Liu, G. Activatable second near-infrared fluorescent probes: A new accurate diagnosis strategy for diseases. *Biosensors* **2021**, *11*, 436. [CrossRef] [PubMed]
10. Zhu, L.; Zhao, Z.; Guo, Z.; Liu, Y.; Chen, H.; Chen, Z.; He, N. Applications of aptamer-bound nanomaterials in cancer therapy. *Biosensors* **2021**, *11*, 344. [CrossRef] [PubMed]

Article

Folic Acid-Modified Fluorescent-Magnetic Nanoparticles for Efficient Isolation and Identification of Circulating Tumor Cells in Ovarian Cancer

Yue Pan [1,2], Zhili Wang [2], Jialing Ma [1,2], Tongping Zhou [1,2], Zeen Wu [1,2], Pi Ding [2], Na Sun [2], Lifen Liu [1], Renjun Pei [2,*] and Weipei Zhu [1,*]

1. Department of Gynecology and Obstetrics, The Second Affiliated Hospital of Soochow University, Suzhou 215004, China; ypan2020@sinano.ac.cn (Y.P.); jlma2021@sinano.ac.cn (J.M.); tpzhou2021@sinano.ac.cn (T.Z.); ezwu2019@sinano.ac.cn (Z.W.); liulifen1981@126.com (L.L.)
2. CAS Key Laboratory for Nano-Bio Interface, Suzhou Institute of Nano-Tech and Nano-Bionics, Chinese Academy of Sciences, Suzhou 215123, China; zlwang2013@sinano.ac.cn (Z.W.); pding2017@sinano.ac.cn (P.D.); nsun2013@sinano.ac.cn (N.S.)
* Correspondence: rjpei2011@sinano.ac.cn (R.P.); zwp333xx@126.com (W.Z.); Tel.: +86-0512-62872776 (R.P.); +86-0512-67784769 (W.Z.)

Abstract: Ovarian cancer (OC) is a lethal disease occurring in women worldwide. Due to the lack of obvious clinical symptoms and sensitivity biomarkers, OC patients are often diagnosed in advanced stages and suffer a poor prognosis. Circulating tumor cells (CTCs), released from tumor sites into the peripheral blood, have been recognized as promising biomarkers in cancer prognosis, treatment monitoring, and metastasis diagnosis. However, the number of CTCs in peripheral blood is low, and it is a technical challenge to isolate, enrich, and identify CTCs from the blood samples of patients. This work develops a simple, effective, and inexpensive strategy to capture and identify CTCs from OC blood samples using the folic acid (FA) and antifouling-hydrogel-modified fluorescent-magnetic nanoparticles. The hydrogel showed a good antifouling property against peripheral blood mononuclear cells (PBMCs). The FA was coupled to the hydrogel surface as the targeting molecule for the CTC isolation, held a good capture efficiency for SK-OV-3 cells (95.58%), and successfully isolated 2–12 CTCs from 10 OC patients' blood samples. The FA-modified fluorescent-magnetic nanoparticles were successfully used for the capture and direct identification of CTCs from the blood samples of OC patients.

Keywords: circulating tumor cells; ovarian cancer; folic acid; fluorescent-magnetic nanoparticles; isolation; identification

1. Introduction

Ovarian cancer (OC) is the major lethal disease occurring in women worldwide [1]. Primary treatment of OC consists of cytoreductive surgery followed by chemotherapy [2]. The surgery cannot completely remove the tumor that has metastasized. Chemotherapy resistance and recurrence are common, resulting in impaired survival [3]. Thus, a high specificity and sensitivity method for cancer monitoring after treatment is desperately needed. Circulating tumor cells (CTCs) are a kind of cancer cell that circulate in the bloodstream after being released from solid tumors. These cells hold high metastatic ability and can cause tumor metastasis in remote locations [4–7]. CTCs have promising potential in monitoring the tumor prognosis, treatment effect, and metastasis condition [8–10]. However, due to the rare existence of CTCs in peripheral blood, the isolation and identification of variable CTCs is still challenging [11–13]. Epithelial to mesenchymal transition (EMT) plays a significant role during the CTCs' metastatic process. It results in a phenotypic change in CTCs, such as transforming of epithelial properties to mesenchymal properties, and poses hurdles in CTC isolation [14,15]. Currently, many approaches have been devoted to isolating or detecting

CTCs from blood samples, such as immunomagnetic separation, microfluidics, label-free CTC capture, electrochemistry, and fluorescence sensor [16–25]. Cancer-specific biomarkers such as Epithelial Cell Adhesion Molecule (EpCAM), N-cadherin, and Human Epidermal Growth Factor Receptor 2 (HER2) are overexpressed on the CTCs' surface. Among them, antibody-dependent CTC enrichment using anti-EpCAM antibody has been mainly employed for CTC separation [16–19]. Our group has fabricated several antibody-modified interfaces and applied them to detect CTCs [26,27]. Still, the high cost of antibodies limits their practical use and underscores low-cost recognition molecules [28–30]. Folic acid (FA), a small molecule affinity agent specifically targeting the folate receptor (FR), attracts our attention due to its low cost, high affinity, and less immunogenicity [29,30]. The FR has been reported to be overexpressed on the cells' surface of 90% of ovarian carcinomas, while few expressed in most normal tissues [31,32]. It was considered a good choice for CTC isolation of OC patients.

The fluorescence probe is a powerful technology for cell biology research because of its simplicity, noninvasiveness, and real time. Herein, CdSe/ZnS quantum dots (QDs) were assembled onto the surface of Fe_3O_4 nanoparticles through electrostatic attraction, zwitterionic poly(sulfobetaine methacrylate) (pSBMA) was applied to overcome the non-specific adhesion of blood cells, and finally, FA was modified on the nanoparticles' surface for the specific capture of CTCs. The clinical blood samples of OC patients were used to evaluate the potential ability to practice the platform for CTC capture and identification. The schematic process of CTC separation and identification using FA-modified fluorescent-magnetic nanoparticles was outlined in Figure 1. Overall, the results demonstrated that FA-modified fluorescent-magnetic nanoparticles offer a cost-effective, reliable, and noninvasive method for the quick detection and identification of CTCs in patients with ovarian cancer.

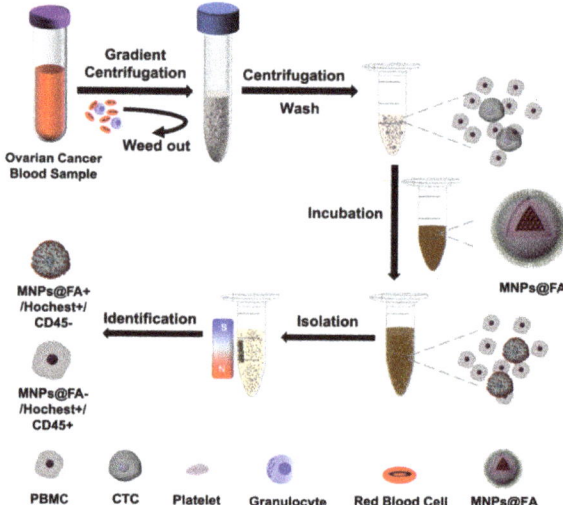

Figure 1. Schematic illustration of the process of CTC separation and identification by folic acid (FA)-modified fluorescent-magnetic nanoparticles (MNPs@FA).

2. Materials and Methods

2.1. Reagents and Cells Culture

Iron chloride hexahydrate ($FeCl_3·6H_2O$), trisodium citrate dihydrate, ammonium acetate, ethylene glycol (EG), ethanol, ammonia aqueous, tetraethyl orthosilicate (TEOS), and acetonitrile were purchased from Sinopharm Chemical Reagent Co., Ltd. (Shanghai, China). The CdSe/ZnS quantum dots (QDs) were obtained from Xingzi New Material Technology Development Co., Ltd. (Shanghai, China). Polyethylenimine (PEI, branched, 25,000 Mw), 3-(trimethoxylsilyl) propyl methacrylate (MPS), N-(3-sulfopropyl)-N-(methacryloxyethyl)-N,N-dimethylammonium betaine (SBMA), methacrylic acid (MAA), N,N′-methylenebis(2-propenamide) (MBA), 2,2-azobisisobutyronitrile (AIBN), 1-ethyl-3-(3′-dimethylaminopropyl) carbodiimide (EDC), N-hydroxysuccinimide (NHS), Histopaque-1077 solution, Hoechst 33,342, 3-3′-dioctadecyloxa-carbocyanine perchlor (DiO), and 1,1′-dioctadecyl-3,3,3′,3′-tetramethylindocarbocyanine perchlo-rate (DiI) were purchased from Sigma-Aldrich (St. Louis, MO, USA). Alexa Fluor 488-modified anti-CD45 (CD45) and Alexa Fluor 555-modified anti-Pan-Keratin (PanCK) antibodies were obtained from Cell Signaling Technology, Inc. (Beverly, MA, USA). Amine-PEG_{5000}-folic acid (NH_2-PEG_{5000}-FA) was prepared by our previous work [30]. RIPA lysis buffer was purchased from Beyotime Biotechnology (Shanghai, China). Protease inhibitor cocktail was obtained from Roche (Mannheim, Germany). BCA protein assay kit was purchased from UU Biotechnology Co., Ltd. (Suzhou, China). FORL1 mouse monoclonal antibody (anti-α-FR), anti-β-actin antibody, and HRP-conjugated secondary antibody (anti-mouse) were purchased from Proteintech (Chicago, IL, USA).

The ovarian adenocarcinoma cell lines (SK-OV-3 and OVCAR-3) and human embryonic kidney cell lines (293T) were provided by the Cell Resource Centre of Life Sciences (Shanghai, China), and the ovarian cancer A2780 cell line was obtained from the Soochow University. The cells culture complied with the instructions of the Cell Resource Centre and the friendship provider.

2.2. Preparation of FA-Modified Fluorescent-Magnetic Nanoparticles

2.2.1. Synthesis of Fe_3O_4 Nanoparticles (MNPs)

Fe_3O_4 magnetic nanoparticles (MNPs) were synthesized according to the reported methods [33]. Briefly, $FeCl_3·6H_2O$ (1.350 g), ammonium acetate (3.854 g), and trisodium citrate dihydrate (0.4 g) were added into EG (70 mL) and stirred until they turned to a brown homogeneous mixture. The solution was heated to 170 °C for 1 h before being transferred to a 100 mL Teflon-lined stainless-steel autoclave. After heating at 200 °C for 16 h, the autoclave was cooled to room temperature and the black Fe_3O_4 magnetic nanoparticles were separated from the solution by using the magnet. After washing with ethanol and deionized water several times, the product was dispersed in ultrapure water and quantified for subsequent use.

2.2.2. Preparation of Fluorescent-Magnetic Nanoparticles (MNPs@QD)

MNPs (200 mg) were dispersed in 40 mL of deionized water containing 200 mg of PEI. The mixture was stirred for 2 h and after washing with water, PEI-modified MNPs (MNPs@PEI) were re-dissolved in 20 mL of deionized water, and then 0.5 mg of quantum dots were added into the solution with stirring overnight. Finally, the fluorescent-magnetic nanoparticles (MNPs@QD) were obtained and protected from light.

2.2.3. Fabrication of Functionalized Hydrogel MNPs (MNPs@hydrogel)

MNPs@QD (50 mg) was dispersed in ethanol (40 mL) and ultrapure water (10 mL) under the ultrasonic condition for 10 min. Next, ammonia solution (0.5 mL) was added into the mixture, and then TEOS (0.25 mL) was dropped into the solution with mechanical stirring at room temperature for 6 h under dark conditions. The obtained product (MNPs@Si) was separated by a magnet and washed with ethanol several times.

The synthesized MNPs@Si (25 mg) was dissolved in the solution consisting of ethanol (50 mL) and MPS (0.5 mL). After reacting for 48 h at room temperature, the product was dried at 40 °C and then re-dissolved in the acetonitrile (80 mL) containing SBMA (1.35 mg), MBA (100 mg), AIBN (10 mg), and MAA (100 mL). Then, the mixture was refluxed at 110 °C for 1 h. After washing, the hydrogel-coated MNPs (MNPs@hydrogel) were resuspended in ultrapure water and quantified for further use.

2.2.4. Synthesis of MNPs@FA

Briefly, the MNPs@hydrogel particles were activated with 0.2 M of EDC and 0.05 M of NHS in 1 mL of MES at room temperature with gentle shaking for 2 h, immediately followed by the reaction with NH_2-PEG_{5000}-FA for at least 24 h in dark conditions at room temperature. After being washed by phosphate buffered solution (PBS) with the help of a magnetic scaffold, unanchored NH_2-PEG_{5000}-FA molecules were removed, and MNPs@FA was obtained for further application.

2.3. Characterization of Nanoparticles and Captured Cells

The morphologies, average size, mono-dispersibility, zeta potential, and fluorescent images of nanoparticles were separately characterized by a transmission electron microscope (TEM, 120 kV, Hitachi-HT7700), a thermal field-emission environmental scanning electron microscope (SEM, 20.0 kV, FEI Quanta 400F), a Zetasizer Nano ZS (Malvern Instruments, Malvern, UK) and a confocal microscope (Olympus FV500-IX81, Tokyo, Japan). Optical density (OD) was measured by microplate reader (Perkin Elmer VICTORTM X4, Waltham, MA, USA). Western blot chemiluminescent signals were detected by the ECL Western blot detection system (Cwbiotech, Beijing, China). The fluorescence signal of the samples was measured by C6 cytometer (BD Biosciences, Ann Arbor, MI, USA). The SEM was also used to image the cells after 4% of paraformaldehyde fixation and gradient ethanol dehydration (30%, 50%, 70%, 85%, 95%, and 100%). The fluorescent images of the captured SK-OV-3 cells were observed by the confocal microscope after fixing with paraformaldehyde (4%) and stained with Hoechst 33,342 and Alexa Fluor 488-modified anti-CD45 or Alexa Fluor 555-modified anti-Pan-Keratin.

2.4. Exploration of Capture Performance

2.4.1. Investigation of Optimum Capture Conditions

The different doses of MNPs@FA (0.04 mg, 0.08 mg, 0.12 mg, 0.16 mg, 0.20 mg, 0.24 mg) were used to incubate with 1.0×10^5 of SK-OV-3 cells in 1 mL of the cell culture medium in 1.5 mL of EP tubes for 30 min at 37 °C to determine the dose-dependent effect of MNPs@FA on their capture efficiency. The cells captured by MNPs@FA were collected and washed by the magnetic scaffold and PBS. Finally, the cell samples were counted by a hemocytometer and a microscope. In addition, the capture efficiency was affected by the modification concentrations of NH_2-PEG_{5000}-FA (0.01, 0.05, 0.1, 0.15, 0.2, 0.25 mg) and the capturing time (5, 10, 15, 20, 25, 30 min). Furthermore, the capture yields of the differently modified magnetic nanoparticles were also researched using the discussed method.

2.4.2. Flow Cytometric Analysis and Western Blotting

For flow cytometric analysis, the SK-OV-3, OVCAR-3, A2780, and HEK293T cells were treated into suspension, then these cells and extracted PBMCs were incubated with FORL1 mouse monoclonal antibody (diluted 1:1000) for 30 min on ice. After washing, the cells were re-suspended in 100 µL of PBS, incubated with FITC-labeled goat anti-mouse antibody (diluted 1:1000) for 30 min on ice. Cells were then washed and analyzed by flow cytometer. The whole cells were lysed with RIPA lysis buffer supplemented with a protease inhibitor cocktail, and the lysates were clarified by centrifugation. The lysates were determined by the BCA protein assay kit, and separated by sodium dodecyl sulfate polyacrylamide gel electrophoresis (SDS-PAGE) at 100 V for 2 h, and then transferred to a polyvinylidene difluoride (PVDF) membrane at 100 V for 1.5 h. After blocking in 5% powdered milk, the membranes were probed with anti-FR antibody and anti-β-actin antibody, followed by HRP-conjugated secondary antibody. The chemiluminescent signals were detected by the ECL Western blot detection system.

2.4.3. Verification of Capture Specificity

The nanoparticles were incubated with 1.0×10^5 of SK-OV-3 cells (high-expressing FR), OVCAR-3 (high-expressing FR), A2780 cells (high-expressing FR), 293T cells (low-expressing FR), and peripheral blood mononuclear cell (PBMCs, separated from healthy blood samples, low-expressing FR) under the optimum capture condition, to determine the captured specificity of DiI pre-stained SK-OV-3 cells (1.0×10^5) and DiO pre-stained PBMCs (1.0×10^5) were mixed in the EP tube. Then, the mixed cells were incubated with the MNPs@FA under the optimum conditions to further verify the capture specificity. The fluorescence microscope enumerated the captured cells.

2.4.4. Cell Viability Analyses of Captured Tumor Cells

The SK-OV-3 cells viabilities captured by MNPs@FA (no modification of QD) and original SK-OV-3 cells (cells before capturing by MNPs@FA) were monitored through staining with 2 μm calcein-AM (green, live cells) and 4.5 μm PI (red, dead cells). The results were counted by fluorescence microscope. Furthermore, the cytotoxicity of MNPs@FA in vitro was analyzed by the CCK-8 assay. Briefly, the SK-OV-3 cells were each seeded into 96-well plates at the density of 5×10^3 cells/well and cultured for 24 h. Washed the cells thrice with PBS, 100 μL fresh medium containing different concentrations of MNPs@FA (0, 0.05, 0.1, 0.2, 0.4 mg/mL) was added per well. After culturing for 24 h, 10 μL CCK-8 was added to each well, and the cells were incubated further for 2 h. Measured optical density (OD) of the cell suspensions at 450 nm by microplate reader (Perkin Elmer VICTORTM X4). The cell viability was calculated as OD sample/OD control $\times 100\%$, where sample refers to the treated cells and control refers to the untreated cells.

2.4.5. Capture of Rare Cells

The artificial samples were prepared by spiking DiI pre-stained SK-OV-3 cells (5, 10, 50, 100, 200) into 1 mL of PBS or pre-treated whole blood from healthy people. The samples were separately incubated with MNP@FA (0.2 mg/mL) for 25 min. The mixtures were then placed on the magnetic separator for 2 min to isolate the captured cells by MNPs@FA, respectively. After washing by PBS, the fluorescence microscope enumerated capture cells to evaluate the capture yields. Meanwhile, the unstained SK-OV-3 cells (5, 10, 50, 100, 200) in the pre-treated whole blood were also captured by MNP@FA (0.2 mg/mL) for 25 min and then counted by the fluorescence microscope under the autofluorescence of magnetic nanoparticles to assess the capture yields.

2.5. CTC Isolation of Ovarian Cancer Patient Peripheral Blood Samples

The whole blood samples (n, 10) from OC patients of the Second Affiliated Hospital of Soochow University were collected and kept in ethylenediaminetetraacetic acid (EDTA) vacutainer tubes. This study was approved by the ethics review board of the Second Affiliated Hospital of Soochow University (Approval # JD-LS-2019-090-01). All blood samples were pretreated via gradient centrifugation to collect PBMCs containing CTCs. The isolation of CTCs was performed according to the same procedure as the artificial samples. After isolating and washing, the captured cells were diluted in 4% paraformaldehyde (PFA) and fixed on an adhesion slide, and then blocked by Triton X-100 (0.3%) and BSA (1%) for 1 h. The samples were stained with Alexa Fluor 488-modified anti-CD45 and/or Alexa Fluor 555-modified anti-Pan-Keratin overnight. Afterward, the samples were mixed with Hoechst 33,342 for 15 min and washed with water several times. Finally, the cells were observed and enumerated by the confocal microscope. Cells that displayed Hoechst 33,342+/MNPs@FA+/CD45− or Hoechst 33,342+/MNPs@FA+/PanCK+/CD45− with morphologically intact were identified as CTCs.

3. Results and Discussion

3.1. Preparation and Characterization of MNPs@FA

The synthesis of MNPs@FA was outlined in Figure S1. In brief, magnetic nanoparticles (MNPs) were synthesized by the solvothermal method. As displayed in Figure S2a, MNPs showed a good uniformity in size with a diameter of about 180 nm. Next, the PEI was modified on the surface of the magnetic nanosphere by electrostatic attraction (MNPs@PEI).

As shown in Figure S2b, MNPs covered by a polymer layer tend to surface smooth of the nanoparticles. As shown in Figures 2a and S2c,d, the QDs (10 nm) were further bedecked on the MNPs@PEI with the help of electrostatic interaction (MNPs@QD). The TEM images showed uniformity QD particles on the surface of MNPs@QD. Figures 2b and S2e showed the MNPs@QD was coated in a silica shell (MNPs@Si). The silica thickness is about 45 nm. Based on our previous method [27], the antifouling hydrogel was formed on the surface MNPs@Si using SBMA, MAA, and MBA to ensure the purity and identification of CTC capture. Figures 2c and S2f showed hydrogel shell (thickness, 12 nm) was successfully coated on the surface MNPs@Si. Meanwhile, the carboxyl group contained in the hydrogel facilitated further NH_2-PEG_{5000}-FA modification to achieve specific and efficient CTC capture. The hydrodynamic size and zeta potential of nanoparticles produced during the process were separately characterized, as shown in Figure 2d,e. The size of MNPs@PEI was significantly increased relative to the MNPs, and the potential of the nanoparticles was changed from a negative potential to a positive potential, which indicated that the PEI was successfully adsorbed on the surface of MNPs. After modifying the QDs, the size slightly increased because the surface of the QDs was modified with carboxyl groups, and the zeta potential of MNPs@QD was lower than that of MNPs@PEI. After further coating of silica shell and hydrogel, due to silicon hydroxyl, sulfonic group, and carboxyl groups, the potential of MNPs@Si and MNPs@hydrogel showed a negative value, and the size increased distinctly. After modifying NH_2-PEG_{5000}-FA, the carboxyl groups on the surface of MNPs@hydrogel were replaced by NH_2-PEG_{5000}-FA; the potential and size of MNPs@FA were visibly increased. The results demonstrated that products from MNPs to MNPs@FA were successfully synthesized.

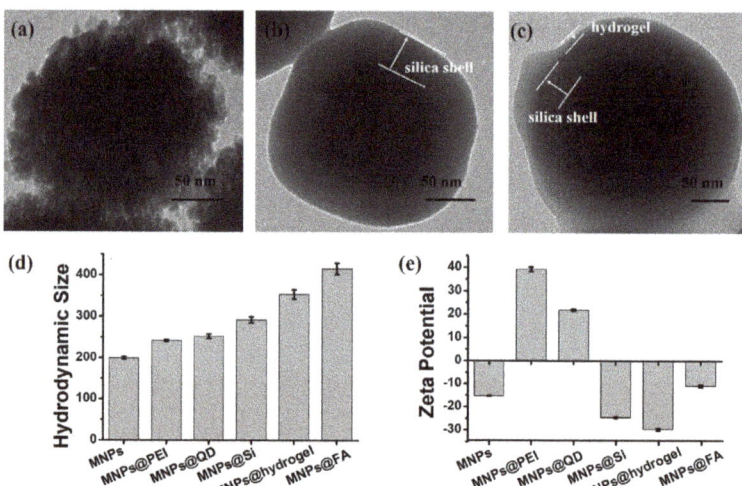

Figure 2. Transmission electron microscope (TEM) images of (**a**) MNPs@QD, (**b**) MNPs@Si, and (**c**) MNPs@hydrogel. Comparison of hydrodynamic size (**d**) and zeta potential (**e**) by the DLS and zeta potential measurement for the nanoparticles with different modifications. All data are expressed as the mean ± stand deviation, n = 3.

To further verify the capacity of MNPs@FA for CTC identification with fluorescence, the MNPs@FA nanoparticles were characterized by a confocal microscope. As shown in Figure 3a, the fluorescent image of MNPs@FA showed that the QDs were well modified into the silica shell, which ensured the fluorescent identification of MNPs@FA for target cells. As shown in Figure S3, SK-OV-3 cells were used as the model cells due to their high expression of folate receptors. SK-OV-3 cells captured by MNPs@FA observed by SEM were shown in Figure 3b,c; compared with original SK-OV-3 cells shown in Figure S4, an abundance of nanoparticles could be seen on the captured cell surface. Additionally, to confirm the feasibility of MNPs@FA for CTC identification, the fluorescence property of MNPs@FA on captured target cells was researched by observing SK-OV-3 cells under the confocal microscope. After staining by Hoechst 33,342 (blue) and anti-PanCK-555 (orange), as shown in Figure 3d, a clear red shell from MNPs@FA could be observed on the surface of the cells, which had a fine fluorescent consistency with immunostaining from anti-PanCK-555. In consideration of the presence of anti-CD45-488 used for PBMC identification in patient blood samples, the target cells were stained with MNPs@FA and anti-CD45-488 at the same time. As shown in Figure 3e, there was no cross-staining between the immunostaining of anti-CD45-488 and MNPs@FA on the captured target cells. The results further indicated that the MNPs@FA could provide the potential for CTC fluorescent identification.

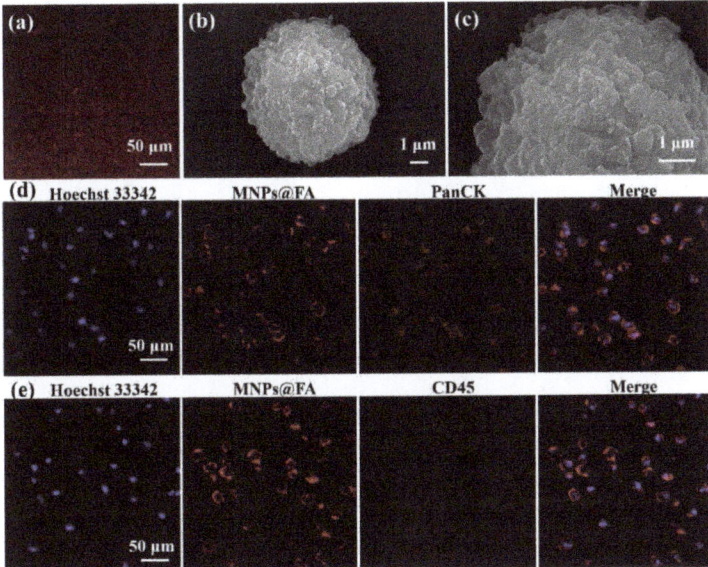

Figure 3. (a) A fluorescent image of MNPs@FA showing a stable fluorescent signal. (b,c) SEM images of an SK-OV-3 cell captured by MNPs@FA with a sufficient number of nanoparticles on the cell surface. (d) Fluorescent images of SK-OV-3 cells captured by MNPs@FA (red) with immunostaining of anti-PanCK-555 (orange) and Hoechst 33,342 (blue). (e) Fluorescent images of SK-OV-3 cells captured by MNPs@FA (red) with immunostaining of anti-CD45-488 (green) and Hoechst 33,342 (blue).

3.2. Cell Capture Performance of MNPs@FA

The experimental conditions are crucial for CTC separation. A series of attempts were conducted to optimize the capture conditions of MNPs@FA with the help of SK-OV-3 cells. Firstly, the experiments were made to optimize the dosages of MNPs@FA, as shown in Figure 4a, 0.04 mg/mL of MNPs@FA showed a 49.92% of capture efficiency, when the dosage increased to 0.2 mg/mL, the capture efficiency increased to 95.58%, while 0.24 mg/mL of MNPs@FA brought a minor change to capture efficiency (95.67%). Therefore, 0.2 mg/mL of the MNPs@FA dosage was chosen for CTC capture. Furthermore, the different modified concentrations (0.01, 0.05, 0.1, 0.15, 0.2, 0.25 mg/mL) of folic acid were also investigated, as shown in Figure 4b, 0.2 mg/mL of FA brought a 96% capture efficiency, while with the addition of FA, the capture yield did not increase. Consequently, 0.2 mg/mL of FA-modified concentration was determined for CTC capture. Moreover, the incubation time was explored based on the above optimal conditions. As shown in Figure 4c, the capture efficiency reached 61.92% in the incubation time of 10 min. However, the capture efficiency rose sharply to 96.08% when the incubation time was 25 min. So, the 25 min incubation time was used for CTC capture.

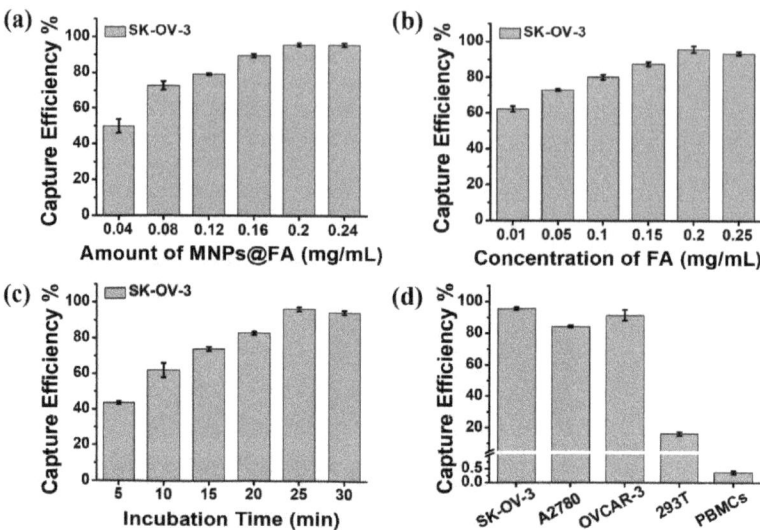

Figure 4. (**a**) Capture efficiency of MNPs@FA for SK-OV-3 cells at different concentrations (0.04, 0.08, 0.12, 0.16, 0.2, 0.24 mg/mL). (**b**) Capture efficiency of MNPs@FA for SK-OV-3 cells at the different modified concentrations of folic acid (0.01, 0.05, 0.1, 0.15, 0.2, 0.25 mg/mL). (**c**) Capture efficiency of MNPs@FA for SK-OV-3 cells at different incubation times (5, 10, 15, 20, 25, 30 min). (**d**) Capture efficiencies of MNPs@FA for different cells. All data are expressed as the mean ± stand deviation, n = 3.

We further explored the capture ability of different modified MNPs (MNPs@Si, MNPs@hydrogel, and MNPs@FA) for SK-OV-3 cells. Figure S5 summarized the capture yields of the different nanoparticles, and the results showed that the hydrogel formed by pSBMA possessed an excellent antifouling property. Meanwhile, the capture specificity of MNPs@FA was evaluated using FR-positive cancer cell lines SK-OV-3, A2780 and OVCAR-3, and FR-negative 293T cells and PBMCs from healthy donors (Figure S3) [34–39]. The capture specificity results shown in Figure 4d. The MNPs@FA had a fine capture efficiency for OVCAR-3 cells (95%) and A2780 cells (85%), and a low adhesion rate for 293T cells (16.17%) and PBMCs (0.39%). The results demonstrated that the FA modified MNPs could capture FR-positive CTCs both efficiently and specifically. Furthermore, the capture sensitivity of MNPs@FA was also investigated by spiking DiI prestained SK-OV-3 cells and DiO prestained PBMCs at a rate of 1:1 into 1 mL of PBS. As shown in Figure S6, the MNPs@FA exhibited a quite different isolation performance for the SK-OV-3 cells (94.25%) and PBMCs (0.41%), indicating an excellent capture specificity of MNPs@FA. In addition, the cell viability of captured SK-OV-3 cells was also assessed by live/dead staining. To ensure the count accuracy of live/dead fluorescence, MNPs without modifying QDs were used in the experiment. As shown in Figure S7, little difference between original cells and captured cells could be observed. The viability percentage of original cells was 97.97% and captured cells was 95.47%. Furthermore, we test the cytotoxicity of MNPs@ FA in the different concentration by the CCK-8 assay. The Figure S8 showed that the viability percentage of the SK-OV-3 cells incubated with 0.2 mg/mL dose of the MNPs@ FA nanoparticles for 24 h was more than 90%. These results indicated that MNPs@FA had little effect on the activity of captured cells.

3.3. Capture Sensitivity Test of MNPs@FA

To demonstrate the capture and identification capability of MNPs@FA for clinical samples, the rare number of DiI pre-stained SK-OV-3 cells (5, 10, 50, 100, 200) were spiked in 1 mL of PBS or pre-treated whole blood from healthy people (PBMCs). Meanwhile, we also spiked the same number of unstained SK-OV-3 cells in 1 mL of PBMCs solution and finished the counting for captured target cells based on the fluorescence performance of MNPs@FA. As shown in Figure 5a–c, the MNPs@FA showed a good capture efficiency for rare target cells, and the counting results by MNPs@FA fluorescence agreed well with the prestained counting group. Moreover, as shown in Figure 5d, the MNPs@FA combined anti-CD45-488 and Hoechst 33,342+ had a fine fluorescence distinction between target cells (Hoechst 33,342+/MNPs@FA+/CD45−) and PBMC (Hoechst 33,342+/CD45−). The results demonstrated that MNPs@FA could provide a dependable identification and enumeration method of CTCs in blood samples.

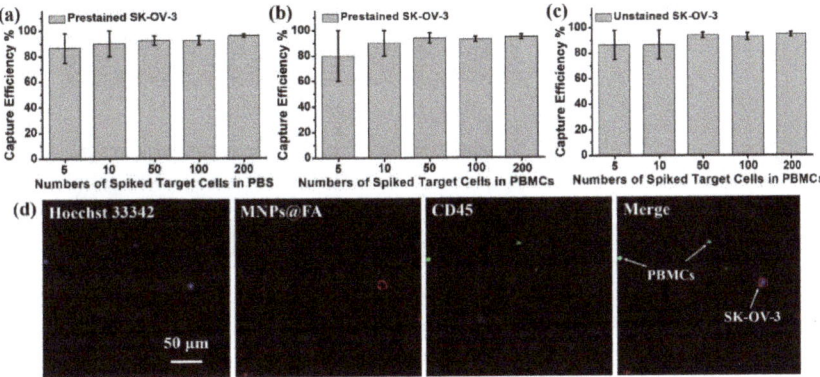

Figure 5. (**a**) The capture efficiency of MNPs@FA for rare DiI prestained SK-OV-3 cells in 1 mL of PBS. (**b**) The capture efficiency of MNPs@FA for rare DiI prestained SK-OV-3 cells in 1 mL of pre-treated PBMCs solution. (**c**) The capture efficiency of MNPs@FA for unstained SK-OV-3 cells in 1 mL of pre-treated PBMCs solution. All data are expressed as the mean ± stand deviation, $n = 3$. (**d**) Fluorescent images of MNPs@FA combined anti-CD45-488 and Hoechst 33,342+ for target cell and PBMCs.

3.4. CTC Detection from OC Patient Blood Samples

The peripheral blood samples of ovarian cancer patients were provided by the Second Affiliated Hospital of Soochow University with patient-informed consent. The optimized capture parameters were applied to detect CTCs in the whole blood samples from 10 OC patients and 10 healthy donors (HD). The clinical characteristics of the blood samples were shown in Table S1. Figure 6a summarizes the CTC enumeration of OC patients and HD people. From 2–12 CTCs were detected from 3 mL of blood samples of 10 OC patients, whereas no CTCs were found in any of the healthy donors' blood samples. Meanwhile, the immunostaining anti-PanCK were applied to identify the isolated CTCs from patient OC01 to validate the identification ability of MNPs@FA, and the same two CTCs were identified by the two identification methods (Figure 6b,c). The fluorescent images of CTCs and PBMCs identified by immunostaining of anti-PanCK-555, anti-CD45-488 and MNPs@FA for the sample OC01 were shown in Figure 6b. Besides, the fluorescent images of the CTCs identified by MNPs@FA and anti-CD45-488 without anti-PanCK-555 for the sample OC01 were displayed in Figure 6c. Cells that displaying Hoechst 33,342 (blue)+/MNPs@FA (red)−/PanCK(orange)−/CD45 (green)+ were counted as PBMCs, and showing Hoechst 33,342+/MNPs@FA+/CD45− or Hoechst 33,342+/PanCK+/MNPs@FA+/CD45− were counted as CTCs. The obtained results support that MNPs@FA hold powerful potential to capture and identify CTCs from patient's blood samples.

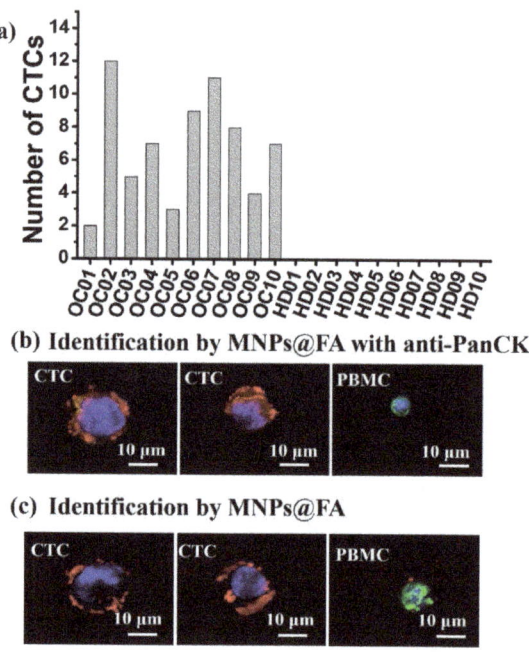

Figure 6. (a) Number of CTCs in the blood of 10 ovarian cancer patients (OC) and 10 healthy donors (HD) detected by the MNPs@FA. (b) Fluorescent images of CTCs and PBMCs identified by immunostaining of anti-PanCK-555 (orange) and anti-CD45-488 (green) as well as MNPs@FA (red) and Hoechst 33,342 (blue). (c) Fluorescent images of CTCs and PBMCs identified by Hoechst 33,342 (blue), anti-CD45-488 (green), and MNPs@FA (red).

4. Conclusions

This work developed a quick and effective method for the nondestructive and rapid capture and identification of CTCs using FA-modified fluorescent-magnetic nanoparticles. The incubation time might be different among different cell lines and different cancer types due to differences in biomarker expression levels. For exploring capture conditions in CTC capture, one cell line was used to be mainly chosen for obtaining the optimal incubation time in many pieces of research [28,40,41]. For example, Ding et al. built a simple and broad-spectrum method to efficiently isolate the heterogeneous CTCs from patient blood samples using tannic acid (TA)-functionalized magnetic nanoparticles (MNPs); the MCF-7 cell line was mainly chosen to explore the optimal capture conditions [28]; and Li et al. synthesized MN@Cys@PEG2k-FA magnetic nanospheres for early-stage cancer diagnosis targeted FR, which only chose the HeLa cell line for the experiment [41]. Therefore, in this work, SK-OV-3 cells with high FR expression were used as representative model cells to explore the relative optimal incubation time. A total of 0.2 mg MNPs@FA was incubated with 1×10^5 SK-OV-3 cells at different incubation time. The optimal incubation time may depend on the targeting of MNPs@FA to the FR on SK-OV-3 cell surface to some extent. The capture efficiency rose to 96.08% when the incubation time was 25 min, and the 25 min incubation time was used for CTC capture in this work. Binding with folic acid, MNPs@FA held a good capture efficiency against FR-positive ovarian cancer cells (higher than 85%), while showing a low capture yield against PBMCs (0.39%). In addition, MNPs@FA exhibited a fine capture sensitivity in capturing four SK-OV-3 cells when five cells were spiked in PBMC solution separated from 1 mL of whole blood. In our study, 2–12 CTCs were detected from 3 mL blood samples of ovarian cancer patients. The large blood sample sizes may provide more chances in the capture number of CTCs from the blood samples of cancer patients. Moreover, the use of multiple trapping agents may increase the capture number for CTCs due to the presence of CTC heterogeneity. For example, the most typical representative and the only system approved by the U.S. Food and Drug Administration (FDA) in 2004, the CellSearch™ system, captures CTCs using the 7.5 mL peripheral blood sample [40]. However, most research detected CTCs in less than 7.5 mL blood samples. The nanocage-featured film successfully detected CTCs and CTC clusters in 2 mL or 4 mL blood taken from 21 cancer patients (stages I-IV) suffering from various types of cancers [24]. A PLGA nanofiber-based microfluidic device was fabricated with dual aptamer-targeting EpCAM and N-cadherin proteins, and 1 to 13 CTCs were successfully detected in 3 mL blood samples of ovarian cancer patients [34]. The FR is highly expressed on the surface of 90% of ovarian cancer cells and FA has a low cost and good specificity, therefore FA was chosen as a capture agent in this work for the CTC isolation of OC patients. However, the application of this method for the early detection of OC may have some limitations due to blood volume. To obtain more CTCs for downstream analysis, many efforts have been devoted to the CTC culture [42,43]. However, the optimization of CTC culture conditions will be needed before this strategy can be incorporated into clinical practice; therefore, more exploration and effort are needed on CTC in vitro culture studies. A few studies have been successful, including the CTC-derived pre-clinical model which consists of 2D cultures, the CTC-derived explant (CDX) model, and the 3D organoid generation; compared with other models, 3D culture has attracted the attention of scientists because of its advantages of stable morphology, gene expression, cell signaling, equal behavior and heterogeneity with cancer cells in the tumor mass, high throughput for drug screening, low cost, and easy operation "in a dish" [43]. In this study, MNPs@FA provided a reliable and noninvasive method for the quick detection and identification CTCs in ovarian cancer patients' blood samples. Moreover, the high viability of captured cancer cells could further be used in in vitro cultures to help obtain more biological information of ovarian cancer patients in clinical applications.

Supplementary Materials: The following supporting information can be downloaded at: https://www.mdpi.com/article/10.3390/bios12030184/s1, Figure S1: The interfacial modification of MNPs@FA for CTC isolation. Figure S2: Transmission electron microscope (TEM) images of (a) Fe_3O_4 nanoparticles (MNPs), (b) PEI-modified Fe_3O_4 nanoparticles (MNPs@PEI), (c) CdSe/ZnS quantum dots (QDs), (d) QDs-modified Fe_3O_4 nanoparticles (MNPs@QD), (e) silica-modified Fe_3O_4 nanoparticles (MNPs@Si) and (f) hydrogel-coated MNPs@Si (MNPs@hydrogel). Figure S3: The expression levels of α-FR protein in A2780, OVCAR-3, SK-OV-3, PBMCs, and HEK 293T cells analyzed by flow cytometry (a-e) and Western blots (f). Figure S4: Scanning electron microscopy (SEM) images of an SK-OV-3 cell without MNPs@FA nanoparticles on the cell surface (a,b). Figure S5: Comparison of capture efficiencies for SK-OV-3 cells by MNPs@Si, MNPs@hydrogel, and MNPs@FA (folic acid-modified MNPs@hydrogel). Figure S6: Capture performance of MNPs@FA for SK-OV-3 cells and PBMCs from the mixture sample at the ratio of 1:1. Figure S7: Cell viability of captured SK-OV-3 cells as well as original SK-OV-3 cells was evaluated by a live/dead staining. (a) Comparison of viability percentage of original and captured cells. Fluorescence imaging of cell viability of (b) original SK-OV-3 cells and (c) captured SK-OV-3 cells using live/dead staining (green: live; red: dead). Figure S8: The viability of SK-OV-3 cells incubated with different concentrations of MNPs@FA for 24 h. Table S1: Clinical information of ovarian cancer (OC) patients and healthy donors (HD) enrolled in this study.

Author Contributions: Conceptualization, Y.P., Z.W. (Zhili Wang), R.P. and W.Z.; methodology, Y.P., Z.W. (Zhili Wang), Z.W. (Zeen Wu), P.D. and N.S.; data analysis, Y.P., J.M., T.Z. and L.L.; writing—original draft preparation, Y.P., J.M., T.Z., Z.W. (Zhili Wang), Z.W. (Zeen Wu), P.D., N.S. and L.L.; writing—review and editing, R.P. and W.Z.; project administration, Z.W. (Zhili Wang), R.P. and W.Z.; resources, R.P. and W.Z.; funding acquisition, R.P. and W.Z.; supervision, R.P. and W.Z. All authors have read and agreed to the published version of the manuscript.

Funding: Funds for this work were provided by the National Natural Science Foundation of China (No. 32000984, 21904135), the Science Foundation of Jiangsu Province (No. BK20180250, BE2021665), the Project of State Key Laboratory of Radiation Medicine and Protection of Soochow University (No. GZK1202106), a project funded by the Priority Academic Program Development of The Second Affiliated Hospital of Soochow University (XKTJ-XK202006).

Institutional Review Board Statement: The study was conducted according to the guidelines of the Second Affiliated Hospital of Soochow University and Suzhou Institute of Nano-Tech and Nano-Bionics (SINANO), Chinese Academy of Science. The use of blood samples was approved by the Ethics Committee of the Second Affiliated Hospital of Soochow University (Approval # JD-LS-2019-090-01).

Informed Consent Statement: Informed consent was obtained from all subjects involved in the study.

Data Availability Statement: Not applicable.

Conflicts of Interest: The authors declare no conflict of interest.

References

1. Asante, D.B.; Calapre, L.; Ziman, M.; Meniawy, T.M.; Gray, E.S. Liquid biopsy in ovarian cancer using circulating tumor DNA and cells: Ready for prime time? *Cancer Lett.* **2020**, *468*, 59–71. [CrossRef] [PubMed]
2. Lheureux, S.; Gourley, C.; Vergote, I.; Oza, A.M. Epithelial ovarian cancer. *Lancet* **2019**, *393*, 1240–1253. [CrossRef]
3. Pokhriyal, R.; Hariprasad, R.; Kumar, L.; Hariprasad, G. Chemotherapy resistance in advanced ovarian cancer patients. *Biomark. Cancer* **2019**, *11*, 1–19. [CrossRef] [PubMed]
4. Shen, Z.; Wu, A.; Chen, X. Current detection technologies for circulating tumor cells. *Chem. Soc. Rev.* **2017**, *46*, 2038–2056. [CrossRef] [PubMed]
5. Bankó, P.; Lee, S.Y.; Nagygyörgy, V.; Zrínyi, M.; Chae, C.H.; Cho, D.H.; Telekes, A. Technologies for circulating tumor cell separation from whole blood. *J. Hematol. Oncol.* **2019**, *12*, 48. [CrossRef]
6. Franses, J.W.; Philipp, J.; Missios, P.; Bhan, I.; Liu, A.; Yashaswini, C.; Tai, E.; Zhu, H.; Ligorio, M.; Nicholson, B.; et al. Pancreatic circulating tumor cell profiling identifies LIN28B as a metastasis driver and drug target. *Nat. Commun.* **2020**, *11*, 3303. [CrossRef]
7. Rao, L.; Meng, Q.F.; Huang, Q.Q.; Wang, Z.X.; Yu, G.T.; Li, A.; Ma, W.J.; Zhang, N.G.; Guo, S.S.; Zhao, X.Z.; et al. Platelet-leukocyte hybrid membrane-coated immunomagnetic beads for highly efficient and highly specific isolation of circulating tumor cells. *Adv. Funct. Mater.* **2018**, *28*, 1803531. [CrossRef]
8. Lux, A.; Bott, H.; Malek, N.P.; Zengerle, R.; Maucher, T.; Hoffmann, J. Real-time detection of tumor cells during capture on a filter element significantly enhancing detection rate. *Biosensors* **2021**, *11*, 312. [CrossRef]
9. Kilgour, E.; Rothwell, D.G.; Brady, G.; Dive, C. Liquid biopsy-based biomarkers of treatment response and resistance. *Cancer Cell* **2020**, *37*, 485–495. [CrossRef]
10. Che, J.; Yu, V.; Garon, E.B.; Goldman, J.W.; Di Carlo, D. Biophysical isolation and identification of circulating tumor cells. *Lab. Chip* **2017**, *17*, 1452–1461. [CrossRef]
11. Li, N.; Xiao, T.; Zhang, Z.; He, R.; Wen, D.; Cao, Y.; Zhang, W.; Chen, Y. A 3D graphene oxide microchip and an Au-enwrapped silica nanocomposite-based supersandwich cytosensor toward capture and analysis of circulating tumor cells. *Nanoscale* **2015**, *7*, 16354–16360. [CrossRef] [PubMed]
12. Xue, T.; Wang, S.Q.; Ou, G.Y.; Li, Y.; Ruan, H.M.; Li, Z.H.; Ma, Y.Y.; Zou, R.F.; Qiu, J.Y.; Shen, Z.Y.; et al. Detection of circulating tumor cells based on improved SERS-active magnetic nanoparticles. *Anal. Methods* **2019**, *11*, 2918–2928. [CrossRef]
13. Kim, T.H.; Wang, Y.; Oliver, C.R.; Thamm, D.H.; Cooling, L.; Paoletti, C.; Smith, K.J.; Nagrath, S.; Hayes, D.F. A temporary indwelling intravascular aphaeretic system for in vivo enrichment of circulating tumor cells. *Nat. Commun.* **2019**, *10*, 1478. [CrossRef] [PubMed]
14. Zhang, Z.; Wuethrich, A.; Wang, J.; Korbie, D.; Lin, L.L.; Trau, M. Dynamic monitoring of EMT in CTCs as an indicator of cancer metastasis. *Anal. Chem.* **2021**, *93*, 16787–16795. [CrossRef]
15. Gao, T.; Ding, P.; Li, W.; Wang, Z.; Lin, Q.; Pei, R. Isolation of DNA aptamers targeting N-cadherin and high-efficiency capture of circulating tumor cells by using dual aptamers. *Nanoscale* **2020**, *12*, 22574–22585. [CrossRef]
16. Park, C.; Abafogi, A.T.; Ponnuvelu, D.V.; Song, I.; Ko, K.; Park, S. Enhanced luminescent detection of circulating tumor cells by a 3D printed immunomagnetic concentrator. *Biosensors* **2021**, *11*, 278. [CrossRef]
17. Shamloo, A.; Ahmad, S.; Momeni, M. Design and parameter study of integrated microfluidic platform for CTC isolation and enquiry; A Numerical Approach. *Biosensors* **2018**, *8*, 56. [CrossRef]
18. Li, W.; Li, R.; Huang, B.; Wang, Z.; Sun, Y.; Wei, X.; Heng, C.; Liu, W.; Yu, M.; Guo, S.S.; et al. TiO_2 nanopillar arrays coated with gelatin film for efficient capture and undamaged release of circulating tumor cells. *Nanotechnology* **2019**, *30*, 335101. [CrossRef]
19. Wang, D.; Ge, C.; Liang, W.; Yang, Q.; Liu, Q.; Ma, W.; Shi, L.; Wu, H.; Zhang, Y.; Wu, Z.; et al. In vivo enrichment and elimination of circulating tumor cells by using a black phosphorus and antibody functionalized intravenous catheter. *Adv. Sci.* **2020**, *7*, 2000940. [CrossRef]

20. Li, C.; Feng, X.; Yang, S.; Xu, H.; Yin, X.; Yu, Y. Capture, detection, and simultaneous identification of rare circulating tumor cells based on a rhodamine 6G-loaded metal-organic framework. *ACS Appl. Mater. Interfaces* **2021**, *13*, 52406–52416. [CrossRef]
21. Qin, W.; Chen, L.; Wang, Z.; Li, Q.; Fan, C.; Wu, M.; Zhang, Y. Bioinspired DNA nanointerface with anisotropic aptamers for accurate capture of circulating tumor cells. *Adv. Sci.* **2020**, *7*, 2000647. [CrossRef] [PubMed]
22. Ribeiro-Samy, S.; Oliveira, M.I.; Pereira-Veiga, T.; Muinelo-Romay, L.; Carvalho, S.; Gaspar, J.; Freitas, P.P.; López-López, R.; Costa, C.; Diéguez, L. Fast and efficient microfluidic cell filter for isolation of circulating tumor cells from unprocessed whole blood of colorectal cancer patients. *Sci. Rep.* **2019**, *9*, 8032. [CrossRef] [PubMed]
23. Salmon, C.; Levermann, J.; Neves, R.P.L.; Liffers, S.T.; Kuhlmann, J.D.; Buderath, P.; Kimmig, R.; Kasimir-Bauer, S. Image-based identification and genomic analysis of single circulating tumor cells in high grade serous ovarian cancer patients. *Cancers* **2021**, *13*, 3748. [CrossRef] [PubMed]
24. Jiang, W.; Han, L.; Yang, L.; Xu, T.; He, J.; Peng, R.; Liu, Z.; Zhang, C.; Yu, X.; Jia, L. Natural Fish Trap-Like Nanocage for Label-Free Capture of Circulating Tumor Cells. *Adv. Sci.* **2020**, *7*, 2002259. [CrossRef] [PubMed]
25. Yang, L.W.; Sun, H.; Jiang, W.N.; Xu, T.; Song, B.; Peng, R.L.; Han, L.L.; Jia, L.Y. A Chemical Method for Specific Capture of Circulating Tumor Cells Using Label-Free Polyphenol-Functionalized Films. *Chem. Mater.* **2018**, *30*, 4372–4382. [CrossRef]
26. Wang, Z.; Sun, N.; Liu, H.; Chen, C.; Ding, P.; Yue, X.; Zou, H.; Xing, C.; Pei, R. High-efficiency isolation and rapid identification of heterogeneous circulating tumor cells (CTCs) using dual-antibody-modified fluorescent-magnetic nanoparticles. *ACS Appl. Mater. Interfaces* **2019**, *11*, 39586–39593. [CrossRef]
27. Wang, Z.; Wu, Z.; Sun, N.; Cao, Y.; Cai, X.; Yuan, F.; Zou, H.; Xing, C.; Pei, R. Antifouling hydrogel-coated magnetic nanoparticles for selective isolation and recovery of circulating tumor cells. *J. Mater. Chem. B* **2021**, *9*, 677–682. [CrossRef] [PubMed]
28. Ding, P.; Wang, Z.; Wu, Z.; Hu, M.; Zhu, W.; Sun, N.; Pei, R. Tannic acid (TA)-functionalized magnetic nanoparticles for EpCAM-independent circulating tumor cell (CTC) isolation from patients with different cancers. *ACS Appl. Mater. Interfaces* **2021**, *13*, 3694–3700. [CrossRef]
29. Nie, L.; Li, F.; Huang, X.; Aguilar, Z.P.; Wang, Y.A.; Xiong, Y.; Fu, F.; Xu, H. Folic acid targeting for efficient isolation and detection of ovarian cancer CTCs from human whole blood based on two-step binding strategy. *ACS Appl. Mater. Interfaces* **2018**, *10*, 14055–14062. [CrossRef]
30. Chen, C.C.; Wang, Z.L.; Zhao, Y.W.; Cao, Y.; Ding, P.; Liu, H.; Su, N.; Pei, R.J. A folic acid modified polystyrene nanosphere surface for circulating tumor cell capture. *Anal. Methods* **2019**, *11*, 5718–5723. [CrossRef]
31. Muller, C. Folate based radiopharmaceuticals for imaging and therapy of cancer and inflammation. *Curr. Pharm. Des.* **2012**, *18*, 1058–1083. [CrossRef] [PubMed]
32. Marchetti, C.; Palaia, I.; Giorgini, M.; De Medici, C.; Iadarola, R.; Vertechy, L.; Domenici, L.; Di Donato, V.; Tomao, F.; Muzii, L.; et al. Targeted drug delivery via folate receptors in recurrent ovarian cancer: A review. *OncoTargets Ther.* **2014**, *7*, 1223–1236. [CrossRef] [PubMed]
33. Zhang, Y.; Li, D.; Yu, M.; Ma, W.; Guo, J.; Wang, C. Fe_3O_4/PVIM-Ni^{2+} magnetic composite microspheres for highly specific separation of histidine-rich proteins. *ACS Appl. Mater. Interfaces* **2014**, *6*, 8836–8844. [CrossRef] [PubMed]
34. Wu, Z.; Pan, Y.; Wang, Z.; Ding, P.; Gao, T.; Li, Q.; Hu, M.; Zhu, W.; Pei, R. A PLGA nanofiber microfluidic device for highly efficient isolation and release of different phenotypic circulating tumor cells based on dual aptamers. *J. Mater. Chem. B* **2021**, *9*, 2212–2220. [CrossRef] [PubMed]
35. Lu, Y.; Low, P.S. Immunotherapy of folate receptor-expressing tumors: Review of recent advances and future prospects. *J. Control Release* **2003**, *91*, 17–29. [CrossRef]
36. Li, L.; Liu, J.; Diao, Z.; Shu, D.; Guo, P.; Shen, G. Evaluation of specific delivery of chimeric phi29 pRNA/siRNA nanoparticles to multiple tumor cells. *Mol. Biosyst.* **2009**, *5*, 1361–1368. [CrossRef]
37. Viola-Villegas, N.; Rabideau, A.E.; Cesnavicious, J.; Zubieta, J.; Doyle, R.P. Targeting the folate receptor (FR): Imaging and cytotoxicity of ReI conjugates in FR-overexpressing cancer cells. *Chem. Med. Chem.* **2008**, *3*, 1387–1394. [CrossRef]
38. Werner, M.E.; Karve, S.; Sukumar, R.; Cummings, N.D.; Copp, J.A.; Chen, R.C.; Zhang, T.; Wang, A.Z. Folate-targeted nanoparticle delivery of chemo- and radiotherapeutics for the treatment of ovarian cancer peritoneal metastasis. *Biomaterials* **2011**, *32*, 8548–8554. [CrossRef] [PubMed]
39. Smith, A.E.; Pinkney, M.; Piggott, N.H.; Calvert, H.; Milton, I.D.; Lunec, J. A novel monoclonal antibody for detection of folate receptor alpha in paraffin-embedded tissues. *Hybridoma* **2007**, *26*, 281–288. [CrossRef]
40. Li, F.L.; Xu, H.Y.; Zhao, Y.F. Magnetic particles as promising circulating tumor cell catchers assisting liquid biopsy in cancer diagnosis: A review. *Trend. Anal. Chem.* **2021**, *145*, 116453. [CrossRef]
41. Li, F.; Wang, M.; Cai, H.; He, Y.; Xu, H.; Liu, Y.; Zhao, Y. Nondestructive capture, release, and detection of circulating tumor cells with cystamine-mediated folic acid decorated magnetic nanospheres. *J. Mater. Chem. B* **2020**, *8*, 9971–9979. [CrossRef] [PubMed]
42. Yu, M.; Bardia, A.; Aceto, N.; Bersani, F.; Madden, M.W.; Donaldson, M.C.; Desai, R.; Zhu, H.; Comaills, V.; Zheng, Z.; et al. Ex vivo culture of circulating breast tumor cells for individualized testing of drug susceptibility. *Science* **2014**, *345*, 216–220. [CrossRef] [PubMed]
43. Yang, G.; Amidi, E.; Zhu, Q. Photoacoustic tomography reconstruction using lag-based delay multiply and sum with a coherence factor improves in vivo ovarian cancer diagnosis. *Biomed. Opt. Express* **2021**, *12*, 2250–2263. [CrossRef] [PubMed]

Article

Exploring Sensitive Label-Free Multiplex Analysis with Raman-Coded Microbeads and SERS-Coded Reporters

Umar Azhar [1], Qazi Ahmed [1], Saira Ishaq [1], Zeyad T. Alwahabi [1,*] and Sheng Dai [1,2,*]

1. School of Chemical Engineering and Advanced Materials, The University of Adelaide, Adelaide, SA 5005, Australia; umar_azhar@hotmail.com (U.A.); engrqaziahmed@hotmail.com (Q.A.); saira_chem@yahoo.com (S.I.)
2. School of Chemical and Process Engineering, University of Leeds, Leeds LS2 9JT, UK
* Correspondence: zeyad.alwahabi@adelaide.edu.au (Z.T.A.); s.dai1@leeds.ac.uk (S.D.)

Abstract: Suspension microsphere immunoassays are rapidly gaining attention in multiplex bioassays. Accurate detection of multiple analytes from a single measurement is critical in modern bioanalysis, which always requires complex encoding systems. In this study, a novel bioassay with Raman-coded antibody supports (polymer microbeads with different Raman signatures) and surface-enhanced Raman scattering (SERS)-coded nanotags (organic thiols on a gold nanoparticle surface with different SERS signatures) was developed as a model fluorescent, label-free, bead-based multiplex immunoassay system. The developed homogeneous immunoassays included two surface-functionalized monodisperse Raman-coded microbeads of polystyrene and poly(4-tert-butylstyrene) as the immune solid supports, and two epitope modified nanotags (self-assembled 4-mercaptobenzoic acid or 3-mercaptopropionic acid on gold nanoparticles) as the SERS-coded reporters. Such multiplex Raman/SERS-based microsphere immunoassays could selectively identify specific paratope–epitope interactions from one mixture sample solution under a single laser illumination, and thus hold great promise in future suspension multiplex analysis for diverse biomedical applications.

Keywords: SERS; Raman; microbeads; multiplex; immunoassays

1. Introduction

Nowadays, accurate diagnosis of human malignant diseases at the earliest stage is vital to curtail the side effects of surgical procedures and radio- and/or chemotherapies, subsequently followed by costly treatment. Typically, infectious or immune-system-related diseases generate various antigens as the first sign of abnormality in organisms. The most representative method for analysis of such disease biomarkers is immunoassays, realized by the immune recognition between antigens and relevant specific antibodies. In clinical applications, diseases as complex as cancers always necessitate multiplex diagnostics; that is, to simultaneously analyze multiple analytes from a single sample measurement. Hence, high sensitivity and high throughput are the desired features in developing novel multiplex immunoassays.

As per signal readout arrangements, immunoassays can be separated into a few classifications, including the enzyme-linked immunosorbent assay (ELISA) [1–3], luminescence [4,5], colorimetric [6,7], fluorescence [8,9], surface plasmon resonance (SPR) [10,11], and surface-enhanced Raman scattering (SERS) [12,13]. To date, traditional biological detection strategies such as ELISA and fluorescence are widely used in various clinical applications, but suffer from various shortcomings, such as high background noise, high spectral overlap, low sensitivity, and photobleaching. Among these readout technologies, SERS-based immunoassays have realized significant application in multiplex analyses. SERS has shown great promise in biomolecular analysis, as it only requires a small amount of body fluids (e.g., blood, urine, ascites, or saliva) for repetitive and timely examination, rather than complicated tissue analysis. In addition, SERS-based immunoassays achieve a low limit of detection (LOD), large dynamic range, and high sensitivity due to the enhanced Raman signals offered by SERS-active molecules physically adsorbed on noble metal nanoparticle surfaces [14–16]. In addition, the full width at half wavelength (FWHW) of a Raman signal is much narrower than that of fluorescence. Therefore, the interference caused by overlapping broad emission spectra will be greatly decreased between different SERS coding elements, which makes it an efficient encoding system for multiplex analysis.

To date, substrate-based SERS immunoassays have been widely utilized in biochemical and biomedical applications for the identification of different biological targets; for example, proteins [17–19], nucleic acids [20,21], infection [22], cells [23,24], and poisons [25]. This type of biomolecular recognition uses substrate-immobilized, immuno-functionalized SERS nanoprobes to recognize various target analytes. In a typical encrypting method, distinct Raman signatures of nanoprobes can be obtained by selecting different SERS-active molecules or their combinations [26,27]. As such, a series of SERS-coded tags have been developed for substrate-based analysis [28–37]. Although this technique provides high-sensitivity multiplex analysis, it is limited by high throughput and a large dynamic detection range. On the other hand, bead-based immunoassays have gained exponential recognition in recent years in developing novel SERS-based multiplex bioanalysis platforms. Compared to these well-established bead-based fluorescence immunoassays, the bead-based Raman and/or SERS immunoassays show obvious advantages such as label-free, narrow vibration spectra; high sensitivity; and a single excitation wavelength, in turn offering higher multiplexing capability. In particular, these Raman beads not only act as a platform for immune bead support, but also as a label-free code [38–40].

In this study, we developed a model bead-based multiplex immunoassay system in the presence of Raman-coded microbeads and SERS-coded nanotags for simultaneous multiplex analysis of epitopes (antigens). As illustrated in Figure 1, the SERS-coded nanotags were simply prepared by the self-assembly of functional SERS-active molecules to gold nanoparticle (AuNP) surfaces, and the SERS-coded reporters obtained from interaction of SERS-coded nanotags to different epitopes. On the other hand, paratopes (antibodies) could be immobilized on the surface of Raman-coded microbeads. In an immunocomplex mixture of various functionalized SERS reporters and Raman microbeads, Raman imaging from one sample measurement allowed us to accurately identify analyte interactions by tailoring different vibrational bands of the Raman/SERS signatures of microbeads and nanotags. Our results showed the abilities of applying Raman spectroscopy and imaging in bead-based multiplex bioanalysis with high sensitivity and high specificity. This holds great promise for future applications in multiplexing, high-throughput screening, and detection of complex human diseases.

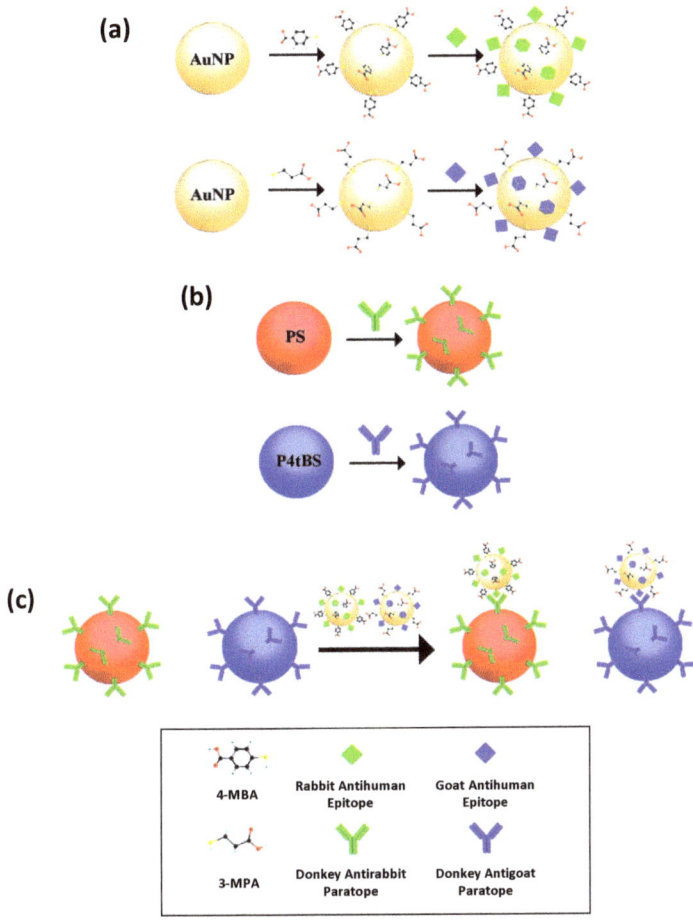

Figure 1. Schematic description of the preparation of: (**a**) SERS-coded reporters; (**b**) surface functionalized Raman-coded polymeric microbeads; (**c**) the resulting bead-based multiplex immunoassays.

2. Materials and Methods

2.1. Materials

Gold chloride trihydrate ($HAuCl_4 \cdot 3H_2O$), 4-mercaptobenzoic acid (4-MBA), 3-mercaptopropionic acid (3-MPA), styrene, 4-tert-butylstyrene, polyvinylpyrrolidone (PVP 360,000) and bovine serum albumin (BSA) were procured from Sigma-Aldrich (St. Louis, MO, USA). Acrylic acid (AA), 2,2′-Azobis(2-methylpropionitrile) (AIBN), N-(3-dimethylaminopropyl)-N'-ethylcarbodiimide hydrochloride (EDC), and N-hydroxysuccinimide (NHS) were procured from Acros Organics (Geel, Belgium). Tri-sodium citrate dehydrate was attained from Prolabo (Phnom Penh, Cambodia). AffiniPure donkey antirabbit IgG (paratope in this study), AffiniPure donkey antigoat IgG (paratope in this study), AffiniPure rabbit antihuman IgG (epitope in this study), and AffiniPure goat antihuman IgG (epitope in this study) were obtained from Jackson ImmunoResearch (West Grove, PA, USA). These lyophilized IgG samples (H + L) were polyclonal in sterile filtered liquid purified from antisera by immunoaffinity chromatography using antigen coupled to agarose beads at a concentration or dilution range of 10–20 µg/mL. Hydroxylamine (1M), phosphate-buffered saline (PBS, 0.01 M sodium phosphate, 0.25 M NaCl, pH 7.4), block solution (50 mM Tris, 0.14 M NaCl, 1% BSA, pH 8), and wash solution (50 mM Tris, 0.14 M NaCl, 0.05% Tween

20, pH 8) were prepared in-house. Deionized water (DI water, 18.2 MΩ·cm^{-1}) was from an Easypure II ultrapure water purification system.

2.2. Preparation of Raman-Coded Microbeads

Monodisperse poly (styrene-co-AA) (PS) and poly (4-tert-butylstyrene-co-AA) (P4tBS) beads were prepared following a two-stage dispersion polymerization method [41]. In the first stage, 6.0 g of styrene or 4-tert-butylstyrene monomer, 0.24 g of AIBN initiator, 0.27 g of PVP 360,000 stabilizer, 0.30 g of 70% Triton X-305 co-stabilizer, and 34 g of 95% ethanol were charged to a 250 mL three-neck flask. The mixture was bubbled by nitrogen at room temperature for 40 min. The flask was then merged into a preheated 70 °C oil bath and stirred mechanically at 100 rpm for 1 h. In the second stage, the preheated comonomer AA solution (2 wt% of AA to the feed monomer mixed with 16 g of 95% ethanol) was added to the reaction flask using a syringe. The reaction was continued for another 24 h. For purification, the synthesized microbeads were washed three times with 95% ethanol and four times with DI water to remove any reaction residuals. The washed microbeads were rescattered in DI water, and the detailed contents of solids were measured by gravity analysis.

The functional carboxylic group on the microbeads' surfaces was used to conjugate IgG (paratope) by EDC chemistry. In a typical experiment, 100 μL of 1.9 wt% purified microbeads and 100 μL of EDC/NHS/PBS (20 mg of EDC and 30 mg of NHS in 1 mL of phosphate buffer saline at pH 7.4) were mixed for 15 min at room temperature, followed by the addition of 10 μg of model paratopes (e.g., donkey antirabbit IgG for PS microbeads and donkey antigoat IgG for P4tBS microbeads). The mixture solution was incubated at room temperature for 2 h with gentle rotation and then quenched by using 1 M hydroxylamine. The conjugated microbeads were washed with PBS buffer and then mixed with 1 mL block solution (50 mM Tris, 0.14 M NaCl, 1% BSA, pH 8) for 30 min at room temperature. Finally, the microbeads were washed four times using wash solution (50 mM Tris, 0.14 M NaCl, 0.05% Tween 20, pH 8), and the paratope-conjugated microbeads were redispersed in DI water.

2.3. Preparation of SERS-Coded Nanotags and SERS-Coded Reporters

AuNPs were synthesized by reduction of gold (III) chloride by sodium citrate [42]. Briefly, 50 mL of 10^{-3} M HAuCl$_4$·3H$_2$O was brought to a boil, followed by rapidly adding 1 wt% sodium citrate at a mixing molar ratio of 1:2 [39]. After continuously stirring at the boiling temperature, a reddish-gold colloidal solution was obtained.

The SERS-coded nanotags were prepared by the formation of different self-assembled monolayers (SAMs) of functional thiols on AuNP surfaces. Here, the solution of 4-MBA or 3-MPA (2.5 μL, 10^{-3} M) was added separately to 1 mL of the above AuNP suspension under vigorous stirring for 2 min at room temperature, and the mixture solution turned purple. The resulting SERS nanotags were washed three times by operating centrifugation and redispersing cycles, and finally redispersed in 100 μL DI water. Hereafter, two different sets of epitopes (rabbit antihuman IgG and goat antihuman IgG) were separately attached to the performed SERS-coded nanotags via ionic and hydrophobic interactions [43]. Then, 5 μg of epitopes in phosphate buffer were separately charged to these 100 μL SERS-coded nanotag suspensions under gentle rotation at room temperature for 2 h. Then, 100 μL of 10% BSA solution was added to the mixtures and rotated for another 30 min, followed by three-time washing using DI water. The final epitope coupled SERS-coded nanotags (aka. SERS-coded reporters) were then redispersed in 100 μL DI water and stored at 4 °C.

2.4. Multiplex Immunoassays

For spectroscopic analysis, 50 μL of the two paratope-conjugated Raman-coded microbead suspensions were separately mixed with their matched and unmatched epitope-coupled, SERS-coded reporters. The mixtures were incubated for 1 h under gentle rotation at room temperature. After washing several times using DI water, the resulting microbeads were redispersed in 50 µL DI water. The air-dried microbeads on a glass substrate were subjected to Raman spectroscopic analysis. For multiplex immunoassays, a 50 µL mixture of two paratope-conjugated, Raman-coded microbeads were incubated with a mixture of two epitope-coupled, SERS-coded reporters for 1 h. After washing using DI water, air-dried microbeads on a glass substrate were subjected to Raman imaging analysis at characteristic wavenumbers of Raman microbeads and SERS nanotags.

2.5. Equipment

The images of microbeads and AuNPs were obtained by using a Quanta 450 scanning electron microscope (SEM) and a Phillips CM200 transmission electron microscope (TEM). The UV–vis absorption spectra were recorded using a Shimadzu UV-1601 UV–vis spectrophotometer. A LabRAM Horiba Raman microscope equipped with LabSpec 6 software was used to measure the Raman spectra and image microbeads, SERS reporters, and immunocomplexes. A 785 nm Xtra II diode laser from Toptica was applied for Raman imaging, with the monochromator comprising 600 grooves per mm grating. The Raman spectra were recorded using an acquisition time of 5 s and an accumulation time of 3 s. The SWIFT mode in LabSpec 6 was used for Raman imaging with an acquisition time of 1 s and a step size of 0.1 micron. The peak and CLS mode in the Raman microscope were used to generate the Raman false-color images to establish selective binding in the immunoassays. The CLS fitting was used to set up and perform the multivariate classical least squares fitting procedure on single spectra and multidimensional spectral arrays using a set of reference component spectra. This method is a supervised multivariate decomposition technique [44].

3. Results and Discussion

3.1. Raman-Coded Microbead Supports

Raman-coded polymeric microbeads were used as the immune solid supports in bead-based immunoassays [45–47]. Dispersion polymerization was used for the synthesis of monodisperse polymer microbeads with average sizes of ~1.5 µm (Figure 2a,b) and narrow size distributions (CV < 1%), where a small amount of AA was introduced as the comonomer to generate functional carboxylic acid groups on the surfaces of microbeads [41]. These surface functional groups not only promoted microbead stability, but also allowed further surface paratope IgG immobilization via the EDC chemistry.

Styrene and 4-tert-butylstyrene monomers have different Raman spectra. Since polymers are long-chain molecules composed of many repeating monomers connected by covalent bonds, the Raman signals of their polymer microbeads are strong and distinguishable, which render them suitable as a Raman-coded support for bioanalysis. As shown in Figure 2c, the Raman spectrum of the PS microbeads showed a distinct vibrational band centered at 1002 cm^{-1} (strong), which was assigned to the υ1 symmetrical ring. Another peak related to the υ18A vibration was located at a wavenumber of 1032 cm^{-1} (medium) [48]. Similarly, the Raman spectrum of the P4tBS microbeads showed distinct vibrational peaks located at 1109 cm^{-1} (strong) and 1611 cm^{-1} (very strong). Due to small amount of surface carboxyl groups, no Raman spectral representation for carboxyl groups before and after EDC conjugation was evident.

3.2. SERS-Coded Nanotags and Reporters

The details of preparing the SERS-coded reporters are illustrated in Figure 1; SERS-coded nanotags were used to report the specific paratope–epitope interaction in the immunoassay analysis. The AuNPs used to prepare SERS-coded nanotags were synthesized at a 1:2 feed molar ratio of $HAuCl_4 \cdot 3H_2O$ and Na_3Ct to achieve sufficient surface citrate ions and good colloidal stability. The localized surface plasmon resonance (LSPR) of the ~25 nm AuNPs showed a characteristic absorption peak at 525 nm (Figure 3a). The same amount of SERS-active molecules of 4-MBA and 3-MPA were separately mixed with AuNPs to produce the SERS-coded nanotags; the thiol groups of 4-MBA and 3-MPA had a strong affinity to the AuNPs, and carboxylic acid groups enhanced colloid stability [49]. The 4-MBA and 3-MPA formed SAM on the AuNPs via the Au–S bond, and an inevitable aggregation took place immediately after the SAM formation due to hydrophobic interaction [50]. The aggregates formed "hotspots" and facilitated strong SERS signals. The successful formation of SAM can be signified by an absorption-peak redshift to 536 nm upon addition of 4-MBA and 3-MPA [51]. The TEM image could be further used to identify the formation of SAM on the AuNPs and hotspots (Figure 3b); each aggregate contained several small AuNPs. Model epitopes of rabbit antihuman IgG and goat antihuman IgG were then mixed with the 4-MBA or 3-MPA SERS-coded nanotags separately to form two distinct SERS-coded reporters, and 10% BSA was used as a blocking solution to avoid future nonspecific bounding in bioanalysis. No redshift in the LSPR of the SERS-coded nanotags after epitope coupling indicated their core–shell–corona structure, in which the hydrophilic corona layer further enhanced the stability of the SERS-coded nanotags in aqueous solution.

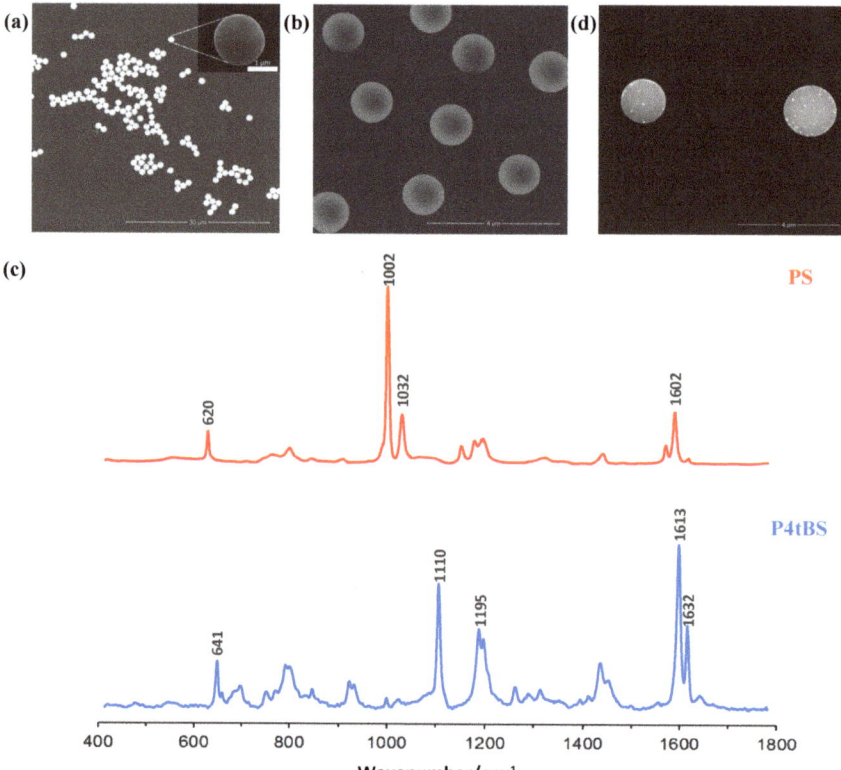

Figure 2. SEM images of monodisperse (**a**) PS and (**b**) P4tBS microbeads. (**c**) Raman spectra of these two Raman-coded microbeads with their characteristic wavenumbers labelled. (**d**) A typical SEM image of the immunocomplexes (SERS-coded reporters on the surface of Raman-coded microbeads formed through specific epitope-paratope interaction).

Figure 3c shows the Raman spectra of bulk 4-MBA and the SERS spectra of 4-MBA on the AuNPs' surfaces before and after epitope introduction. Both the Raman and SERS (with and without IgG) spectra revealed the characteristic vibrational peaks of the υ(CC) ring-breathing at 1074 and 1583 cm^{-1}. However, other less intense peaks at the δ(CH) (1132 and 1173 cm^{-1}) and υs(COO-) (1375 cm^{-1}) were observed in the SERS spectra, but not the Raman spectrum of bulk 4-MBA [52–54]. Similarly, the Raman spectra of bulk 3-MPA, as well as the SERS spectra of 3-MPA on the AuNPs surfaces with and without IgG, are shown in Figure 3d. All spectra were dominated by three distinct peaks at 674, 867, and 1430 cm^{-1}, which were assigned to the υ(CS)G, υ(SH), and υ(CH$_2$) characteristic vibrational modes [55]. We also observed that the SERS signals before and after IgG (epitope) coupling were identical (Figure 3c,d), which agreed with the LSPR results, indicating no direct interaction between the AuNPs and IgGs.

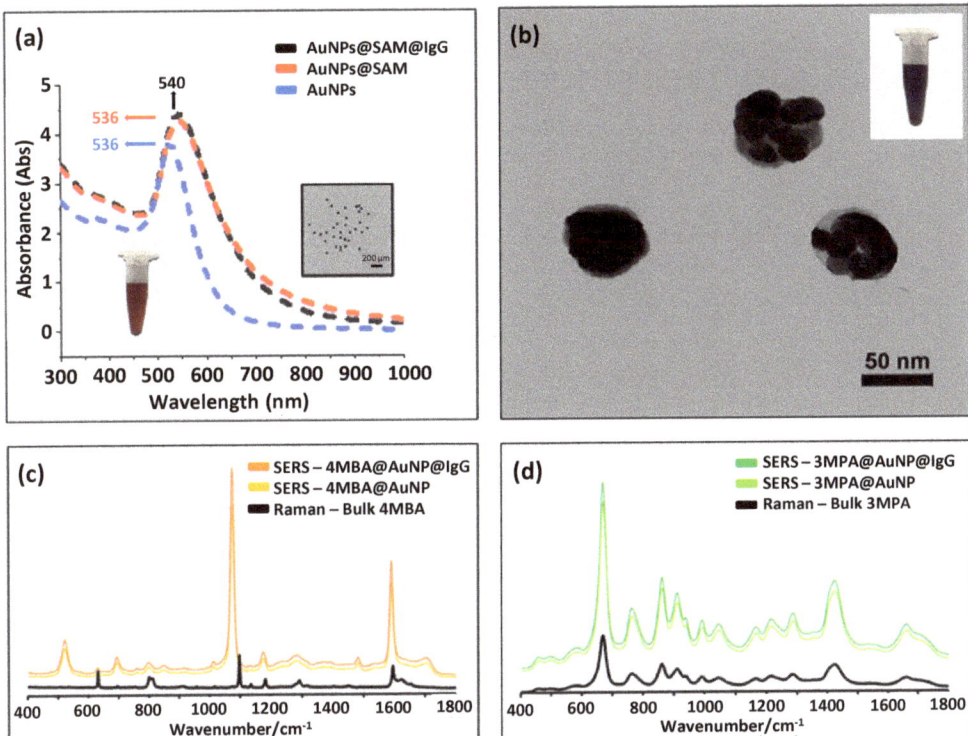

Figure 3. (a) UV–vis absorption spectra of AuNPs, AuNP@SAM, and AuNP@SAM@IgG; inset shows a TEM image of the AuNPs. (b) TEM image of SERS-coded nanotags. (c,d) Raman spectra of bulk 4 MBA/3 MPA, SERS-coded nanotags before and after epitope IgG loading.

The most intensive peaks of 4-MBA and 3-MPA were distinguishable with narrow bandwidths, resulting in their SERS-coded nanotags and reporters having different SERS signatures. Moreover, as the Raman scattering cross-sections of nanotags were much larger than bulk small molecules (4-MBA and 3-MPA), the intensities of Raman signals enhanced by AuNPs could reach a level of 10^6. Such a great increase in the signal-to-noise ratio was favorable for highly sensitive analysis. The formed self-assembled gold colloid helped enhance the Raman signals of the 4-MBA and 3-MPA. Due to this enhancement effect of the SERS-coded nanotags, it was possible to lower the concentration limit for analyte detection. The increase in the intensity of SERS signal to Raman signal was quantified by the apparent effective enhancement factor (*EEF*) using the following expression:

$$EEF = \frac{I_{SERS} N_{bulk}}{I_{bulk} N_{SERS}}$$

where I_{SERS} and I_{bulk} are the intensities of the same band for the SERS and bulk Raman spectra, while N_{bulk} and N_{SERS} are the number of molecules for the bulk and SERS sample, respectively [56,57]. Under the experimental conditions, 2.5 µL 10^{-3} M of 4-MBA or 3-MPA was charged separately to 1 mL of 10^{-3} M AuNP aqueous solution to prepare the SAMs. Using the band of 1074 cm^{-1} for 4-MBA, the apparent EEF was calculated to be 1.08×10^{6}. Similarly, using the band of 674 cm^{-1} for 3-MPA, the EEF was calculated to be 1.25×10^{7}.

3.3. Immunocomplexes and Multiplex Immunoassays

We aimed to develop a novel bioassay for multiplex detection using Raman-coded microbeads and SERS-coded reporters; details are shown in Figure 1. In this study, donkey antirabbit IgG and donkey antigoat IgG were considered as model paratopes, while rabbit antihuman IgG and goat antihuman IgG were considered as model epitopes. The PS and P4tBS microbeads were separately immobilized with donkey antirabbit and donkey antigoat IgGs. On the other hand, rabbit antihuman IgG was coupled to the 4-MBA-coded SERS nanotag as the epitope of the donkey antirabbit paratope, and the 3-MPA-encoded SERS nanotag was decorated with the goat antihuman IgG as the epitope of the donkey antigoat paratope. Due to the specific recognition between these paired paratopes and epitopes, the SERS-coded nanotags were used to report the matched immune-interaction on the microbeads' surfaces. In other words, in the presence of matched paratope–epitope pairs, both codes of the Raman bands from the microbeads and the SERS signals from the nanotags could be read simultaneously due to the formation of immunocomplexes. Otherwise, only the Raman bands of microbeads could be read in the presence of unmatched paratope–epitope pairs.

After mixing paired paratope-conjugated, Raman-coded microbeads and epitope-coupled, SERS-coded nanotags, the SEM image of the immunocomplexes in Figure 2d clearly showed the presence of SERS-coded reporters on the Raman-coded microbeads. To further conclude the specific biorecognition of the paratope-conjugated Raman microbeads to the counterpart epitope loaded to SERS-coded nanotags, we performed a Raman spectrum analysis (Figure 4a,b). As mentioned previously, PS microbeads displayed two strong Raman vibrational bands at 1002 and 1032 cm^{-1}, whereas the SERS vibrational bands for 4-MBA were located at 1074 and 1583 cm^{-1}. After the specific binding of the matched paratope and epitope, the individual Raman signals of the PS microbeads and the SERS signals of 4-MBA could be detected in the dry bead sample. Similarly, the P4tBS microbeads showed typical Raman strong vibrational bands at 1110 and 1613 cm^{-1}, whereas the SERS vibrational bands of 4-MPA were located at 674 and 2576 cm^{-1}. Again, due to the specific binding of the matched paratope and epitope, both the Raman signals of the P4tBS microbeads and the SERS signals of 3-MPA could be observed for the resulting dry microbeads. In presence of unmatched paratope–epitope pairs, only the Raman signals of PS or P4tBS could be observed. These results demonstrated that SERS reporters could selectively bind to microbeads through the specific matched paratope–epitope interaction with high sensitivity and selectivity. Experimentally, Raman shifts were reproducible in repeated experiments for Raman-coded microbeads and SERS-coded nanotags, as well as the paratope-conjugated microbeads with the matched or unmatched SERS-coded reporters. However, the intensity ratios of Raman-coded microbeads and SERS-coded reporters varied due to the inhomogeneous distribution of the antibody on the microbeads' surfaces and difficulty in controlling the amount of loaded SERS-coded nanotags.

The specific recognition of paratopes and epitopes in the above immunoassays could be further examined by Raman mapping to explore future multiplex analysis through spectroscopic imaging of SERS nanotags on the Raman microbead surfaces. After mixing the two SERS-coded reporters with the two epitope-conjugated Raman-coded microbeads, the Raman images clearly showed the selective and specific interaction of paratopes and epitopes from one sample measurement. Figure 5f shows the optical image of dry microbead mixtures from this immunoassay, and no immunoprecipitation or aggregation was observed. That was important for the bead-based multiplex analysis to ensure high reproducibility and reliability. Raman mapping of this area was then carried out with a scanning step size of 0.1 micron. The false-colored Raman images were achieved based on the distinct spectral peaks of the Raman and SERS codes, as well as by using the classical least squares (CLS) algorithm mode. By selecting and deselecting distinct Raman signatures of the two polymer microbeads and the two SERS nanotags, different Raman mapping images were acquired (Figure 5a–d). From that, we could distinguish the microbeads (and their relevant paratopes) with or without SERS nanotags (and their relevant epitopes) (Figure 5e) simultaneously. Raman imaging revealed that there was no cross-interaction of the SERS nanotag signals on the microbead surfaces due to the presence of specific paratope and epitope interactions. Therefore, the false-colored Raman imaging analysis reinforced the Raman and SERS immunoassay results, and showed that the interaction was highly selective and highly specific.

The unabridged Raman spectrum and Raman imaging analysis indicated the specific recognition between matched paratopes and epitopes with high selectivity. This study combined the signatures of Raman-coded microbeads (support) and SERS-coded nanotags (reporter) for multiplex analysis. Since a variety of Raman-coded microbeads and SERS nanotags can be easily prepared, this Raman and SERS bioassay using vibrational information of microbead supports and SERS reporters can be expanded to simultaneous multiplex analysis from one homogeneous immunoassay. The capacity of the Raman dual-encoding system will be far more than that of a fluorescent coding system [58], qualifying its high analyte throughput.

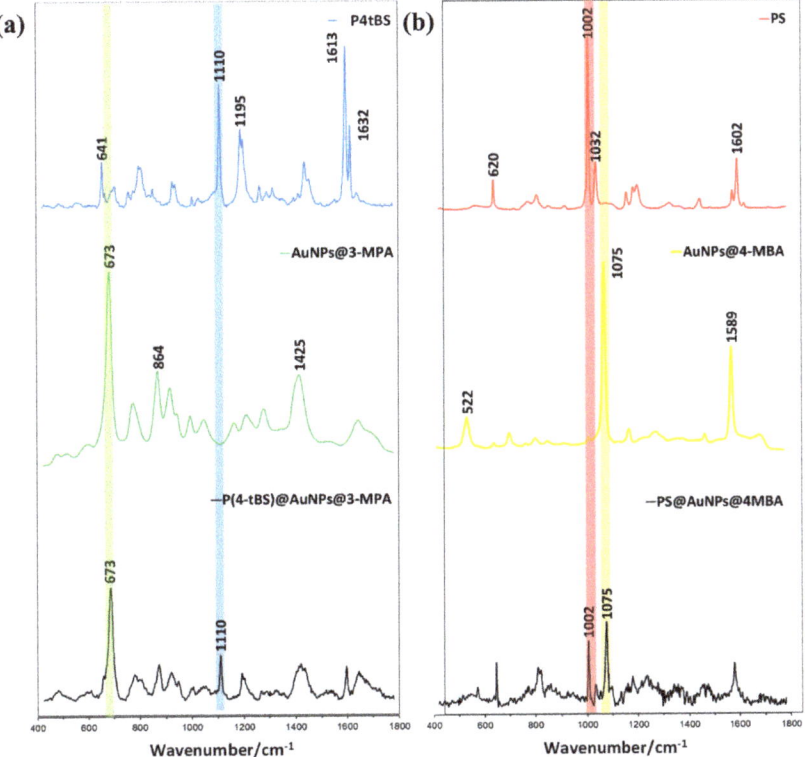

Figure 4. (a) Raman spectra of paratope-conjugated P4tBS microbeads, epitope-coupled AuNP@3-MPA SERS-nanotags, and the resulting immunocomplexes after the specific paratope and epitope interactions. (b) Raman spectra of paratope-conjugated PS microbeads, epitope-coupled AuNPs@4-MBA SERS-nanotags, and resulting immunocomplexes after the specific paratope and epitope interactions.

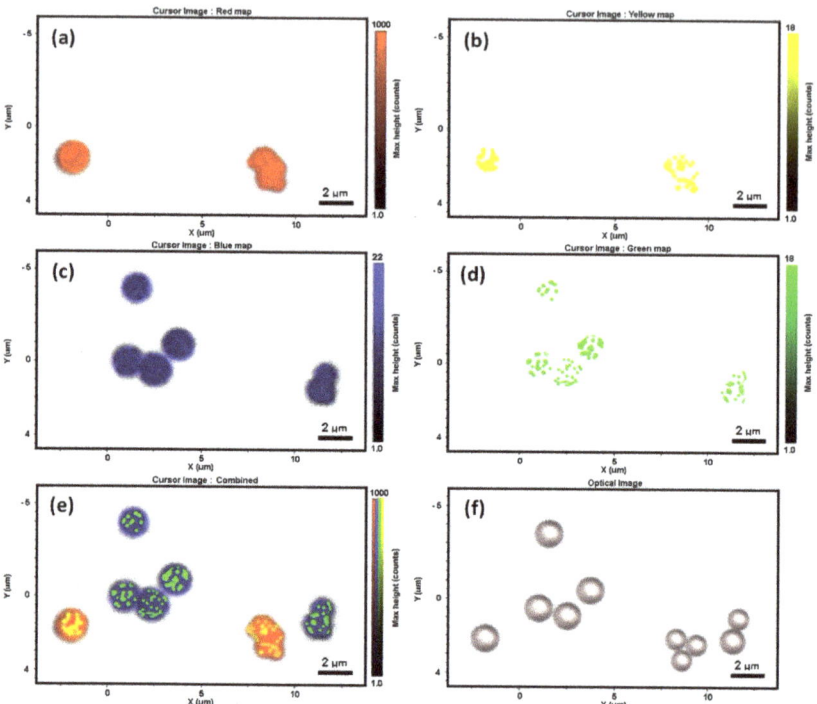

Figure 5. Raman imaging of the multiplex assays. (**a–d**) Raman images of PS microbeads (red) and 4-MBA SERS-reporters on PS microbead surface (yellow); P4tBS microbeads (blue) and 3-MPA SERS reporters on P4tBS microbead surface (green). (**e,f**) Combined false-colored spectroscopic and optical images of dry PS and P4tBS microbeads with their specific SERS reporters from immunoassays.

4. Conclusions

A novel bioanalytical technique was demonstrated for multiplex analyte detection based on a Raman and SERS suspension immunoassay. Surface-functionalized, Raman-coded microbeads could be used as the support to immobilize various paratopes for bead-based immunoassays. AuNPs, SERS-active molecules, and epitopes could be used to prepare the core–shell–corona structured SERS-coded reporters. In a homogeneous immunoassay of mixed microbeads and SERS reporters, both Raman spectroscopic and Raman imaging analysis demonstrated that the SERS (from reporter) and Raman (from microbead) signatures could only be successfully read in the presence of specific paratope–epitope interactions. Such a system offered high selectivity, high specificity, high multiplex, no photobleaching, narrow spectra, and a single excitation wavelength when compared to a traditional bead-based fluorescent bioanalysis. This Raman and SERS immunoassay has great potential for future high-throughput multiplexing analysis of many target analytes from a single sample measurement.

Author Contributions: Experiments, investigation, data analysis and writing, U.A.; analysis and discussion, Q.A. and S.I.; conceptualization and supervision, Z.T.A. and S.D. All authors have read and agreed to the published version of the manuscript.

Funding: This research was funded by the Australian Research Council (ARC) under grant number DP110102877.

Institutional Review Board Statement: Not applicable.

Informed Consent Statement: Not applicable.

Acknowledgments: U.A. would like to express appreciation for the APA scholarship provided by the University of Adelaide.

Conflicts of Interest: The authors declare no conflict of interest.

References

1. Hurley, I.P.; Coleman, R.C.; Ireland, H.E.; Williams, J.H. Measurement of bovine IgG by indirect competitive ELISA as a means of detecting milk adulteration. *J. Dairy Sci.* **2004**, *87*, 543–549. [CrossRef]
2. Hurley, I.P.; Coleman, R.C.; Ireland, H.E.; Williams, J.H. Use of sandwich IgG ELISA for the detection and quantification of adulteration of milk and soft cheese. *Int. Dairy J.* **2006**, *16*, 805–812. [CrossRef]
3. Berg, B.; Cortazar, B.; Tseng, D.; Ozkan, H.; Feng, S.; Wei, Q.; Chan, R.Y.-L.; Burbano, J.; Farooqui, Q.; Lewinski, M. Design, Synthesis, and Characterization of Small-Molecule Reagents that Cooperatively Provide Dual Readouts for Triaging and, When Necessary, Quantifying Point-of-Need Enzyme Assays. *ACS Nano* **2015**, *9*, 7857–7866. [CrossRef]
4. Algaar, F.; Eltzov, E.; Vdovenko, M.M.; Sakharov, I.Y.; Fajs, L.; Weidmann, M.; Mirazimi, A.; Marks, R.S. Fiber-optic immunosensor for detection of Crimean-Congo hemorrhagic fever IgG antibodies in patients. *Anal. Chem.* **2015**, *87*, 8394–8398. [CrossRef]
5. Li, H.; Zhao, M.; Liu, W.; Chu, W.; Guo, Y. Polydimethylsiloxane microfluidic chemiluminescence immunodevice with the signal amplification strategy for sensitive detection of human immunoglobin G. *Talanta* **2016**, *147*, 430–436. [CrossRef]
6. Shi, H.; Yuan, L.; Wu, Y.; Liu, S. Colorimetric immunosensing via protein functionalized gold nanoparticle probe combined with atom transfer radical polymerization. *Biosens. Bioelectron.* **2011**, *26*, 3788–3793. [CrossRef]
7. Li, H.; Wang, C.; Ma, Z.; Su, Z. Colorimetric detection of immunoglobulin G by use of functionalized gold nanoparticles on polyethylenimine film. *Anal. Bioanal. Chem.* **2006**, *384*, 1518–1524. [CrossRef]
8. Linder, V.; Verpoorte, E.; Thormann, W.; de Rooij, N.F.; Sigrist, H. Surface Biopassivation of Replicated Poly (dimethylsiloxane) Microfluidic Channels and Application to Heterogeneous Immunoreaction with On-Chip Fluorescence Detection. *Anal. Chem.* **2001**, *73*, 4181–4189. [CrossRef]
9. Wang, M.; Hou, W.; Mi, C.-C.; Wang, W.-X.; Xu, Z.-R.; Teng, H.-H.; Mao, C.-B.; Xu, S.-K. Immunoassay of Goat Antihuman Immunoglobulin G Antibody Based on Luminescence Resonance Energy Transfer between Near-Infrared Responsive NaYF4:Yb, Er Upconversion Fluorescent Nanoparticles and Gold Nanoparticles. *Anal. Chem.* **2009**, *81*, 8783–8789. [CrossRef]
10. Lyon, L.A.; Musick, M.D.; Smith, P.C.; Reiss, B.D.; Pena, D.J.; Natan, M.J. Surface plasmon resonance of colloidal Au-modified gold films. *Sens. Actuators B Chem.* **1999**, *54*, 118–124. [CrossRef]
11. Cao, C.; Sim, S.J. Signal enhancement of surface plasmon resonance immunoassay using enzyme precipitation-functionalized gold nanoparticles: A femto molar level measurement of anti-glutamic acid decarboxylase antibody. *Biosens. Bioelectron.* **2007**, *22*, 1874–1880. [CrossRef]
12. Song, C.; Wang, Z.; Zhang, R.; Yang, J.; Tan, X.; Cui, Y. Highly sensitive immunoassay based on Raman reporter-labeled immuno-Au aggregates and SERS-active immune substrate. *Biosens. Bioelectron.* **2009**, *25*, 826–831. [CrossRef]
13. Chon, H.; Lim, C.; Ha, S.-M.; Ahn, Y.; Lee, E.K.; Chang, S.-I.; Seong, G.H.; Choo, J. On-Chip Immunoassay Using Surface-Enhanced Raman Scattering of Hollow Gold Nanospheres. *Anal. Chem.* **2010**, *82*, 5290–5295. [CrossRef]
14. Kim, K.; Lee, Y.M.; Lee, H.B.; Shin, K.S. Silver-coated silica beads applicable as core materials of dual-tagging sensors operating via SERS and MEF. *ACS Appl. Mater. Interfaces* **2009**, *1*, 2174–2180. [CrossRef]
15. Shin, M.H.; Hong, W.; Sa, Y.; Chen, L.; Jung, Y.-J.; Wang, X.; Zhao, B.; Jung, Y.M. Multiple detection of proteins by SERS-based immunoassay with core shell magnetic gold nanoparticles. *Vib. Spectrosc.* **2014**, *72*, 44–49. [CrossRef]
16. Song, C.; Min, L.; Zhou, N.; Yang, Y.; Su, S.; Huang, W.; Wang, L. Synthesis of novel gold mesoflowers as SERS tags for immunoassay with improved sensitivity. *ACS Appl. Mater. Interfaces* **2014**, *6*, 21842–21850. [CrossRef]
17. Zhang, X.; Du, X. Carbon Nanodot-Decorated Ag@SiO2 Nanoparticles for Fluorescence and Surface-Enhanced Raman Scattering Immunoassays. *ACS Appl. Mater. Interfaces* **2015**, *8*, 1033–1040. [CrossRef]
18. Han, X.X.; Chen, L.; Ji, W.; Xie, Y.; Zhao, B.; Ozaki, Y. Label-Free Indirect Immunoassay Using an Avidin-Induced Surface-Enhanced Raman Scattering Substrate. *Small* **2011**, *7*, 316–320. [CrossRef]
19. Israelsen, N.D.; Wooley, D.; Hanson, C.; Vargis, E. Rational Design of Raman Labeled Nanoparticles for a Dual-Modality, Light Scattering Immunoassay on a Polystyrene Substrate. *J. Biol. Eng.* **2016**, *10*, 2. [CrossRef]
20. Kang, T.; Yoo, S.M.; Yoon, I.; Lee, S.Y.; Kim, B. Patterned Multiplex Pathogen DNA Detection by Au Particle-on-Wire SERS Sensor. *Nano Lett.* **2010**, *10*, 1189–1193. [CrossRef]
21. Zheng, J.; Hu, Y.; Bai, J.; Ma, C.; Li, J.; Li, Y.; Shi, M.; Tan, W.; Yang, R. Universal surface-enhanced Raman scattering amplification detector for ultrasensitive detection of multiple target analytes. *Anal. Chem.* **2014**, *86*, 2205–2212. [CrossRef]
22. Valdez, J.; Bawage, S.; Gomez, I.; Singh, S.R. Facile and rapid detection of respiratory syncytial virus using metallic nanoparticles. *J. Nanobiotechnol.* **2016**, *14*, 13. [CrossRef] [PubMed]
23. Wu, X.; Xia, Y.; Huang, Y.; Li, J.; Ruan, H.; Chen, T.; Luo, L.; Shen, Z.; Wu, A. Improved SERS-active nanoparticles with various shapes for CTC detection without enrichment process with supersensitivity and high specificity. *ACS Appl. Mater. Interfaces* **2016**, *8*, 19928–19938. [CrossRef]

24. Li, J.; Zhu, Z.; Zhu, B.; Ma, Y.; Lin, B.; Liu, R.; Song, Y.; Lin, H.; Tu, S.; Yang, C. Surface-Enhanced Raman Scattering Active Plasmonic Nanoparticles with Ultrasmall Interior Nanogap for Multiplex Quantitative Detection and Cancer Cell Imaging. *Anal. Chem.* **2016**, *88*, 7828–7836. [CrossRef] [PubMed]
25. Feng, J.; Xu, L.; Cui, G.; Wu, X.; Ma, W.; Kuang, H.; Xu, C. Building SERS-active heteroassemblies for ultrasensitive Bisphenol A detection. *Biosens. Bioelectron.* **2016**, *81*, 138–142. [CrossRef] [PubMed]
26. Gellner, M.; Kömpe, K.; Schlücker, S. Multiplexing with SERS labels using mixed SAMs of Raman reporter molecule. *Anal. Bioanal. Chem.* **2009**, *394*, 1839–1844. [CrossRef]
27. Hassanain, W.A.; Izake, E.L.; Schmidt, M.S.; Ayoko, G.A. Gold nanomaterials for the selective capturing and SERS diagnosis of toxins in aqueous and biological fluids. *Biosens. Bioelectron.* **2017**, *91*, 664–672. [CrossRef]
28. Doering, W.E.; Piotti, M.E.; Natan, M.J.; Freeman, R.G. SERS as a Foundation for Nanoscale, Optically Detected Biological Labels. *Adv. Mater.* **2007**, *19*, 3100–3108. [CrossRef]
29. Wu, L.; Wang, Z.; Fan, K.; Zong, S.; Cui, Y. A SERS-Assisted 3D Barcode Chip for High-Throughput Biosensing. *Small* **2015**, *11*, 2798–2806. [CrossRef]
30. Li, M.; Kang, J.W.; Sukumar, S.; Dasari, R.R.; Barman, I. Multiplexed detection of serological cancer markers with plasmon-enhanced Raman spectro-immunoassay. *Chem. Sci.* **2015**, *6*, 3906–3914. [CrossRef]
31. Kang, H.; Jeong, S.; Park, Y.; Yim, J.; Jun, B.H.; Kyeong, S.; Yang, J.K.; Kim, G.; Hong, S.; Lee, L.P. Near-Infrared SERS Nanoprobes with Plasmonic Au/Ag Hollow-Shell Assemblies for In Vivo Multiplex Detection. *Adv. Funct. Mater.* **2013**, *23*, 3719–3727. [CrossRef]
32. Lai, Y.; Sun, S.; He, T.; Schlücker, S.; Wang, Y. Raman-encoded microbeads for spectral multiplexing with SERS detection. *RSC Adv.* **2015**, *5*, 13762–13767. [CrossRef]
33. Zheng, P.; Li, M.; Jurevic, R.; Cushing, S.K.; Liu, Y.; Wu, N. A gold nanohole array based surface-enhanced Raman scattering biosensor for detection of silver (I) and mercury (II) in human saliva. *Nanoscale* **2015**, *7*, 11005–11012. [CrossRef] [PubMed]
34. Dinish, U.; Balasundaram, G.; Chang, Y.-T.; Olivo, M. Actively targeted in vivo multiplex detection of intrinsic cancer biomarkers using biocompatible SERS nanotags. *Sci. Rep.* **2014**, *4*, 4075. [CrossRef] [PubMed]
35. Li, M.; Yang, H.; Li, S.; Zhao, K.; Li, J.; Jiang, D.; Sun, L.; Deng, A. Ultrasensitive and Quantitative Detection of a New β-Agonist Phenylethanolamine A by a Novel Immunochromatographic Assay Based on Surface-Enhanced Raman Scattering (SERS). *J. Agric. Food Chem.* **2014**, *62*, 10896–10902. [CrossRef] [PubMed]
36. Xu, L.; Yan, W.; Ma, W.; Kuang, H.; Wu, X.; Liu, L.; Zhao, Y.; Wang, L.; Xu, C. SERS encoded silver pyramids for attomolar detection of multiplexed disease biomarkers. *Adv. Mater.* **2015**, *27*, 1706–1711. [CrossRef]
37. Deng, D.; Yang, H.; Liu, C.; Zhao, K.; Li, J.; Deng, A. Ultrasensitive detection of diclofenac in water samples by a novel surface-enhanced Raman scattering (SERS)-based immunochromatographic assay using AgMBA@SiO$_2$-Ab as immunoprobe. *Sens. Actuators B Chem.* **2019**, *283*, 563–570. [CrossRef]
38. Raez, J.; Blais, D.R.; Zhang, Y.; Alvarez-Puebla, R.A.; Bravo-Vasquez, J.P.; Pezacki, J.P.; Fenniri, H. Spectroscopically encoded microspheres for antigen biosensing. *Langmuir* **2007**, *23*, 6482–6485. [CrossRef]
39. Wei, J.; Jin, B.; Dai, S. Polymer microbead-based surface enhanced Raman scattering immunoassays. *J. Phys. Chem. C* **2012**, *116*, 17174–17181. [CrossRef]
40. Azhar, U.; Ahmed, Q.; Alwahabi, Z.; Dai, S. Synthesis and spectroscopic study of dual self-encoded polymer microbeads with Raman scattering and surface-enhanced Raman scattering. *J. Raman Spectrosc.* **2020**, *51*, 910–918. [CrossRef]
41. Song, J.-S.; Chagal, L.; Winnik, M.A. Monodisperse micrometer-size carboxyl-functionalized polystyrene particles obtained by two-stage dispersion polymerization. *Macromolecules* **2006**, *39*, 5729–5737. [CrossRef]
42. Turkevich, J.; Stevenson, P.C.; Hillier, J. A study of the nucleation and growth processes in the synthesis of colloidal gold. *Discuss. Faraday Soc.* **1951**, *11*, 55–75. [CrossRef]
43. Hayat, M.A. *Colloidal Gold, Principles, Methods and Applications*; Academic Press, Inc.: San Diego, CA, USA, 1989.
44. Li, J.; Qin, J.; Zhang, X.; Wang, R.; Liang, Z.; He, Q.; Wang, Z.; Wang, K.; Wang, S. Label-free Raman imaging of live osteosarcoma cells with multivariate analysis. *Appl. Microbiol. Biotechnol.* **2019**, *103*, 6759–6769. [CrossRef] [PubMed]
45. Jun, B.-H.; Kim, J.-H.; Park, H.; Kim, J.-S.; Yu, K.-N.; Lee, S.-M.; Choi, H.; Kwak, S.-Y.; Kim, Y.-K.; Jeong, D.H. Surface-enhanced Raman spectroscopic-encoded beads for multiplex immunoassay. *J. Comb. Chem.* **2007**, *9*, 237–244. [CrossRef]
46. McCabe, A.F.; Eliasson, C.; Prasath, R.A.; Hernandez-Santana, A.; Stevenson, L.; Apple, I.; Cormack, P.A.; Graham, D.; Smith, W.E.; Corish, P. SERRS labelled beads for multiplex detection. *Faraday Discuss.* **2006**, *132*, 303–308. [CrossRef]
47. Abdelrahman, A.I.; Ornatsky, O.; Bandura, D.; Baranov, V.; Kinach, R.; Dai, S.; Thickett, S.C.; Tanner, S.; Winnik, M.A. Metal-containing polystyrene beads as standards for mass cytometry. *J. Anal. At. Spectrom.* **2010**, *25*, 260–268. [CrossRef]
48. Liang, C.; Krimm, S. Infrared spectra of high polymers. VI. Polystyrene. *J. Polym. Sci.* **1958**, *27*, 241–254. [CrossRef]
49. Weisbecker, C.S.; Merritt, M.V.; Whitesides, G.M. Molecular self-assembly of aliphatic thiols on gold colloids. *Langmuir* **1996**, *12*, 3763–3772. [CrossRef]
50. Hung, Y.-L.; Hsiung, T.-M.; Chen, Y.-Y.; Huang, Y.-F.; Huang, C.-C. Colorimetric detection of heavy metal ions using label-free gold nanoparticles and alkanethiols. *J. Phys. Chem. C* **2010**, *114*, 16329–16334. [CrossRef]
51. Neng, J.; Harpster, M.H.; Wilson, W.C.; Johnson, P.A. Surface-enhanced Raman scattering (SERS) detection of multiple viral antigens using magnetic capture of SERS-active nanoparticles. *Biosens. Bioelectron.* **2013**, *41*, 316–321. [CrossRef]

52. Park, H.; Lee, S.B.; Kim, K.; Kim, M.S. Surface-enhanced Raman scattering of p-aminobenzoic acid at silver electrode. *J. Phys. Chem.* **1990**, *94*, 7576–7580. [CrossRef]
53. Michota, A.; Bukowska, J. Surface-enhanced Raman scattering (SERS) of 4-mercaptobenzoic acid on silver and gold substrates. *J. Raman Spectrosc.* **2003**, *34*, 21–25. [CrossRef]
54. Lin-Vien, D.; Colthup, N.B.; Fateley, W.G.; Grasselli, J.G. *The Handbook of Infrared and Raman Characteristic Frequencies of Organic Molecules*; Academic Press, Inc.: London, UK, 1991.
55. Castro, J.; López-Ramírez, M.; Arenas, J.; Otero, J. Surface-enhanced Raman scattering of 3-mercaptopropionic acid adsorbed on a colloidal silver surface. *J. Raman Spectrosc.* **2004**, *35*, 997–1000. [CrossRef]
56. Camargo, P.H.; Cobley, C.M.; Rycenga, M.; Xia, Y. Measuring the surface-enhanced Raman scattering enhancement factors of hot spots formed between an individual Ag nanowire and a single Ag nanocube. *Nanotechnology* **2009**, *20*, 434020. [CrossRef]
57. Le Ru, E.; Blackie, E.; Meyer, M.; Etchegoin, P.G. Surface Enhanced Raman Scattering Enhancement Factors: A Comprehensive Study. *J. Phys. Chem. C* **2007**, *111*, 13794–13803. [CrossRef]
58. Zhao, X.; Zhao, Y.; Gu, Z. Surface enhanced Raman scattering enhancement factors: A comprehensive study. *Sci. China Chem.* **2011**, *54*, 1185. [CrossRef] [PubMed]

Communication

Effective CpG Delivery Using Zwitterion-Functionalized Dendrimer-Entrapped Gold Nanoparticles to Promote T Cell-Mediated Immunotherapy of Cancer Cells

Huan Chen, Yiming Zhang, Lulu Li, Rui Guo, Xiangyang Shi * and Xueyan Cao *

State Key Laboratory for Modification of Chemical Fibers and Polymer Materials, Shanghai Engineering Research Center of Nano-Biomaterials and Regenerative Medicine, College of Chemistry, Chemical Engineering and Biotechnology, Donghua University, Shanghai 201620, China; chenhuan800@gmail.com (H.C.); 2200737@mail.dhu.edu.cn (Y.Z.); 2200744@mail.dhu.edu.cn (L.L.); ruiguo@dhu.edu.cn (R.G.)
* Correspondence: xshi@dhu.edu.cn (X.S.); caoxy_116@dhu.edu.cn (X.C.)

Abstract: Recently, cell-based immunotherapy has become one of the most promising ways to completely eliminate cancer. The major challenge is to effectively promote a proper immune response to kill the cancer cells by activated T cells. This study investigated the effect of T cell-mediated immunotherapy trigged by Au DENPs-MPC (zwitterion 2-methacryloyloxyethyl phosphorylcholine (MPC)-functionalized dendrimer-entrapped gold nanoparticles) loading oli-godeoxynucleotides (ODN) of unmethylated cytosine guanine dinucleotide (CPG). Here, we first synthesized Au DENPs-MPC, evaluated their capability to compress and transfect CpG-ODN to bone marrow dendritic cells (BMDCs), and investigated the potential to use T cells stimulated by matured BMDCs to inhibit the growth of tumor cells. The developed Au DENPs-MPC could apparently reduce the toxicity of Au DENPs, and enhanced transfer CpG-ODN to the BMDCs for the maturation as demonstrated by the 44.41–48.53% increase in different surface maturation markers. The transwell experiments certificated that ex vivo activated T cells display excellent anti-tumor ability, which could effectively inhibit the growth of tumor cells. These results suggest that Au DENPs-MPC can deliver CpG-ODN efficiently to enhance the antigen presentation ability of BMDCs to activate T cells, indicating that T cells-based immunotherapy mediated by Au DENPs-MPC loaded with CpG-ODN may become the most promising treatment of cancer.

Keywords: dendrimers; gold nanoparticles; CpG-ODN; immunotherapy; T cells

1. Introduction

In the last few decades, immunotherapy has become an important and promising treatment for cancer [1,2], which can stimulate or boost our immune system to defend cancer cells in a much more robust and smarter manner [3]. The immune system detects abnormal cells and prevents the growth of many cancers with the help of tumor-infiltrating lymphocytes (TILs). However, cancer cells have ways to avoid destruction by the immune system. Immunotherapy can strengthen multiple antitumor capabilities to kill cancer cells in-site and achieve anti-metastasis and anti-relapse effects by strengthening the response of the immune system [4]. Recently, cell-based immunotherapy has been believed to be one of the most effective clinical therapy modalities for tumor therapy [5,6]. The patient's immune system can eliminate tumor cells mostly through cytotoxic T lymphocyte (CTL) mediation [7,8]. While some of the stimulators are unable to reach high levels of immune response owing to the phenomenon of immune escape [9], it is necessary to find a much more efficient stimulator for immune cells.

Dendritic cells (DCs) are the unique and mighty antigen presenting cells (APCs), where only APCs could stimulate primary T lymphocytes and trigger the immunologic responses of cytotoxic T lymphocytes [10]. The immature DCs (iDCs) will become mature

DCs (mDCs) in vitro with the procedure of stimulation by some stimulators, and then will stimulate T cells, especially CD8+ killer T cells and CD4+ helper T cells, to boost different antitumor immune reactions [11,12]. The immune stimulating DNA containing unmethylated cytosine-guanine (CpG) motifs have been successfully used as adjuvants to increase immune responses [13,14]. By MyD88-dependent nuclear factor-KB (NF-κB) and mitogen-activated protein kinase (MAPK) signaling pathways [15], the Toll-like receptor 9 (TLR9) on the surfaces of DCs can directly activate iDCs to mDCs when it detects CpG-ODN.

However, owing to various systemic and intracellular obstacles in gene therapy, containing fast degradation, poor cellular uptake, and inefficient endosomal escape, some special transport ways may be needed to effectively deliver genes to the cell coleus or cytosol [16,17]. Therefore, it is a challenge to find an outstanding gene carrier that could efficiently transport the exogenous nucleic acids into cells. Nonviral delivery systems have attracted extensive attention thanks to their properties of easy preparation and modification, high safety and genetic loading capacity, and perfect delivery efficiency [18,19]. Fifth-generation (G5) polyamidoamine (PAMAM) dendrimers are a kind of highly branched and symmetrical polymer with a precise molecular structure and abundant surface functional groups, which make it useful for gene delivery [20,21]. Modifications for dendrimers are necessary to overcome a series of shortcomings (e.g., high cytotoxicity and low gene transfection efficiency) [22,23]. Metallic and metal oxide NPs have been widely used in cell-based immunotherapy and other biomedical fields in recently years [24–26]. In our previous study, dendrimer-entrapped Au NPs (Au DENPs) showed more high gene delivery efficiency and lower cytotoxicity than it alone [27,28]. This is because of the truth that entrapped Au NPs can help to maintain the 3-dimensional spherical morphology of dendrimers, allowing more efficient compaction of DNA [29,30]. Meanwhile, the cytotoxicity of the gene vector was also reduced after dendrimer entrapment of Au NPs. After surface modification, it can further decrease the cytotoxicity and strengthen the efficiency of gene delivery upon Au DENPs [31,32].

To reduce the nonspecific adsorption and improve the transport capacity to immune cells, the antifouling effect of the delivery vectors is also worthy of attention. Zwitterionic materials have attracted more attention in the field of biomaterials than other materials owing to a range of advantages (e.g., ultralow nonspecific protein adsorption, bacterial adhesion, and biofilm formation) in recent years [33]. Zwitterions can form stronger hydration shells through ion–dipole interaction with denser and tighter adsorbed water [34], making them preferable substitutes for other materials, such as PEG [35]. To enhance the efficiency of gene delivery, it is needed to remove the existence of serum protein in the medium. This is because of the strong interaction between serum protein and positively charged vector/gene complexes, leading to low efficiency of cellular uptake and gene delivery [36]. However, it is very conflicting that the cells will become weak owing to lack of adequate nutritional in the serum-free medium, which also affect the gene delivery efficiency. Hence, it would be significant to discover a new protein-resistant carrier program to retain the formed positively charged vector/gene polyplexes to be intact in a serum culture environment for highly efficient gene delivery. Liu et al. and Xong et al. modified G5.NH$_2$ with the zwitterion: MPC and carboxy betaine acrylamide (CBAA) to bear the G5.NH$_2$ with outstanding compatibility with protein and cells [37,38].

In our study, we explored a novel T cell-based tumor immunotherapy induced by CpG-loaded zwitterion-functionalized Au DENPs. G5.NH$_2$ was used as a template to synthesize Au DENPs-MPC (Au DENPs-MPC) (Scheme 1). The cytocompatibility of Au DENPs-MPC on BMDCs was assessed by the MTT kit. Au DENPs-MPC were completely characterized via different techniques. Then, the surface antigens on BMDCs were detected by flow cytometry to certificate that they were successfully stimulated to maturation. The activation of T cells was detected by allogeneic mixed lymphocyte reaction (MLR) setting. The in vitro anti-tumor effect was detected by the transwell system and it indicated that T cells activated in vitro have a satisfactory anti-tumor effect. Therefore, Au DENPs-MPC can be served as an efficient gene delivery vector and DCs stimulator, and potentially enhance T cell-mediated tumor immunotherapy after loading with CpG-ODN. It also provides a good reference for the design of biosensors, and broadens the application field of biosensor.

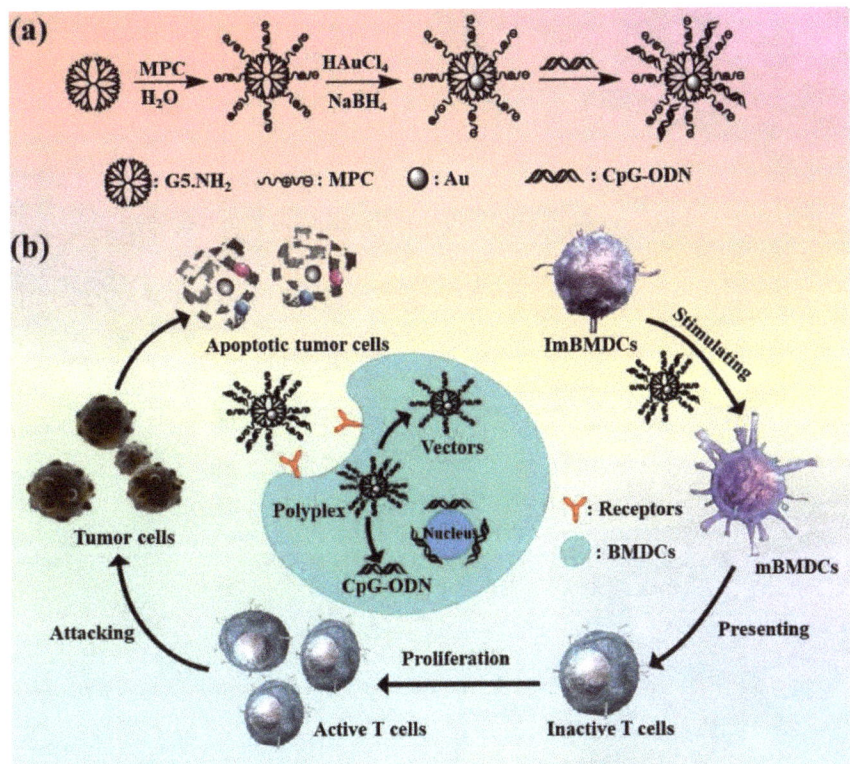

Scheme 1. Synthesis of Au DENPs-MPC (a) and the anti-tumor immunotherapy route (b).

2. Results and Discussion

2.1. Synthesis and Characterization of Au DENPs-MPC

Based on the ^{1}H-NMR (Figure 1a), there are about 21.2 MPCs attached to each G5.NH$_2$ by integrating the areas of G5.NH$_2$ proton peaks (range from 2.2 to 3.4 ppm) and MPC proton peaks (about −1.0 ppm). Figure 1b indicates the existence of Au NPs inside G5-NH$_2$ due to the absorption peak at 520 nm. The Au NPs could also be visualized by transmission electron microscopy (TEM) (Figure 1c). It was shown that the Au NPs were spherical, and the average diameter was about 1.7 nm. In addition, the calculated Mws and mean numbers of primary amines of the G5.NH$_2$ and Au DENPs-MPC vectors are shown in Table S1, respectively. The gel retardation results demonstrated that the Au DENPs-MPC showed the best ability to compress CpG-ODN, and the mobility of CpG-ODN could be retarded at N/P ratios of 1 or greater (Figure 1d).

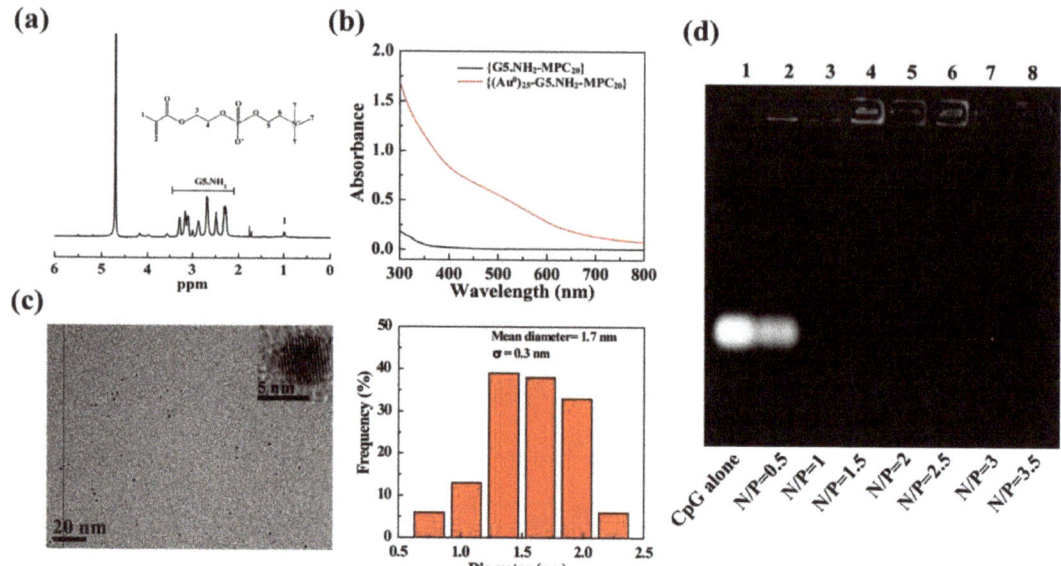

Figure 1. (a) ^{1}H-NMR spectrum of Au DENPs-MPC. (b) UV/Vis absorption spectra of {G5.NH$_2$-MPC$_{20}$} and Au DENPs-MPC. (c) TEM image and size distribution histogram of Au DENPs-MPC. (d) Agarose gel electrophoresis of Au DENPs-MPC/CpG-ODN complexes at different N/P ratios (range from 0.5~3.5).

Based on the above results, Au DENPs-MPC/CpG complexes with five different nitrogen phosphorus (N/P) ratios (1, 2, 4, 6, and 8) were selected to evaluate the hydrodynamic size and zeta potential. The average hydrodynamic sizes of the Au DENPs-MPC/CpG-ODN complexes were about 200 nm, and their average zeta potentials were about 15 mV, as shown in Table S2. Considering the optimal endocytosis efficiency and gene delivery efficiency [39,40], we chose the complexes with an N/P ratio of 2 for the subsequent studies.

2.2. In Vitro Cytotoxicity and Cellular Uptake Assays

Next, the cytocompatibility of the tested nanomaterials was evaluated, which was significant for the following experiments. According to the MTT assay results, Au DENPs-MPC showed lower cytotoxicity to BMDCs after being modified with MPC, indicating the outstanding cytocompatibility (Figure 2a). Meanwhile, the Au DENPs-MPC loaded with CpG-ODN could further decrease the cytotoxicity compared with Au DENPs-MPC alone, which is due to the decreased positive potential of the complexes after complexation of negatively charged CpG ODN [41], thereby improving the cytocompatibility of Au DENPs-MPC/CpG-ODN. It is noted that, when the concentration is 10 μg/mL, the cell viability is higher than 100%, which should be due to the lower cytotoxicity of the complexes under the lower concentrations. Even when the concentration of Au DENPs-MPC/CpG-ODN complexes was up to 50 μg/mL, the cell viability of BMDCs could still reach higher than 90%, indicating that the Au DENPs-MPC/CpG-ODN complexes were safe for subsequent experiments.

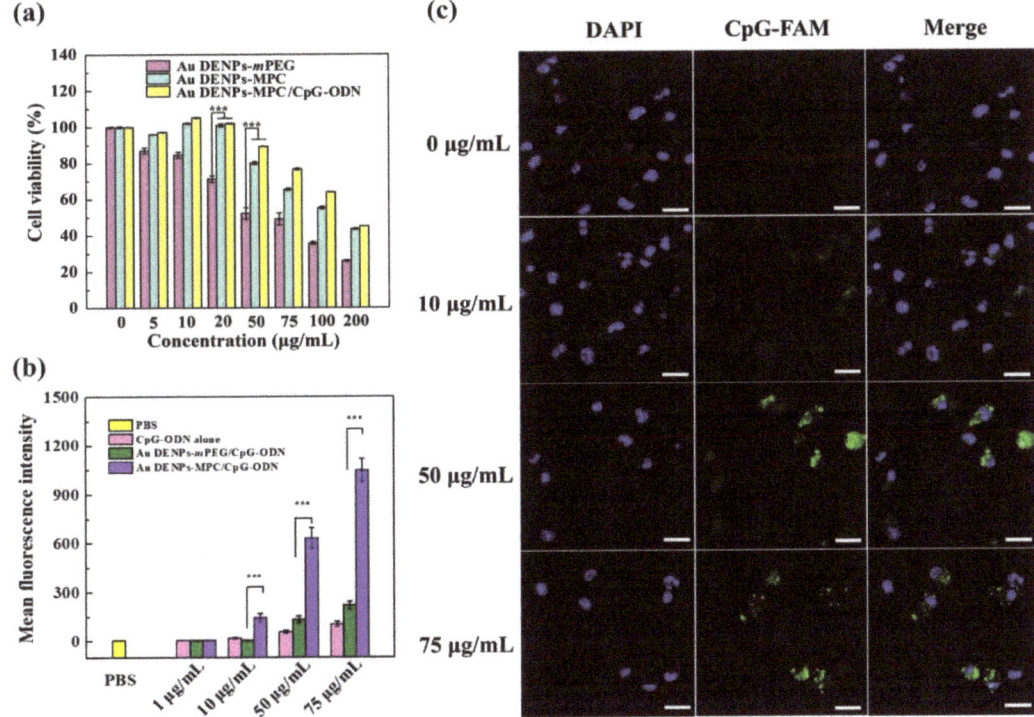

Figure 2. (a) MTT assay results of the BMDCs cultured with Au DENPs-mPEG, Au DENPs-MPC alone, and Au DENPs-MPC/CpG-ODN complexes at different concentrations for 24 h. (b) Flow cytometry results of BMDCs cultured with CpG-ODN alone, Au DENPs-mPEG/CpG-ODN, and Au DENPs-MPC/CpG-ODN complexes at different concentrations for 4 h. (c) Confocal fluorescence images results of BMDCs cultured with Au DENPs-MPC/CpG-ODN complexes at concentration of 50 μg/mL. The scale bar is 20 μm. *** $p < 0.001$.

As shown in Figure 2b, the mean fluorescence intensity of CpG-ODN labeled with FAM (carboxyfluorescein) was increased obviously with the concentration of Au DENPs-MPC/CpG-ODN complexes and Au DENPs-*m*PEG/CpG-ODN complexes, respectively, while the intensity did not change obviously in the group of CpG-ODN alone (Figure 2b and Figure S1). This is because of the negative charge and easy degradation of CpG-ODN, which are not beneficial for uptake by cells. What is more, the decoration of zwitterion (MPC) onto the Au DENPs' surface is able to efficiently degrade the adsorption of protein to Au DENPs-MPC/CpG-ODN complexes, thereby greatly reducing the influences of the presence of FBS [37]. Thus, Au DENPs-MPC/CpG-ODN showed about 4.7 times higher cellular uptake ability of BMDCs than Au DENPs-*m*PEG/CpG-ODN. In addition, the flow cytometry results were consistent with the results of confocal fluorescence detection (Figure 2c).

2.3. Maturation of BMDCs

Maturation of BMDCs is the first step in immune response for the anti-tumor effect. Only mature BMDCs can present antigens and effectively activate T cells. In Figure S2, the expression of CD11c on the surfaces of BMDCs reached more than 75%, suggesting the high purity of BMDCs. The results from Figure S3 presented the extracted cell morphology, and were similar to those in the previous reports [42]. Compared with the group without any stimulation, the expression of BMDCs' maturation markers, CD80, CD86 and MHC-II, were remarkably raised to 51.23%, 45.44%, and 44.41%, respectively, indicating the BMDCs were matured after incubated with 50 μg/mL Au DENPs-MPC/CpG-ODN (Figure 3). To further confirm the stimulation effect of the Au DENPs-MPC/CpG-ODN, liposome was also used as a positive control. In the liposome group, the expressions of CD86 and MHC-II were just 29.40% and 20.08%, respectively. This means that more CpG-ODN are transferred into cells and then stimulate the maturation of DCs by Au DENPs-MPC rather than liposome. It is notable that, although Au DENPs-MPC/CpG-ODN at 75 μg/mL showed much better cellular uptake efficiency, the expression levels of related cytokines are lower than those of Au DENPs-MPC/CpG-ODN at 50 μg/mL. This may be due to the increasing cytotoxicity of Au DENPs-MPC/CpG-ODN at high concentrations, which limits the function of CpG-ODN. According to the above results, the concentration of Au DENPs-MPC at 50 μg/mL was selected for the following tests.

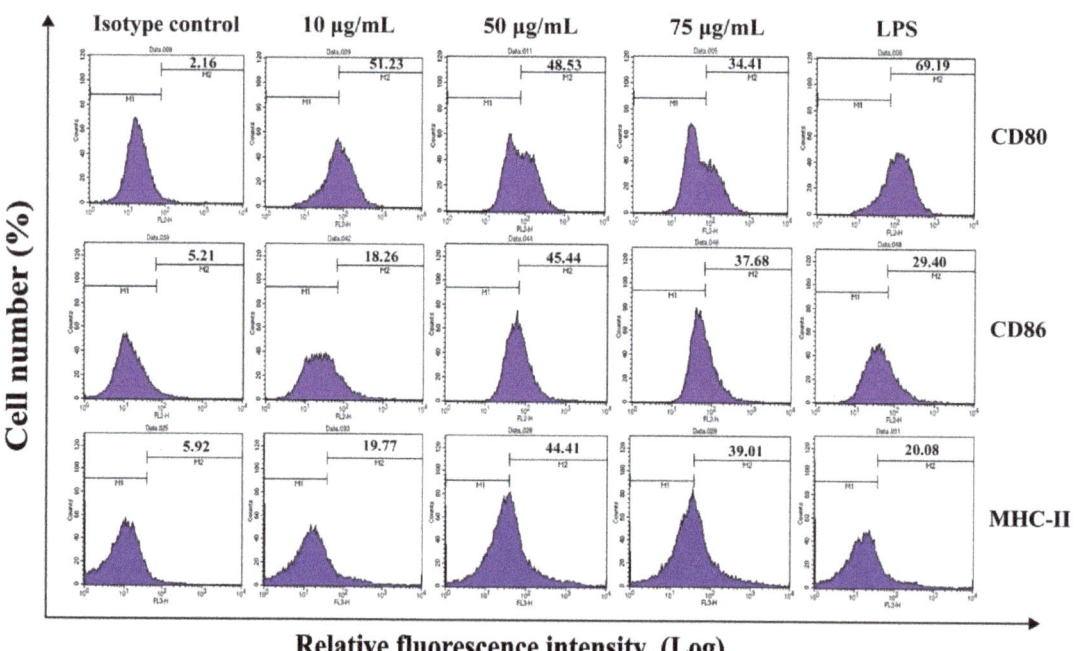

Figure 3. The expression detection results of different maturation markers were detected by flow cytometry and the isotype controls in panels are related to the determination background.

2.4. T Cells' Activation

Next, the activation of T cells in vitro through mature BMDCs (mBMDCs) was checked. Firstly, we checked the purity of the extracted T cells according to our previous reports [43]. The flow cytometry detection results (Figure S4) showed that the expression of CD3 reached above 90%, suggesting high purity of the extracted T cells [43]. Then, the proliferation of T cells was detected by the allogeneic MLR setting [44]. It is clear that BMDCs matured by the Au DENPs-MPC/CpG-ODN complexes could activate the proliferation of allogeneic T cells at a DC/T ratio higher than 1:200, and the highest stimulus index was 2.8 times higher than that of the control group (T cells without stimulation) (Figure 4).

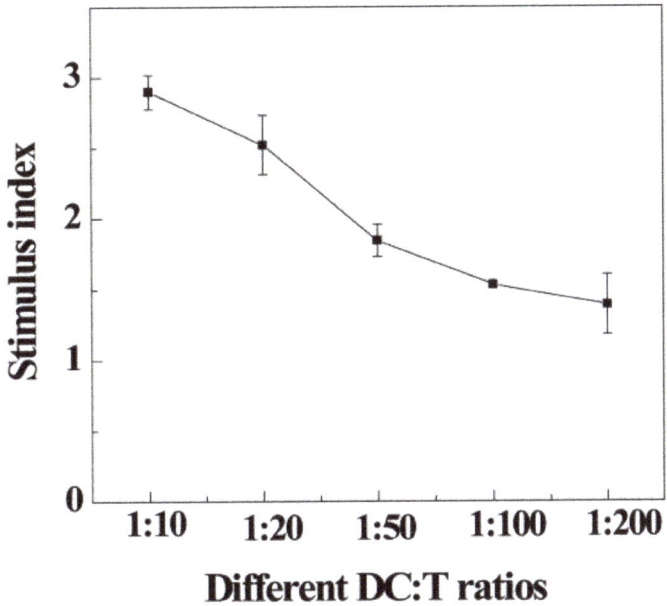

Figure 4. The proliferation result of T cells after co-culture with different ratio of mature BMDCs for 3 days by MTT assay.

Meanwhile, the flow cytometry results also showed that, after co-culturing T cells with mBMDCs, the expressions of CD4 and CD8 on the surface of T cells increased (Figure S5). Thus, Au DENPs-MPC/CpG-ODN complexes-treated BMDCs exhibited a remarkably enhanced capacity for presenting antigen to T cells for the stimulation of cell proliferation, showing that Au DENPs-MPC/CpG-ODN complexes were able to induce functional maturation of BMDCs.

2.5. In Vitro Anti-Tumor Effect of T Cells

Considering the remarkable activation of T cells stimulated by Au DENPs-MPC/CpG-treated BMDCs, the in vitro antitumor effect of T cells was detected by the transwell system (Figure 5a). Based on our results, the activated T cells showed the most powerful and good inhibition effect of 4T1 cells, compared with inactivated T cells and 4T1 alone. When it was co-cultured with activated T cells for 24 h, the cell viability of 4T1 cells dropped to 37% (Figure 5b). It is believed that activated T cells in the upper chamber could secrete some abundant related cytokines, such as interferon-γ (IFN-γ) and tumor necrosis factor-α (TNF-α), to inhibit the growth of 4T1 cells in the lower chamber [45]. Meanwhile, the inactivated T cells group also showed a low inhibition effect on the growth of 4T1 cells, which may because the inactivated T cells also secrete a small amount of related cytokines to kill the 4T1 cells [45]. This showed that T cells activated by mBMDCs could be able to inhibit the growth of tumor cells by triggering the immune response.

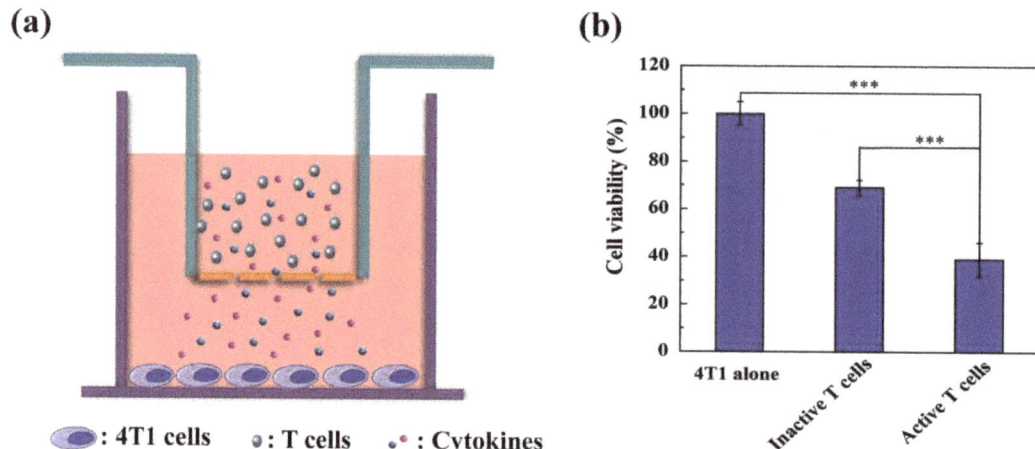

Figure 5. (a) Schematic illustration of T cells (upper) and tumor cell (lower) co-cultured using the transwell plates. (b) Cell viability of 4T1 after co-culture with different T cells. *** $p < 0.001$.

3. Conclusions

In summary, we synthesized a kind of non-viral carrier Au DENPs-MPC with improved cytocompatibility and delivery efficiency of CpG-ODN for enhancing the antigen presentation ability of BMDCs to activate T cells. The results showed that Au DENPs-MPC could reduce the toxicity of Au DENPs and increase the efficiency of delivery of CpG-ODN to BMDCs. The expression of specific surface proteins detected by flow cytometry demonstrated the maturation of BMDCs and the activation of T cells. The transwell experiments certificated that ex vivo activated T cells display excellent anti-tumor ability, which could effectively inhibit the growth of tumor cells. These results show that the T cells-based immunotherapy mediated by Au DENPs-MPC loaded with CpG-ODN may become the most promising treatment of cancer.

Supplementary Materials: The following supporting information can be downloaded at https://www.mdpi.com/article/10.3390/bios12020071/s1, Table S1: Physicochemical parameters of the $G_5.NH_2$, {$(Au^0)_{25}$-$G_5.NH_2$-mPEG$_{20}$}, and {$(Au^0)_{25}$-$G_5.NH_2$-MPC$_{20}$} vectors. Table S2: Zeta potentials and hydrodynamic diameters of Au DENPs-MPC alone and the Au DENPs-MPC/CpG complexes at five different N/P ratios. Figure S1: Cellular uptake results of BMDCs detected by flow cytometry. Figure S2: The expression of CD11c on the surfaces of BMDCs related the purity of BMDCs was detected by flow cytometry. Figure S3: The inverted microscope images of mouse bone marrow-derived dendritic cells after induction for 1, 3, 5, and 7 days. Figure S4: The expression of CD3 on the surfaces of T cells related to the purity of T cells was detected by flow cytometry. Figure S5: The flow cytometry results of the expression of CD4 and CD8 on the surfaces of T cells. (a,b) T cells cultured for 3 days without stimulation. (c,d) T cells cultured with mBMDCs for 3 days. CD4 and CD8 were labeled with FITC (F1 channel).

Author Contributions: Conceptualization, X.S. and X.C.; methodology, H.C. and R.G.; software, R.G., Y.Z. and L.L.; validation, H.C.; formal analysis, H.C.; investigation, Y.Z. and L.L.; resources, X.S. and X.C.; writing—original draft preparation, H.C.; writing—review and editing, X.S. and X.C.; project administration, X.C.; funding acquisition, X.S. All authors have read and agreed to the published version of the manuscript.

Funding: This research was financially supported by the Science and Technology Commission of Shanghai Municipality (19XD1400100, 21490711500, and 20DZ2254900), the National Natural Science Foundation of China (81761148028 and 21875031), and Shanghai Education Commission through the Shanghai Leading Talents Program.

Institutional Review Board Statement: Not applicable.

Informed Consent Statement: Not applicable.

Data Availability Statement: Data are contained within the article.

Conflicts of Interest: The authors declare no conflict of interest.

References

1. Xu, C.N.; Tian, H.Y.; Chen, X.S. Recent progress in cationic polymeric gene carriers for cancer therapy. *Sci. China Chem.* **2017**, *60*, 19–28. [CrossRef]
2. Chen, W.; Zheng, R.; Baade, P.D.; Zhang, S.; Zeng, H.; Bray, F.; Jemal, A.; Yu, X.Q.; He, J. Cancer Statistics in China, 2015. *CA A Cancer J. Clin.* **2016**, *66*, 115–132. [CrossRef] [PubMed]
3. Apetoh, L.; Locher, C.; Ghiringhelli, F.; Kroemer, G.; Zitvogel, L. Harnessing Dendritic Cells in Cancer. *Semin. Immunol.* **2011**, *23*, 42–49. [CrossRef]
4. Binnewies, M.; Roberts, E.W.; Kersten, K.; Chan, V.; Fearon, D.F.; Merad, M.; Coussens, L.M.; Gabrilovich, D.I.; Ostrand-Rosenberg, S.; Hedrick, C.C.; et al. Understanding the Tumor Immune Microenvironment (TIME) for Effective Therapy. *Nat. Med.* **2018**, *24*, 541–550. [CrossRef] [PubMed]
5. Bol, K.F.; Schreibelt, G.; Gerritsen, W.R.; de Vries, I.J.M.; Figdor, C.G. Dendritic Cell-Based Immunotherapy: State of the Art and Beyond. *Clin. Cancer Res.* **2016**, *22*, 1897–1906. [CrossRef] [PubMed]
6. Fridman, W.H.; Zitvogel, L.; Sautes-Fridman, C.; Kroemer, G. The Immune Contexture in Cancer Prognosis and Treatment. *Nat. Rev. Clin. Oncol.* **2017**, *14*, 717–734. [CrossRef] [PubMed]
7. June, C.H.; Sadelain, M. Chimeric Antigen Receptor Therapy. *N. Engl. J. Med.* **2018**, *379*, 64–73. [CrossRef]
8. June, C.H.; O'Connor, R.S.; Kawalekar, O.U.; Ghassemi, S.; Milone, M.C. CAR T Cell Immunotherapy for Human Cancer. *Science* **2018**, *359*, 1361–1365. [CrossRef]
9. Wilson, D.R.; Sen, R.; Sunshine, J.C.; Pardoll, D.M.; Green, J.J.; Kim, Y.J. Biodegradable STING Agonist Nanoparticles for Enhanced Cancer Immunotherapy. *Nanomedicine* **2018**, *14*, 237–246. [CrossRef]
10. Banchereau, J.; Steinman, R.M. Dendritic Cells and the Control of Immunity. *Nature* **1998**, *392*, 245–252. [CrossRef]
11. Wang, M.J.; Yin, B.N.; Wang, H.Y.; Wang, R.F. Current Advances in T-cell-Based Cancer Immunotherapy. *Immunotherapy* **2014**, *6*, 1265–1278. [CrossRef] [PubMed]
12. Li, K.; Zhang, Q.; Zhang, Y.; Yang, J.; Zheng, J.N.A. T-Cell-Associated Cellular Immunotherapy for Lung Cancer. *J. Cancer Res. Clin. Oncol.* **2015**, *141*, 1249–1258. [CrossRef] [PubMed]
13. Rosenberg, S.A.; Yang, J.C.; Restifo, N.P. Cancer Immunotherapy: Moving Beyond Current Vaccines. *Nat. Med.* **2004**, *10*, 909–915. [CrossRef]
14. Baines, J.; Celis, E. Immune-Mediated Tumor Regression Induced by CpG-Containing Oligodeoxynucleotides. *Clin. Cancer Res.* **2003**, *9*, 2693–2700. [PubMed]
15. Hanagata, N. Structure-Dependent Immunostimulatory Effect of CpG Oligodeoxynucleotides and Their Delivery System. *Int. J. Nanomed.* **2012**, *7*, 2181–2195. [CrossRef] [PubMed]
16. Markowicz, S.; Niedzielska, J.; Kruszewski, M.; Ołdak, T.; Gajkowska, A.; Machaj, E.K.; Skurzak, H.; Pojda, Z. Nonviral Transfection of Human Umbilical Cord Blood Dendritic Cells is Feasible, but the Yield of Dendritic Cells with Transgene Expression Limits the Application of this Method in Cancer Immunotherapy. *Acta Biochim. Pol.* **2006**, *53*, 203–211. [CrossRef]
17. Li, A.; Zhou, B.; Alves, C.S.; Xu, B.; Guo, R.; Shi, X.; Cao, X. Mechanistic Studies of Enhanced PCR Using PEGylated PEI-Entrapped Gold Nanoparticles. *ACS Appl. Mater. Interfaces* **2016**, *8*, 25808–25817. [CrossRef]
18. Li, S.D.; Huang, L. Non-Viral is Superior to Viral Gene Delivery. *J. Control. Release* **2007**, *123*, 181–183. [CrossRef]
19. Xiao, T.Y.; Hou, W.X.; Cao, X.Y.; Wen, S.H.; Shen, M.W.; Shi, X.Y. Dendrimer-Entrapped Gold Nanoparticles Modified with Folic Acid for Targeted Gene Delivery Applications. *Biomater. Sci.* **2013**, *1*, 1172–1180. [CrossRef]
20. Braun, C.S.; Vetro, J.A.; Tomalia, D.A.; Koe, G.S.; Koe, J.G.; Middaugh, C.R. Structure/Function Relationships of Polyamidoamine/DNA Dendrimers as Gene Delivery Vehicles. *J. Pharm. Sci.* **2005**, *94*, 423–436. [CrossRef]
21. Almasian, A.; Olya, M.E.; Mahmoodi, N.M. Synthesis of polyacrylonitrile/polyamidoamine composite nanofibers using electrospinning technique and their dye removal capacity. *J. Taiwan Inst. Chem. Eng.* **2015**, *49*, 119–128. [CrossRef]
22. Qi, R.; Gao, Y.; Tang, Y.; He, R.R.; Liu, T.L.; He, Y.; Sun, S.; Li, B.Y.; Li, Y.B.; Liu, G. PEG-Conjugated PAMAM Dendrimers Mediate Efficient Intramuscular Gene Expression. *AAPS J.* **2009**, *11*, 395–405. [CrossRef] [PubMed]
23. Majoros, I.J.; Keszler, B.; Woehler, S.; Bull, T.; Baker, J.R. Acetylation of Poly(amidoamine) Dendrimers. *Macromolecules* **2003**, *36*, 5526–5529. [CrossRef]
24. Elahi, N.; Kamali, M.; Baghersad, M.H. Recent biomedical applications of gold nanoparticles: A review. *Talanta* **2018**, *184*, 537–556. [CrossRef] [PubMed]
25. Rkrk, A.; Mos, B.; Chc, A. A Review of Recent Advances in Non-Enzymatic Electrochemical Creatinine Biosensing. *Anal. Chim. Acta* **2021**, *1183*, 1–29.
26. Agarwal, H.; Nakara, A.; Shanmugam, V.K. Anti-inflammatory mechanism of various metal and metal oxide nanoparticles synthesized using plant extracts: A review. *Biomed. Pharmacother.* **2019**, *109*, 2561–2572. [CrossRef]

27. Hou, W.X.; Wei, P.; Kong, L.D.; Guo, R.; Wang, S.G.; Shi, X.Y. Partially PEGylated Dendrimer-Entrapped Gold Nanoparticles: A Promising Nanoplatform for Highly Efficient DNA and siRNA Delivery. *J. Mat. Chem. B* **2016**, *4*, 2933–2943. [CrossRef]
28. Shan, Y.; Luo, T.; Peng, C.; Sheng, R.; Cao, A.; Cao, X.; Shen, M.; Guo, R.; Tomás, H.; Shi, X. Gene Delivery Using Dendrimer-Entrapped Gold Nanoparticles as Nonviral Vectors. *Biomaterials* **2012**, *33*, 3025–3035. [CrossRef]
29. Lin, L.; Fan, Y.; Gao, F.; Jin, L.; Li, D.; Sun, W.; Li, F.; Qin, P.; Shi, Q.; Shi, X.; et al. UTMD-Promoted Co-Delivery of Gemcitabine and miR-21 Inhibitor by Dendrimer-Entrapped Gold Nanoparticles for Pancreatic Cancer Therapy. *Theranostics* **2018**, *8*, 1923–1939. [CrossRef]
30. Shi, X.Y.; Wang, S.H.; Lee, I.; Shen, M.W.; Baker, J.R. Comparison of the Internalization of Targeted Dendrimers and Dendrimer-Entrapped Gold Nanoparticles into Cancer Cells. *Biopolymers* **2009**, *91*, 936–942. [CrossRef]
31. Yu, F.; Wenjie, S.; Xiangyang, S. Design and Biomedical Applications of Poly(amidoamine)-Dendrimer-Based Hybrid Nanoarchitectures. *Small Methods* **2017**, *1*, 1700224.
32. Xu, B.; Li, A.J.; Hao, X.X.; Guo, R.; Shi, X.Y.; Cao, X.Y. PEGylated Dendrimer-Entrapped Gold Nanoparticles with Low Immunogenicity for Targeted Gene Delivery. *RSC Adv.* **2018**, *8*, 1265–1273. [CrossRef]
33. He, M.; Gao, K.; Zhou, L.; Jiao, Z.; Wu, M.; Cao, J.; You, X.; Cai, Z.; Su, Y.; Jiang, Z. Zwitterionic Materials for Antifouling Membrane Surface Construction. *Acta Biomater.* **2016**, *40*, 142–152. [CrossRef] [PubMed]
34. He, Y.; Hower, J.; Chen, S.; Bernards, M.T.; Chang, Y.; Jiang, S. Molecular Simulation Studies of Protein Interactions with Zwitterionic Phosphorylcholine Self-assembled Monolayers in the Presence of Water. *Langmuir ACS J. Surf. Colloids* **2008**, *24*, 10358. [CrossRef]
35. Schlenoff, J.B. Zwitteration: Coating Surfaces with Zwitterionic Functionality to Reduce Nonspecific Adsorption. *Langmuir ACS J. Surf. Colloids* **2014**, *30*, 9625–9636. [CrossRef]
36. Zhang, T.; Huang, Y.; Ma, X.; Gong, N.; Liu, X.; Liu, L.; Ye, X.; Hu, B.; Li, C.; Tian, J.H.; et al. Fluorinated Oligoethylenimine Nanoassemblies for Efficient siRNA-Mediated Gene Silencing in Serum-Containing Media by Effective Endosomal Escape. *Nano Lett.* **2018**, *18*, 6301–6311. [CrossRef]
37. Liu, J.; Xiong, Z.; Zhang, J.; Peng, C.; Klajnert-Maculewicz, B.; Shen, M.; Shi, X. Zwitterionic Gadolinium(III)-Complexed Dendrimer-Entrapped Gold Nanoparticles for Enhanced Computed Tomography/Magnetic Resonance Imaging of Lung Cancer Metastasis. *ACS Appl. Mater. Interfaces* **2019**, *11*, 15212–15221. [CrossRef]
38. Xiong, Z.; Alves, C.S.; Wang, J.; Li, A.; Liu, J.; Shen, M.; Rodrigues, J.; Tomás, H.; Shi, X. Zwitterion-Functionalized Dendrimer-Entrapped Gold Nanoparticles for Serum-Enhanced Gene Delivery to Inhibit Cancer Cell Metastasis. *Acta Biomater.* **2019**, *99*, 320–329. [CrossRef]
39. Peng, C.; Zheng, L.; Chen, Q.; Shen, M.; Guo, R.; Wang, H.; Cao, X.; Zhang, G.; Shi, X. PEGylated dendrimer-entrapped gold nanoparticles for in vivo blood pool and tumor imaging by computed tomography. *Biomaterials* **2012**, *33*, 1107–1119. [CrossRef]
40. Conner, S.D.; Schmid, S.L. Regulated portals of entry into the cell. *Nature* **2003**, *422*, 37–44. [CrossRef]
41. Li, J.; Chen, L.; Xu, X.; Fan, Y.; Xue, X.; Shen, M.; Shi, X. Targeted Combination of Antioxidative and Anti-Inflammatory Therapy of Rheumatoid Arthritis using Multifunctional Dendrimer-Entrapped Gold Nanoparticles as a Platform. *Small* **2020**, *16*, 1–11. [CrossRef] [PubMed]
42. Lutz, M.B.; Kukutsch, N.; Ogilvie, A.L.; Rößner, S.; Koch, F.; Romani, N.; Schuler, G. An Advanced Culture Method for Generating Large Quantities of Highly Pure Dendritic Cells from Mouse Bone Marrow. *J. Immunol. Methods* **1999**, *223*, 77–92. [CrossRef]
43. Liu, C.; Xiang, Y.; Qin, X.; Liu, H.; Ju, X.; Zhang, X. A Kind of Method Concurrently Separating Peripheral Blood T, Bone-Marrow-Derived Lymphocyte. CN Patent 106085955-A, 9 November 2016.
44. Yang, D.; Zhao, Y.; Guo, H.; Li, Y.; Tewary, P.; Xing, G.; Hou, W.; Oppenheim, J.J.; Zhang, N. Gd@C-82(OH)(22) (n) Nanoparticles Induce Dendritic Cell Maturation and Activate Th1 Immune Responses. *ACS Nano* **2010**, *4*, 1178–1186. [CrossRef] [PubMed]
45. de Waard, M. Efficient Neutralization of Deadly Toxins in vivo by DNA Oligonucleotides. *Toxicon* **2018**, *149*, 88. [CrossRef]

Article

Mechanism Study of Thermally Induced Anti-Tumor Drug Loading to Engineered Human Heavy-Chain Ferritin Nanocages Aided by Computational Analysis

Shuang Yin [1], Yongdong Liu [2], Sheng Dai [3], Bingyang Zhang [1], Yiran Qu [1], Yao Zhang [2], Woo-Seok Choe [4] and Jingxiu Bi [1],*

[1] School of Chemical Engineering and Advanced Materials, The University of Adelaide, Adelaide, SA 5005, Australia; shuang.yin@adelaide.edu.au (S.Y.); bingyang.zhang@adelaide.edu.au (B.Z.); yiran.qu@adelaide.edu.au (Y.Q.)
[2] State Key Laboratory of Biochemistry Engineering, Institute of Process Engineering, Chinese Academy of Sciences, Beijing 100190, China; ydliu@ipe.ac.cn (Y.L.); zhangyao@ipe.ac.cn (Y.Z.)
[3] Department of Chemical Engineering, Brunel University London, London UB8 3PH, UK; sheng.dai@brunel.ac.uk
[4] School of Chemical Engineering), Sungkyunkwan University (SKKU), Suwon 16419, Korea; checws@skku.edu
* Correspondence: jingxiu.bi@adelaide.edu.au

Citation: Yin, S.; Liu, Y.; Dai, S.; Zhang, B.; Qu, Y.; Zhang, Y.; Choe, W.-S.; Bi, J. Mechanism Study of Thermally Induced Anti-Tumor Drug Loading to Engineered Human Heavy-Chain Ferritin Nanocages Aided by Computational Analysis. *Biosensors* **2021**, *11*, 444. https://doi.org/10.3390/bios11110444

Received: 13 October 2021
Accepted: 9 November 2021
Published: 11 November 2021

Publisher's Note: MDPI stays neutral with regard to jurisdictional claims in published maps and institutional affiliations.

Copyright: © 2021 by the authors. Licensee MDPI, Basel, Switzerland. This article is an open access article distributed under the terms and conditions of the Creative Commons Attribution (CC BY) license (https:// creativecommons.org/licenses/by/ 4.0/).

Abstract: Diverse drug loading approaches for human heavy-chain ferritin (HFn), a promising drug nanocarrier, have been established. However, anti-tumor drug loading ratio and protein carrier recovery yield are bottlenecks for future clinical application. Mechanisms behind drug loading have not been elaborated. In this work, a thermally induced drug loading approach was introduced to load anti-tumor drug doxorubicin hydrochloride (DOX) into HFn, and 2 functionalized HFns, HFn-PAS-RGDK, and HFn-PAS. Optimal conditions were obtained through orthogonal tests. All 3 HFn-based proteins achieved high protein recovery yield and drug loading ratio. Size exclusion chromatography (SEC) and transmission electron microscopy (TEM) results showed the majority of DOX loaded protein (protein/DOX) remained its nanocage conformation. Computational analysis, molecular docking followed by molecular dynamic (MD) simulation, revealed mechanisms of DOX loading and formation of by-product by investigating non-covalent interactions between DOX with HFn subunit and possible binding modes of DOX and HFn after drug loading. In in vitro tests, DOX in protein/DOX entered tumor cell nucleus and inhibited tumor cell growth.

Keywords: ferritin; drug delivery; thermally induced drug loading; computational analysis

1. Introduction

Mammalian ferritin is a 12 nm symmetrical protein cage consisting of 24 subunits. Each subunit contains a 4-helix bundle (helix A, B, C, and D) and a fifth short helix (helix E). Three N-terminals of subunits gather and form 8 hydrophilic channels in each ferritin shell to allow iron ion penetration [1]. Residues from 4 helices E make another 6 ferritin hydrophobic channels. All 14 channels on each ferritin shell are around 0.3–0.5 nm wide [2]. Ferritin's unique structure and high biocompatibility have made it a potential drug nanocarrier [3]. Especially, human heavy-chain ferritin (HFn) has shown an intrinsic active tumor targeting ability because it can recognize and bind to human transferrin receptor 1 (TfR1) [4].

Through decades of efforts, research have explored diverse drug loading approaches. Disassembly/reassembly and passive diffusion are 2 mainstream drug loading approaches. A disassembly/reassembly approach involves a dissociation/re-association of HFn assembly induced by pH or 8 M urea. This approach suits drugs either smaller or larger than ferritin channels. However, the disassembly/reassembly process in pH-induced pathway has been criticized. Kim et al. has proven that the process damaged ferritin structure and led to random aggregation of ferritin and drug, in which small aggregates were soluble and huge ones became precipitates [5]. The damage results in 2 problems in drug loading performance: (1) precipitation causes the loss of ferritin and an unsatisfactory protein recovery yield; (2) soluble ferritin-drug aggregates with different sizes can affect drug performance in vitro and in vivo. Condition optimization in drug loading was often required to mitigate these problems. For example, Mehmet et al. and Ruozi et al. critically investigated the pH adjustment course in drug loading and used a stepwise pH adjustment or optimization of final pH to boost protein recovery yield to 55% [6,7]. The 8 M urea-based approach is less frequently used in contrast with the pH-induced one. In two studies, it showed a DOX loading ratio comparable with that of optimized pH-induced approach (around 33 DOX per HFn nanocage), but the protein recovery yield was still undesirable, around 64.8% [8,9].

Passive diffusion approach loads drugs through the hydrophobic or hydrophilic channels on ferritin shell, through incubating ferritin and drugs together under suitable mixing conditions. Different stressors, such as high temperature, additives, and pressure, have been introduced to expand the channels and facilitate drug loading. This approach poses minor effects on ferritin structure and causes relatively low ferritin aggregation and loss compared with the disassembly–reassembly approach. However, the loading efficiency is low [10]. In the study of Yang et al., soybean ferritin was heated with Rutin at 60 °C for 1 h, resulting in a loading ratio of 10.5 Rutin molecules per ferritin [11]. They used chaotropic chemicals, urea, and guanidine chloride, to expand soybean ferritin channels and load molecules in 2 other studies [12,13]. To boost passive loading ratio, Wang et al., have successfully applied high hydrostatic pressure, explored different levels of variables, such as: pressure values, buffer pH, and additives, to finally achieve a 99% of HFn recovery and high DOX loading ratio (32 DOX per HFn nanocage) [14]. However, the high pressurized device is expensive and possesses a number of potential safety risks in operation. Therefore, it is challenging to achieve concomitantly a desirable drug loading ratio and a protein recovery yield in ferritin drug loading process.

In theory, after a drug enters ferritin, it retains in ferritin either by physical entrapment or chemical interaction, or both. For small molecule drugs, such as DOX (molecular weight < 600 Da), chemical interaction dominates. The chemical interaction type and strength are critical to the stability of drug loaded ferritin to prevent drug leakage from ferritin channel. Currently, the chemical interactions between ferritin and DOX have not been investigated in detail. An investigation on these interactions can help understand the drug loading mechanism, interpret the findings in drug loading and lead to an improvement of drug loading performance. In the investigation of protein-ligand binding mechanism, computational tools, molecular docking and molecular dynamic (MD) simulation are significantly regarded and widely used. Molecular docking provides multiple reliable modes of protein-ligand complexes, based on a searching algorithm, whilst MD simulation can assess the validity of these complexes by stability evaluation [15,16]. Shahwan et al. used AutoDock Vina, a molecular docking service, to find the most possible human ferritin (PDB ID: 3AJO)-enzyme inhibitor Donepezil complex, and ran a MD simulation of the complex to assess its stability [17]. These 2 tools are potentially capable of analyzing the chemical interactions between ferritin and drug in loading process.

In this study, a thermally induced passive diffusion was introduced to load DOX to HFn and 2 functionalized HFns, HFn-PAS, and HFn-PAS-RGDK. It is expected to obtain desirable loading results. HFn-PAS was constructed by fusing PAS peptide to HFn C-terminal. HFn-PAS-RGDK was constructed by fusing PAS and RGDK peptide onto the HFn subunit C-terminus. PAS peptide enlarges hydrodynamic volume and RGDK improves inhibition of tumor cell growth through specific affinity with integrin $\alpha v \beta 3/5$ and neuropilin-1, which are overexpressed by a wide range of tumor cells [18,19]. Three purified HFn-based proteins were characterized by transmission electron microscopy (TEM) before drug loading. Condition optimization in thermally induced drug loading for HFn and HFn-GFLG-PAS-RGDK were conducted. Size exclusion chromatography (SEC) and TEM were used to detect the structures of proteins after drug loading. DOX loaded proteins (protein/DOX) stability test was performed to check drug leakage profile during storage. For the first time, computational analysis, molecular docking followed by MD simulation, was adopted to analyze chemical interactions contributing to drug loading and aggregation in thermally induced DOX loading process. Finally, in vitro evaluations, intracellular distribution, and cytotoxicity assays, compared 3 HFn-based proteins in vitro performances after thermally induced drug loading.

2. Materials and Methods

2.1. Materials

Three HFn-based proteins, HFn, HFn-PAS, and HFn-PAS-RGDK were designed as in a previous work [20]. *Escherichia coli* (*E. coli*) BL21 (DE3) (Tiangen Biotech, Beijing, China) was the expression host. MDA-MB-231 cell line was purchased from Cellbank Australia (Sydney, NSW, Australia). Cell culture related reagents were purchased from Thermo Scientific (Massachusetts, MA, USA). All other chemicals of analytical grade except for Doxorubicin hydrochloride (DOX) (Dalian Meilun Biotechnology, Dalian, China), were bought from Chem-Supply (Gillman, SA, Australia). All chromatography columns used in this work were bought from GE healthcare (Waukesha, WI, USA). Millipore purification system (Merck, Melbourne, VIC, Australia) was used throughout the experiments.

2.2. Preparation and Characterization of HFn and Functionalized HFns

E. coli strains expressing HFn or functionalized HFns were fermented in LB medium at 37 °C and target proteins were expressed by 0.5 mM isopropyl β-D-thiogalactoside (IPTG) 4 h induction. Harvested cell pellets were re-suspended, subjected to ultra-sonication for cell disruption. Lysis supernatants were collected and stored at −20 °C before purification. HFn was purified through the procedure established in a previous work [21]. Harvested *E. coli* lysis supernatants containing HFn-PAS and HFn-PAS-RGDK first underwent 50 °C, pH 5.0, 5 min heat-acidic precipitation to remove host cell proteins, buffer exchange using Hiprep X26/10 G25 desalting column (GE Healthcare, Waukesha, WI, USA), and then pH 7.0 mono Q ion-exchange chromatography (GE Healthcare, Waukesha, WI, USA) for polishing. The 12% reducing SDS-PAGE and TEM were adopted for purity and conformation integrity characterization, respectively. In TEM analysis, a FEI Tecnai G2 Spirit TEM (Eindhoven, NB, The Netherlands) was employed. Operating voltage was 100 kV. Three purified proteins were diluted to 0.1 mg mL^{-1}, spread on TEM support grids, air dried, and then negatively stained with 2% uranyl acetate before micrography capture.

2.3. Thermally-Induced Passive Loading of DOX into HFn, HFn- PAS-RGDK, and HFn-PAS

DOX was loaded to HFn-based nanocages through thermally induced passive diffusion. Temperature, buffer pH and incubation time are the main factors affecting drug loading. An orthogonal test was designed to optimize thermally induced drug loading condition for HFn and HFn-PAS-RGDK. Variables and levels tested are listed in Table 1.

Table 1. Levels of variables used in the orthogonal tests for optimization of thermally induced DOX loading to HFn and HFn-PAS-RGDK.

Variables	Level
Temperature	45, 50, 60 °C
Phosphate buffer pH	7.0, 7.5
Incubation time	2, 4, 6 h

Initial protein concentration (1 mg mL^{-1}) and DOX concentration (0.2 mg mL^{-1}) were used in all conditions. Sample buffer was 20 mM phosphate buffer (PB) with 5 mM guanidinium chloride, pH 7.0 or 7.5. After thermal incubation of DOX and HFn-based nanocages, samples were at 1000 rpm 10 min at 4 °C to remove precipitates. Concentrations of the supernatants after centrifugation were measured using Bradford assay (Bio-Rad,

Gladesville, NSW, Australia) for calculation of protein recoveries yields. Unloaded DOX was removed using Hitrap G25 desalting column (GE healthcare, Waukesha, WI, USA) and DOX loaded HFn-based protein (protein/DOX) were collected. All protein/DOX peaks then underwent SEC by Superose 6 increase 10/300 GL column (GE Healthcare, Waukesha, WI, USA) to detect if any soluble HFn-DOX aggregates existed.

SEC can separate DOX loaded in HFn-based nanocages (DOX loaded in nanocage) from soluble HFn-DOX aggregates. Peak areas (absorbance at 480 nm) can be used to determine the proportion of DOX loaded in nanocage, using Equation (1). Drug loading ratio, protein recovery yield, and the proportion of DOX loaded in nanocages under various conditions were compared to find the optimal condition. For HFn-PAS, DOX loading was conducted at the optimal loading condition of HFn-PAS-RGDK.

$$\text{Proportion of DOX loaded in nanocage (\%)} = \frac{\text{Peak area of DOX loaded in nanocage}}{\text{Peak area of DOX loaded in nanocage} + \text{Peak area of protein–DOX aggregates}} * 100\% \quad (1)$$

Drug loading ratio, which is the number of DOX per HFn or functionalized HFn nanocage (N), was determined using Equation (2). C_{DOX} represents DOX concentration in protein/DOX samples collected from Hitrap G25 desalting chromatography. $C_{nanocage}$ represents the concentration of HFn-based proteins in protein/DOX samples. DOX has absorbance at 280 and 480 nm, and protein has absorbance at 280 nm. Therefore, we assume: (1) $OD480_{nanocage/DOX} = OD480_{DOX}$; (2) $OD280_{nanocage/DOX} = OD280_{DOX} + OD280_{nanocage}$. Five standard OD vs. C linear curves were established, including $OD480_{DOX}$ vs. C_{DOX}, $OD280_{DOX}$ vs. C_{DOX}, $OD280_{HFn}$ vs. C_{HFn}, $OD280_{HFn-PAS}$ vs. $C_{HFn-PAS}$, $OD280_{HFn-PAS-RGDK}$ vs. $C_{HFn-PAS-RGDK}$. DOX concentration range for standard curves was 1–40 µg mL^{-1}, and concentration range of proteins for standard curves was 0.1–1.2 mg mL^{-1}.

$$N = \frac{Number\ of\ DOX}{Number\ of\ nanocage} = \frac{C_{DOX} \bullet Mw_{nanocage}}{C_{nanocage} \bullet Mw_{DOX}} \quad (2)$$

2.4. TEM Characterization of DOX Loaded HFn-Based Proteins and HFn-DOX Aggregate

Three protein/DOX samples under the optimal thermally induced drug loading conditions were analyzed using TEM analysis. A HFn-DOX aggregate sample collected from Superose 6 increase SEC also underwent TEM analysis. Sample treatment and device setting in TEM analysis were the same as in Section 2.2.

2.5. Stability of DOX Loaded HFn and Functionalized HFns

After drug loading, the buffer of protein/DOX samples obtained from 50 °C, 6 h, pH 7.5 were exchanged into either phosphate-buffered saline (PBS) pH 7.4. Buffer exchanged samples were placed at 37 and 4 °C. Aliquots were taken from samples at certain time points (0, 2, 4, 8, 24, 72, 120, 168, 336 h) and desalted using Hitrap G25 desalting column (GE Healthcare, USA) to remove leaked DOX, followed by N value calculation.

2.6. Computational Study of Interactions of HFn and DOX in Thermally-Induced Drug Loading

Molecular docking and Gromacs MD simulation analysis were used to identify the potential HFn and DOX interactions to explain the formation of HFn/DOX and soluble HFn-DOX aggregates. Molecular docking was performed to analyze the possible poses of HFn subunit and DOX interactions. MD simulation of the docking complexes aimed to find out the most stable HFn subunit-DOX complex structures.

Computational analysis was based on 2 prerequisites: (1) we assume that HFn subunit can be a representative of HFn assembly. This is because the assembly was theoretically 24 repetitions of the subunit. DOX is smaller than HFn channels, which makes it unlikely to simultaneously interact with more than one subunit of the same HFn assembly. (2) The computational analysis focus was on the interactions between DOX and the residues located on HFn assembly outer surface and inner surface. All interactions with interface residues of HFn assembly were ignored.

In molecular docking analysis, PyRx software was used. DOX 3D structure was from Pub Chem and HFn subunit structure file (PDB file) from RCSB PDB (ID: 2FHA). DOX was energy minimized before conducting docking. Top 9 docking HFn-DOX complexes were obtained and saved as PDB files. PDB files from docking results underwent Gromacs MD simulation using Gromacs 2018.

In MD simulation, CHARMM36 force field was used. The HFn-DOX complex structure was solvated in a dodecahedral box of size 460.73 nm^3 with water molecules and the box was charge neutralized by replacing eight water molecules with 8 Na+ ions. Energy minimization was conducted using the steepest descent integrator for 50,000 steps, until a tolerance of 10 kJ mol^{-1}. After this, temperature (NVT) and pressure equilibration (NPT) of the full system were performed at 323 K (approximate 50 °C). Finally, 10 ns 323 K simulation were conducted with 5,000,000 steps and 2 fs each step. Lincs constraint algorithm, Verlet cut-off scheme, Particle Mesh Ewald coulomb type were used in this MD simulation. Root-mean-square deviation (RMSD) and short-range non-bonded interaction energy of each complex in MD simulation were analyzed for the stability assessment. Three-dimensional structures of 9 complexes after MD simulation were saved and the interactions of HFn subunit and DOX within were visualized by Discovery Studio Visualizer. Interactions analyzed include hydrogen bond, salt bridge, and Pi (π) effects. Possible hydrophobic interaction was evaluated by analyzing the residue hydrophobicity in DOX binding area.

2.7. In Vitro Anti-Tumor Assessments of DOX Loaded HFn-Based Proteins

MDA-MB-231 is a human breast tumor cell line and has been proven to overexpress human TfR1, neuropilin 1, and integrin $\alpha v \beta 3/5$ [22,23]. MDA-MB-231 cells were cultured in L-15 medium with 10% FBS and 1% PS. Intracellular distribution and MTT assay of all 3 protein/DOX were conducted.

MDA-MB-231 cells in exponential growth phase were utilized in intracellular distribution analysis and cytotoxicity assay. Procedures of these two assays were the same as in a previous work using another tumor cell line [20]. A_{well} and cell viability were calculated using the following equations. IC$_{50}$ values of DOX and three protein/DOX were calculated in Origin 9.0 software. Unpaired T test was employed for statistical assessment.

$$A_{well} = A595 - A630 \qquad (3)$$

$$\text{Cell viability (\%)} = (A_{well} - A_{blank})/(A_{cell} - A_{blank}) \times 100\ (\%) \qquad (4)$$

3. Results

3.1. Characterizations of Purified HFn-Based Proteins

Figure 1 shows the SDS-PAGE and TEM images of 3 purified HFn-based proteins. In Figure 1A of 12% reducing SDS-PAGE, HFn subunit showed a single band with around 21 kDa. However, 2 functionalized HFns (HFn-PAS and HFn-PAS-RGDK) showed higher apparent molecular weights in SDS-PAGE gel than their theoretical 26 kDa and 26.5 kDa. The discrepancies in molecular weights probably result from PAS peptides, which has the tendency of binding to surrounding water molecules to increase the hydrodynamic radius. Other researchers discovered similar molecular weight increase in PAS modified proteins and PEG-conjugated proteins in SDS-PAGE analysis [18,24].

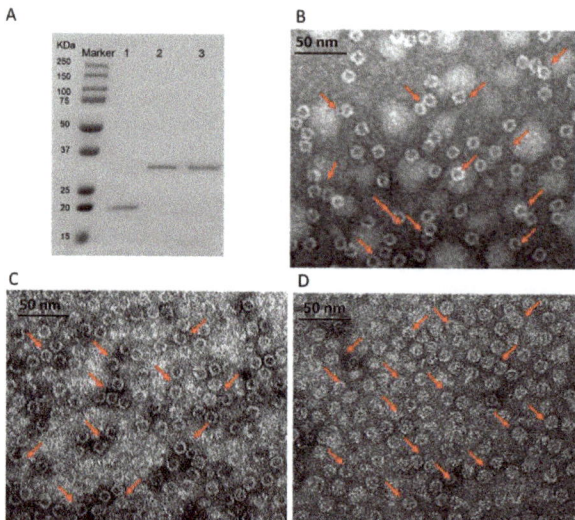

Figure 1. Characterizations of purified HFn-based proteins. (**A**), 12% reducing SDS-PAGE results of purified 3 HFn-based proteins. Lane 1: HFn, 2: HFn-PAS, 3: HFn-PAS-RGDK. (**B**), TEM image of purified HFn. (**C**), TEM image of purified HFn-PAS. (**D**), TEM image of purified HFn-PAS. E, TEM image of purified HFn-PAS-RGDK. Red arrows indicate some spheres.

TEM images in Figure 1B–D demonstrate that both functionalized HFns were assembled hollow spheres, same as HFn. Cages of all proteins were around 12 nm in diameter regardless of functionalization. This is because the inserted functional peptides at the C-terminus did not constitute the ferritin nanocage, while under TEM, the size of the nanocage was visualized.

3.2. Optimization of HFn Thermally Induced Passive Loading to Increase Drug Loading

Thermally induced strategy takes advantage of the thermal energy mediated structural perturbation of selective hydrophilic pore areas. Theil and co-workers used Circular Dichroism to analyze the α-helix content change of HFn following heat treatment at different temperatures, and found that a small amount of secondary structure began to transition into random coil when temperature is greater than 45 °C, and it is very likely to take place in pore areas and expand pores [25]. Heating also accelerates Brownian motion of proteins and drug molecules so that greater efficiency could be achieved than in non-heated passive diffusion. In this work, pH 7.0 and 7.5 were chosen to ensure that DOX carries positive charge (DOX pKa 8.3) and HFn inner surface has the opposite charge (HFn pI 4.8). Temperature conditions were selected based on thermal stability of HFn.

Standard curves for determination of drug loading ratio (N) are in Figure S1. Figure 2 summarizes the changes of drug loading ratios, proportions of DOX loaded in nanocage and HFn recovery yields with varying thermal induction time, temperature, and buffer pH. Table S1 lists all drug loading ratios (Ns), proportions of DOX loaded into nanocage and protein recovery yields for HFn. As is shown in Figure 2A,C,E, with the increase in thermal induction duration from 2 to 6 h, N increased at all tested temperatures. At 45 °C, 50 °C, and 60 °C, the highest N was 30.3, 41.6, and 56.7. Ns at pH 7.5 were slightly higher than those at pH 7.0 in most of the time regardless of temperature.

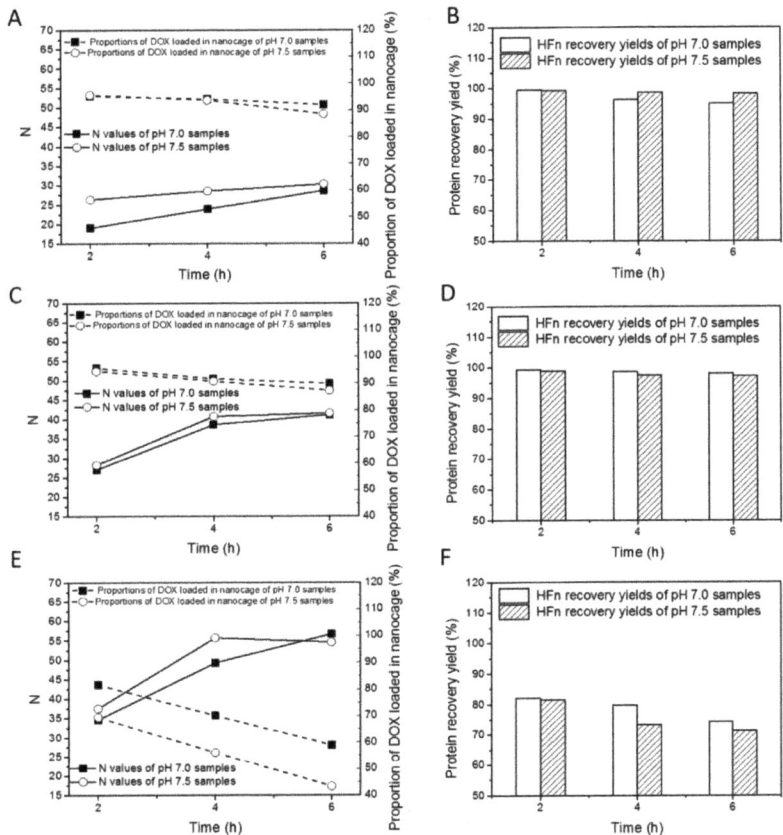

Figure 2. Thermally induced DOX loading results of HFn under different experimental conditions. Loading ratios (Ns) and proportions of DOX loaded in nanocage at 45 °C (**A**), 50 °C (**C**), and 60 °C (**E**). HFn recovery yields at 45 °C (**B**), 50 °C (**D**) and 60 °C (**F**).

In terms of proportions of DOX loading in nanocage, at 45 and 50 °C, they were at least 85% whilst at 60 °C they were below 85%. At all 3 temperatures, the proportion of DOX loaded into nanocage decreased with the duration of thermal induction. The proportions were largely pH-dependent, and the extent of pH-dependency was subject to temperature, hence they decreased by 0.4–3.5% at 45 °C and 50 °C, and by 10–15% at 60 °C as pH increased from 7.0 to 7.5.

For the HFn recovery yields, in Figure 2B,D,F, at 45 °C and 50 °C, they were above 90%, and at 60 °C, they were mostly below 85%. These results suggest that in the thermally induced drug loading process, DOX loaded in individual HFn nanocages, soluble HFn-DOX aggregates, and HFn-DOX precipitates were simultaneously produced as in previous research using disassembly/reassembly drug loading approaches. At 45 °C, proportion of drug loaded in nanocage and HFn recovery yield decreased slowly, and N increased slowly over time, whilst at 60 °C, proportion of drug loaded in nanocage, HFn recovery yield and N behaved in the opposite manner. These results confirm that 45 °C may not be effective to accelerate drug loading. In addition, at 60 °C, the local structures of HFn nanocages undergo excessive changes, resulting in massive formation of aggregates of HFn with DOX. Considering N, proportion of DOX loaded in nanocage and HFn recovery yield together, 50 °C, pH 7.5, 6 h is the best drug loading condition (N of 41.6, proportion of DOX loaded in nanocage of 87.2% and HFn recovery yield of 97.2%).

3.3. Optimization of HFn-PAS-RGDK Thermally Induced Passive Loading

Figure 3 and Table S2 show the DOX loading optimization results of HFn-PAS-RGDK. The relations between drug loading performance indicators (N, proportion of DOX loaded in nanocage and HFn-PAS-RGDK recovery yield) and experimental variables (induction time, pH and temperature) are similar to those in HFn.

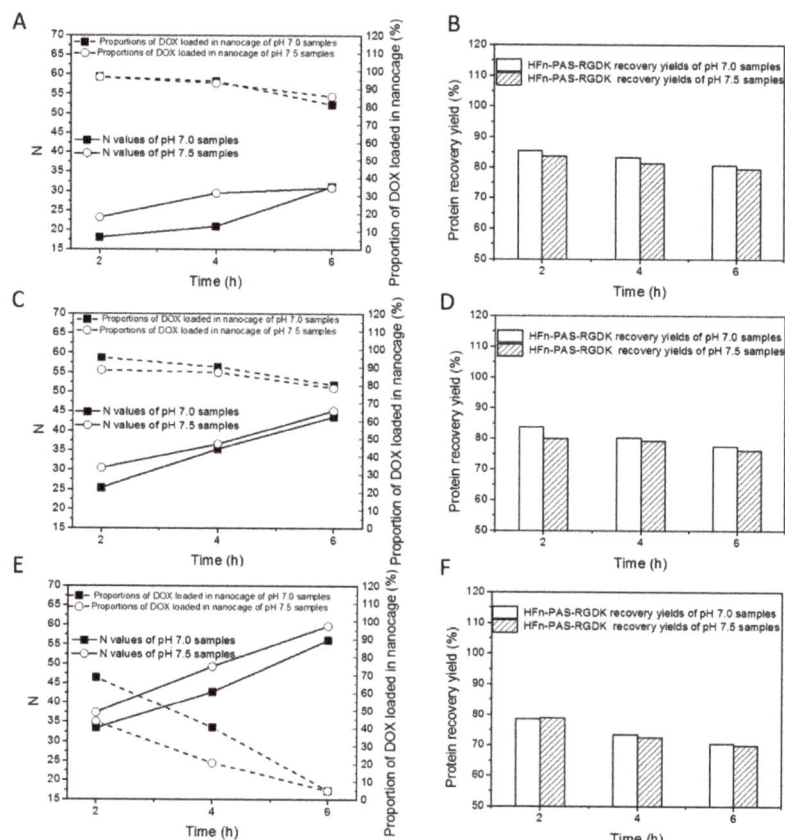

Figure 3. Thermally-induced DOX loading results of HFn-PAS-RGDK under different experimental conditions. Loading ratios (N) and proportions of DOX loaded in nanocage at 45 °C (**A**), 50 °C (**C**), and 60 °C (**E**). HFn-PAS-RGDK recovery yields at 45 °C (**B**), 50 °C (**D**), and 60 °C (**F**).

In Figure 3A,C,E, N positively responded to thermal induction duration and temperature. pH 7.5 showed greater Ns than pH 7.0 in most of time. Proportion of DOX loaded in nanocage was negatively related to temperature and incubation time. At 45 °C and 50 °C, proportions of DOX loaded in nanocage were greater than 75%. At 60 °C, they were lower than 70%. As in Figure 3B,D,E, HFn-PAS-RGDK recovery yields were greater than 75% except at 60 °C 4 h and 6 h. The best DOX loading condition was obtained at 50 °C, pH 7.5 and 6 h, with an N of 45.2, proportion of DOX loaded in nanocage of 78.5% and HFn-PAS-RGDK recovery yield of 76.0%. HFn-PAS DOX loading ratio was 38.4, proportion of DOX loaded in nanocage was 73.4% and protein recovery was 75.1% under the same condition.

3.4. Conformation of DOX Loaded HFn and Functionalized HFns

SEC analysis was performed to prove the success of DOX loading and separate DOX loaded nanocages from HFn-DOX soluble aggregates according to hydrodynamic volume differences. HFn-based proteins have absorbance at 280 nm but not at 480 nm. DOX has absorbance at both wavelengths. Protein/DOX in theory has absorbance at both wavelengths and peak retention time should be similar to HFn-based proteins. Figure 4A–C show the chromatograms of HFn/DOX, HFn-PAS/DOX, and HFn-PAS-RGDK/DOX prepared under the condition of 50 °C, 6 h, pH 7.5. Two peaks, P1 and P2, were observed in all 3 samples. The larger P2 had retention volumes of 13–15 mL in Superose 6 increase column, and absorbance at both 280 and 480 nm. This means it was the DOX loaded HFn-based nanocage. The smaller P1 eluted before P2 was the protein-DOX soluble aggregates, the DOX amount of which accounted for below 27% of the total DOX in the SEC loading samples. The heating process did not affect most of the ferritin nanocage, as are shown in Figure 4D–F. Most of the protein/DOX were hollow spheres. Nanocage sizes were still around 12 nm, the same as before thermally induced passive drug loading process.

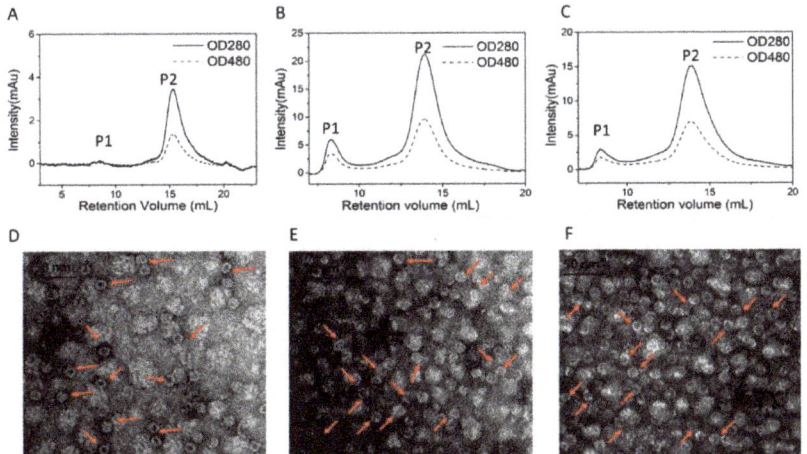

Figure 4. Size-exclusion chromatograms and TEM images of optimal protein/DOX. SEC HFn/DOX (**A,D**), HFn-PAS/DOX (**B,E**), HFn-PAS-RGDK/DOX (**C,F**). Red arrows indicate some spheres.

3.5. DOX Loaded HFn and Functionalized HFns Stability

Protein/DOX stability test was designed to reflect the stability of protein/DOX in storage (4 °C) and blood circulation (37 °C). In storage, drug leakage profiles for all protein/DOX were consistent, where around 20% of loaded drug leaked over 2 weeks (Figure 5). At 37 °C, protein/DOX were less stable in contrast with 4 °C, with around 30% of drug loss detected in HFn/DOX and 35% of drug leaking in other 2 groups for 2 weeks. In all 3 protein/DOX, drug leaked fast during the initial 12 h, then slowed down. Perhaps some of the loaded drugs were just loosely attached or physically trapped inside protein nanocages. Hence, these drugs were more prone to dissociation from protein, while drugs strongly interacted with HFn remained within ferritin. Functionalized HFns showed lower protein/DOX stabilities compared with HFn, which probably result from the insertion of foreign peptides.

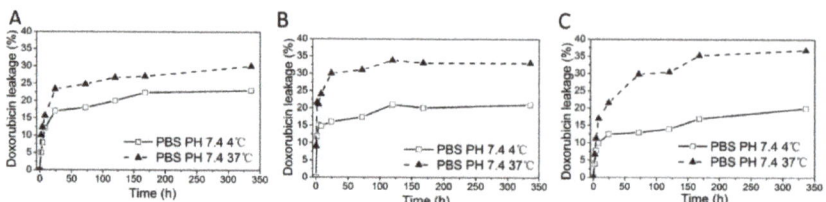

Figure 5. DOX leakage over time at different conditions. (**A**), HFn/DOX. (**B**), HFn-PAS/DOX. (**C**), HFn-PAS-RGDK/DOX.

3.6. Interactions between HFn and DOX in Thermally Induced Drug Loading by Computational Analysis

From molecular docking results, 9 different HFn subunit-DOX complexes were obtained. Complexes underwent 10 ns 50 °C MD simulation for stability assessment. Three-dimensional structures of 9 complexes after simulation are shown in Figure 6. Among them, only in Complex 1, the location DOX binds to was the inner surface in HFn assembly. In the other 8 complexes, DOX bound to areas corresponding to the outer surface in HFn assembly. This implies that Complex 1 is very likely to be the structure of DOX loaded in HFn nanocage, while the interaction ways in the other complexes could form drug loading, soluble HFn-DOX aggregates, and HFn-DOX precipitates in thermally induced drug loading process.

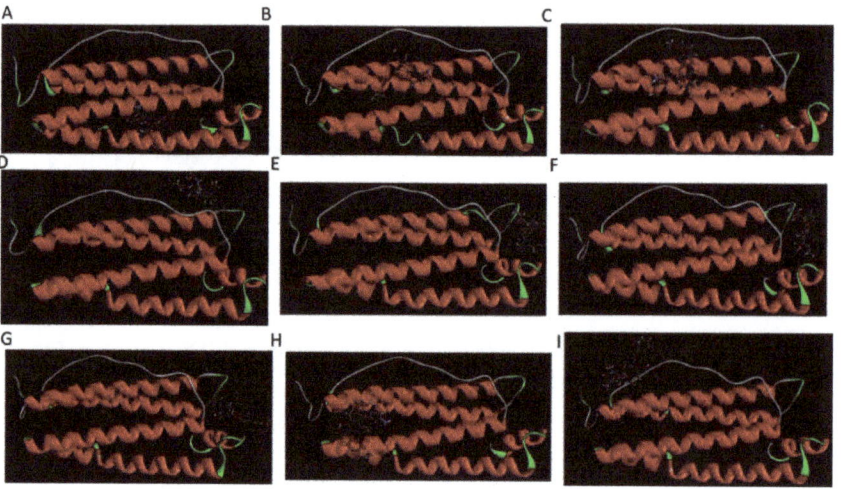

Figure 6. Three-dimensional structures of 9 complexes after 10 ns MD simulation. (**A–I**) are complex 1–9.

To evaluate the stabilities of these structures, the RMSD and the short-range non-bonded interaction energy of HFn subunit and DOX molecule in 9 complexes during simulation were monitored and are presented in Figures 7 and 8. The smaller the RMSD and the lower the energy is, the more stable the complex structure is and the more reliable the complex structure is. The stability orders of structures demonstrated in RMSD and energy are consistent.

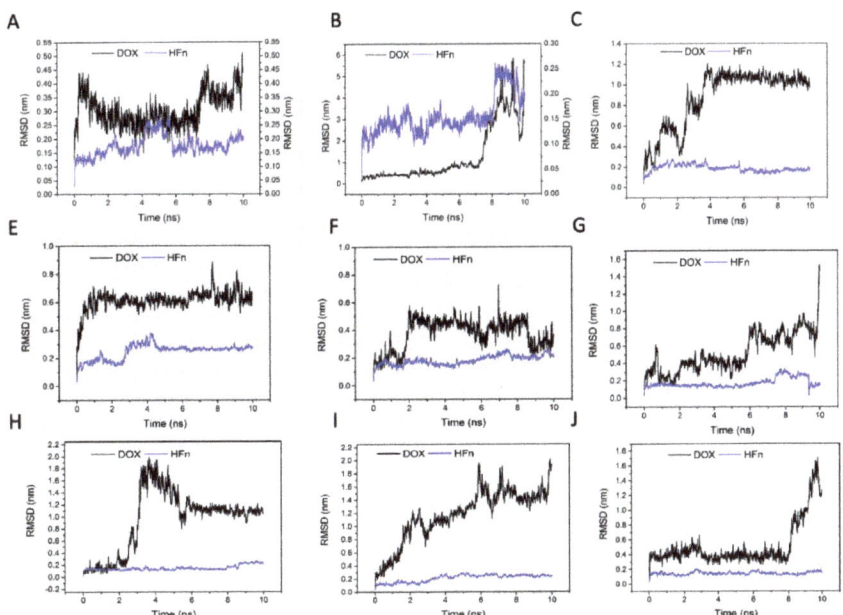

Figure 7. RMSD of HFn subunit and DOX in complexes 1–9 during MD simulation. (**A–I**) are complex 1–9.

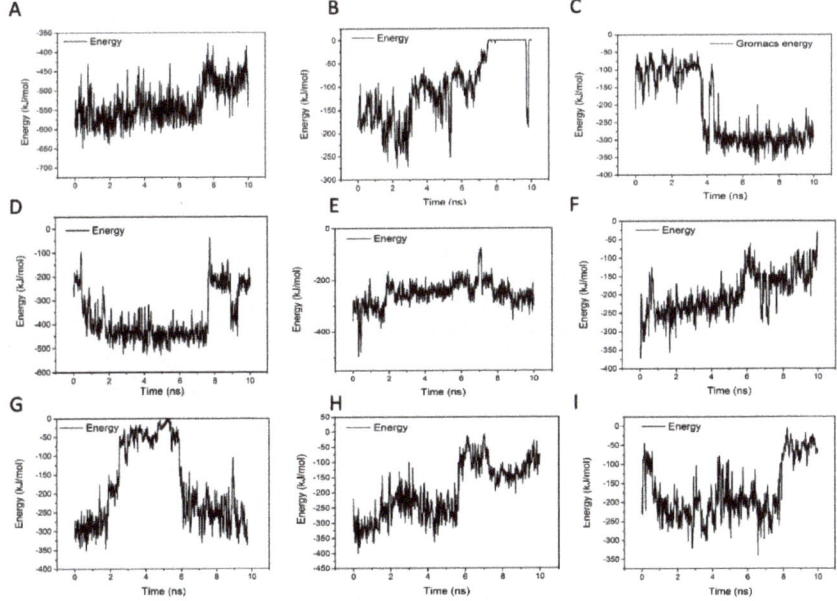

Figure 8. Short-range non-bonded interaction energy of HFn subunit and DOX in complexes 1–9 during MD simulation. (**A–I**) are complex 1–9.

Complex 1 was the most stable structure in 50 °C MD simulation, with RMSD lower than 1 nm and energy below -350 kJ mol^{-1}. This result is in accordance with the experiment result conducted under 50 °C, that more DOX was being loaded to HFn nanocage than forming soluble aggregates and precipitates. Complex 4 and 5 were the second most stable structures, of which the RMSD were below 1 nm and the energies were below -200 kJ mol^{-1} most of the time. Complex 3 was the third most stable structure. Its RMSD was lower than 1.2 nm. RMSD of Complex 7, 6, 8, 9, and 2 were greater than 1.5 nm and energies of them were above -50 kJ mol^{-1}, indicating relatively unstable structures and possibly weak interactions. Based on the stability results, Complex 1, 4, 5, and 3 were the focus in interaction analysis.

Table 2 lists the hydrogen bond, salt bridge and Pi effect interactions between HFn subunit and DOX in Complex 1, 4, 5, and 3 at 10 ns of the simulation. Figure S4 lists the 2D diagrams of 9 complexes of HFn subunit with DOX at 10 ns of the MD simulation. Hydrogen bonds and salt bridges are strong non-covalent bonds, in contrast with van der Waals interaction, such as Pi effects. The more of them in the complex, the more stable the complex structure is. Complex 1 had the most hydrogen bonds and salt bridges.

Table 2. Interactions between HFn subunit and DOX in complexes.

Complex	Residues Forming Hydrogen Bond with DOX	Residues Forming Salt Bridge/Attractive Charge with DOX	Residues Have Pi Effects with DOX
1	GLN58, GLU62, HIS65, GLN141	GLU27, GLU62 (2) [1], GLU107 (2) [1]	HIS57 (Pi-Pi stacked), TYR54 (Pi-alkyl)
4	ARG43, ASP91	ASP91	TYR39 (Pi-Pi stacked), TYR39 and PRO88 (Pi-alkyl)
5	TYR40, ASP45, GLU94, GLU167	ASP45	/
3	/	/	TYR29 (Pi-Pi T shaped), LEU26 (Pi-alkyl)

[1] Numbers in the brackets after the residue are the number of the interactions involving the residue.

Regarding the possible hydrophobic interactions between HFn subunit and DOX in Complex 1, 4, 5, and 3, a 5 residue average hydrophobicity was used to reflect the hydrophobicity level of residues in DOX binding area. This is because in a protein, the hydrophobicity of residues can be affected by the nearby residues. Local area hydrophobicity reflects the possibly of hydrophobic interaction better than considering individual residue hydrophobicity. The calculation of 5 residue average hydrophobicity has considered the impact of nearby residues and its value demonstrates how hydrophobic the local area of the residue is. Hydrophobicity values in Table 3 were calculated using Discovery Studio Visualizer. The greater the value is, the more likely it would interact with the hydrophobic DOX molecule. In Complex 1, 4, 5, and 3, there were at least 3 residues at the binding pocket available for hydrophobic interactions with DOX.

Table 3. Five residue average hydrophobicity of residues in DOX binding area in complex 1, 4, 5, and 3.

Complex	5 Residue Average Hydrophobicity Values of Hydrophobic Residues in DOX Binding Pocket
1	TYR34 (0.92) [1], TYR54 (0.64) [1], LYS143 (0.62) [1], ALA144 (0.54) [1], GLU147 (0.1) [1].
4	TYR32 (0.52) [1], SER36 (0.56) [1], TYR39 (0.26) [1].
5	VAL46 (0.56) [1], ALA47 (0.48) [1], LEU48 (0.48) [1].
3	LEU26 (1.02) [1], GLN83 (0.82) [1], GLN112 (0.04) [1], GLU116 (0.94) [1].

[1] Numbers in the brackets after the residues are the 5 residue average hydrophobicity.

According to the computational analysis, in DOX loaded HFn nanocage, DOX was mostly bound to HFn subunit as in Complex 1. Relatively weak binding ways found in Complex 4, 5, and 3 and physically trapped DOX also existed. Therefore, the loading ratio could reach above 24. However, DOX remained in HFn in these weaker ways are more prone to dissociation during storage. Physically trapped DOX probably accounts for the burst release of DOX in the initial 12 h, and the weakly bounded DOX on HFn surface in Complex 4, 5, and 3 would gradually be released, as is shown in Figure 5.

Because no aggregates nor precipitates were found in 50 °C 6 h heated HFn, it is reasonable to infer that the interaction of DOX and HFn assembly has led to HFn and DOX aggregation. Small aggregates are still soluble while huge ones turn into HFn-DOX precipitates. TEM image in Figure 9A demonstrates that the HFn in soluble HFn-DOX aggregates were still intact spheres but clumped into a large particle. Interaction ways in Complex 4, 5, and 3 and others, except Complex 1, are theoretically possible to cause aggregation in a way that DOX works as a cross linker (Figure 9B). In each HFn assembly, there are 24 subunits for DOX to bind to, and the HFn-DOX aggregates contain multiple DOX molecules.

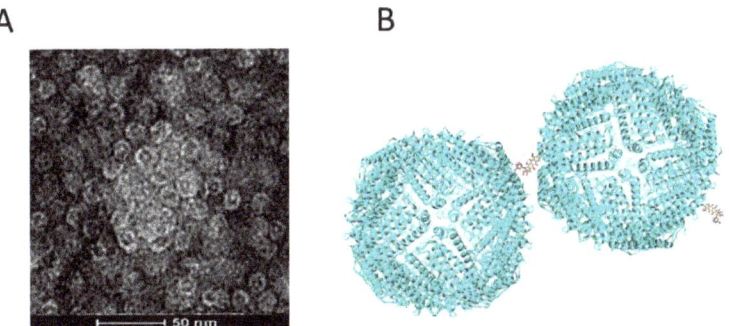

Figure 9. TEM image of DOX loaded in aggregates (**A**) and schematic of conformation of DOX loaded in aggregates (**B**). Cyan part is HFn assembly and brown part is DOX molecule.

3.7. Intracellular Distribution and Cytotoxicity of DOX Loaded HFn-Based Proteins

DOX has been proven to be able to diffuse into cell nucleus and disrupt cell division [26]. However, in theory, the DOX loaded on HFn and functionalized HFns need to be released from protein prior to exerting its function. Intracellular distribution test aimed to check whether the release of DOX form protein/DOX occurred. Cell nucleus locations were visualized as blue dots under cell imager after Hoechst 33,258 staining (Figure 10A). Due to the intrinsic fluorescence of DOX molecules, under the excitation of 480 nm light, DOX molecule accumulation could be observed as green dots. In the merged images, the color of dots in all four groups changed to light cyan, indicating that DOX molecules loaded on proteins through thermally induced loading approach were released and accumulated inside cell nucleus.

MTT assay was designed to compare the inhibition effects of DOX loaded on HFn, two functionalized HFns, and free DOX. Figure 10B shows the cell viabilities at different concentrations of equivalent DOX. Table 4 lists the IC_{50} values of all four groups. Free DOX possessed the lowest IC_{50}. However, it does not indicate free DOX has the greatest anti-proliferation effect. This is because in in vitro assays, the direct incubation of free DOX with cells has maximized the internalization efficiency of free DOX. On the contrary, the uptake efficiency of DOX in in vivo tests and in real practice would be greatly hampered by the blood circulation and metabolism system. Except for free DOX, DOX loaded on HFn-PAS-RGDK had the lowest IC_{50} value, followed by HFn-PAS/DOX. HFn/DOX had the greatest IC_{50}.

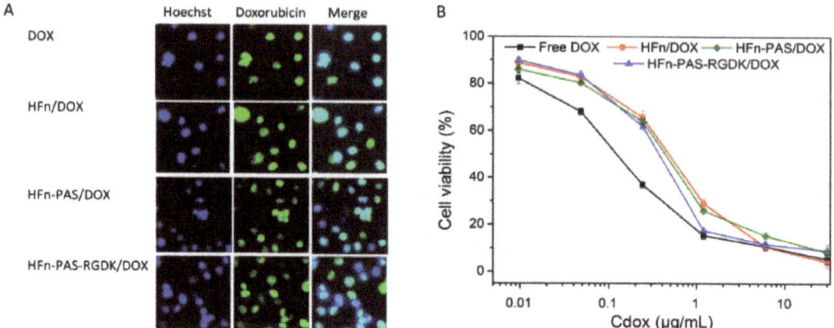

Figure 10. DOX intracelluar distribution photos and cytoxocity comparison of protein/DOX and DOX. (**A**), DOX distribution inside cells shown under cell imager. Blue dots show the locations of cell nucleus. Green dots represent the accumulation of DOX molecules. The light cyan dots in merge photos indicate the DOX molecules accumulated at cell nucleus. (**B**), Cytotoxocity effects of protein/DOX and DOX on MDA-MB-231 cells.

Table 4. IC50 values of all groups.

Group	IC$_{50}$ (µg mL^{-1})
DOX	0.15 ± 0.01
HFn/DOX	0.57 ± 0.02
HFn-PAS/DOX	0.46 ± 0.01
HFn-PAS-RGDK/DOX	0.34 ± 0.01

No statistical significance was found between anti-proliferation abilities of HFn-PAS-RGDK/DOX group and free DOX group ($p > 0.05$). Anti-proliferation effect of HFn-PAS-RGDK/DOX was significantly higher than the other two HFn-based protein/DOX groups ($p < 0.05$). This is because of the tumor targeting ability of the inserted RGDK in HFn-PAS-RGDK.

4. Discussion

In this study, the thermally induced passive diffusion approach succeeded in loading DOX into HFn and 2 functionalized HFns. 50 °C, pH 7.5 and 6 h was found to be the optimal condition for HFn and functionalized HFns. Temperature and incubation time showed a great impact on DOX loading performance. Although HFn and DOX have outstanding thermal stabilities, in the thermally induced drug loading process, both drug loading and irreversible HFn-DOX aggregation occurred under all selected conditions. With the same incubation time, as incubation temperature increased, N value increased whilst the proportion of DOX loaded in nanocage declined. At the same incubation temperature, N value increased and the proportion of DOX loaded in nanocage decreased over incubation time, especially at 60 °C.

Table 5 compares HFn drug loading performance of this work with some previously published studies. In this study, N of HFn (41.6) is greater than previously studies, which adopted 8 M urea or optimized stepwise pH-induced disassembly-reassembly approaches. Recovery yield of HFn in this study, 97.2%, is similar to high hydrostatic pressure passive diffusion approach (99%) and significantly greater than the pH-induced (25%, 55%) or 8 M urea-based approach (64.8%). Disassembly/reassembly approach has been questioned to be not fully reversible because 2 holes were detected by synchrotron small-angle X-ray scattering, and the authors argued that this structural damage may result in protein loss and aggregation in the drug loading process [5]. To the contrary, at 50 °C, HFn nanocage remains intact throughout the thermally induced drug loading process, which involves less structural changes.

Table 5. Comparison on DOX loading to HFn in this work and previous studies.

Protein	Loading Approach	N	Protein Recovery (%)	Reference
HFn	Thermal induction	41.6	97.2	This study
Horse spleen ferritin	Step-wise pH induction	28	55 ± 7	[6]
HFn	pH induction	29 ± 3	40 ± 4	[18]
Equine spleen ferritin	pH induction	22 ± 1	25	[27]
HFn	pH induction	28.3	/	[28]
Ferritin	Urea-based	32.5	64.8	[8]
HFn	Urea-based	33	/	[9]
HFn	High hydrostatic pressure	32 ± 2	99	[14]

'/' means no data.

Compared with HFn, under most experimental conditions, especially at 50 °C and 60 °C, the functionalized HFn, HFn-PAS-RGDK, had relatively low protein recovery yields and low proportions of DOX loaded in nanocage. HFn-PAS also demonstrated reduced proportions of DOX loaded in nanocage and protein recovery yields. Two functionalized HFns were more prone to aggregation in the heating process, suggesting slightly decreased thermal stabilities. This could be ascribed to the 'flip to flop' phenomenon in functionalized HFns, where E-helix with inserted functional peptide are extruded outside HFn nanocage, as was discovered in our previous work [20]. Hydrophobic interactions of 4 helices E around each hydrophobic channel in natural 'flip' HFn have been proven to contribute to HFn stability [29,30]. The turnover of E-helix has hampered helices E interactions and thus declined thermal stability.

Combining the results from molecular docking, MD simulation, and experiments, hydrogen bond and salt bridges between DOX and HFn residues in Complex 1 probably account for most of the loading of DOX. Physical entrapment of DOX in HFn assembly and interactions in other complexes may also contribute but they suffer from a rapid DOX leakage during storage, as shown in Figure 5. In the process of thermally induced DOX loading, DOX may undergo unexpected interactions with multiple HFn assemblies through hydrogen bonds and salt bridges to form HFn-DOX aggregates (Figure 9B).

In vitro tests demonstrate that DOX loaded through thermally induced passive diffusion could exert anti-cancer function as free DOX.

5. Conclusions

A thermally induced drug loading approach has improved DOX loading ratios and protein recovery yields of HFn and functionalized HFns, HFn-PAS and HFn-PAS-RGDK. This mild and efficient strategy can become an alternative to produce HFn-based nanocages with various drugs. According to molecular docking and MD simulation analysis, hydrogen bond, salt bridges and other non-covalent interactions between HFn and DOX molecules contribute to DOX loading and by-product formation. The combination of molecular docking and MD simulation analyses can be a useful tool to shed light on ferritin drug loading mechanism. In vitro tests show that thermally-induced DOX loaded HFn-based proteins can exert tumor inhibition of DOX. RGDK has promoted DOX internalization to tumor cells and enhanced HFn anti-tumor efficacy.

Supplementary Materials: The following are available online at https://www.mdpi.com/article/10.3390/bios11110444/s1, Figure S1: Standard linear curves of correlations between drug or HFn-based protein nanocages concentrations and optical densities. Table S1: Loading ratios (Ns), proportions of DOX loaded in nanocage and protein recovery percentages in HFn thermally induced drug loading optimization. Table S2: Loading ratios (Ns), proportions of DOX loaded in nanocage and protein recovery percentages in HFn-PAS-RGDK thermally induced drug loading optimization. Figure S2: Size exclusion chromatograms of all HFn/DOX samples under 18 conditions in thermally induced drug loading optimization. Figure S3: Size exclusion chromatograms of all HFn-GFLG-PAS-RGDK/DOX samples under 18 conditions in thermally induced drug loading optimization.

Figure S4: Hydrogen bond, salt bridge, and Pi effect interactions between HFn subunit and DOX in Complex 1–9 after 10 ns MD simulation.

Author Contributions: Conceptualization, J.B. and Y.L.; methodology, S.Y.; software, B.Z. and Y.Q.; validation, S.Y.; formal analysis, S.Y.; investigation, Y.Z.; resources, J.B., Y.L.; writing—original draft preparation, S.Y.; writing—review and editing, W.-S.C., S.D., J.B.; project administration, J.B.; funding acquisition, J.B. and Y.L. All authors have read and agreed to the published version of the manuscript.

Funding: This research was funded by joint Ph.D. Scholarship Scheme of the University of Adelaide and Institute of Process Engineering, Chinese Academy of Sciences, the National Natural Science Foundation of China [Grant No. 21576267], and Beijing Natural Science Foundation [Grant Number 2162041].

Institutional Review Board Statement: Not applicable.

Informed Consent Statement: Not applicable.

Data Availability Statement: The data presented in this study are available on request from the corresponding author.

Acknowledgments: Great appreciations to Fabien Voisin and the Phoenix High Performance Computer team for their support on computational analysis. Many thanks to Anton Middelberg from the University of Adelaide for his valuable advice on this study.

Conflicts of Interest: The authors declare no conflict of interest.

References

1. Yin, S.; Davey, K.; Dai, S.; Liu, Y.; Bi, J. A critical review of ferritin as a drug nanocarrier: Structure, properties, comparative advantages and challenges. *Particuology* **2021**, in press. [CrossRef]
2. Jutz, G.; van Rijn, P.; Santos Miranda, B.; Boker, A. Ferritin: A versatile building block for bionanotechnology. *Chem. Rev.* **2015**, *115*, 1653–1701. [CrossRef]
3. Truffi, M.; Fiandra, L.; Sorrentino, L.; Monieri, M.; Corsi, F.; Mazzucchelli, S. Ferritin nanocages: A biological platform for drug delivery, imaging and theranostics in cancer. *Pharmacol. Res.* **2016**, *107*, 57–65. [CrossRef] [PubMed]
4. Li, L.; Fang, C.J.; Ryan, J.C.; Niemi, E.C.; Lebron, J.A.; Bjorkman, P.J.; Arase, H.; Torti, F.M.; Torti, S.V.; Nakamura, M.C.; et al. Binding and uptake of H-ferritin are mediated by human transferrin receptor-1. *Proc. Natl. Acad. Sci. USA* **2010**, *107*, 3505–3510. [CrossRef]
5. Kim, M.; Rho, Y.; Jin, K.S.; Ahn, B.; Jung, S.; Kim, H.; Ree, M. pH-dependent structures of ferritin and apoferritin in solution: Disassembly and reassembly. *Biomacromolecules* **2011**, *12*, 1629–1640. [CrossRef] [PubMed]
6. Mehmet, A.; Kilic, E.O.; Calis, S. A novel protein-based anticancer drug encapsulating nanosphere: Apoferritin-doxorubicin complex. *J. Biomed. Nanotechnol.* **2012**, *8*, 508–514.
7. Ruozi, B.; Veratti, P.; Vandelli, M.A.; Tombesi, A.; Tonelli, M.; Forni, F.; Pederzoli, F.; Belletti, D.; Tosi, G. Apoferritin nanocage as streptomycin drug reservoir: Technological optimization of a new drug delivery system. *Int. J. Pharm.* **2017**, *518*, 281–288. [CrossRef]
8. Lei, Y.; Hamada, Y.; Li, J.; Cong, L.; Wang, N.; Li, Y.; Zheng, W.; Jiang, X. Targeted tumor delivery and controlled release of neuronal drugs with ferritin nanoparticles to regulate pancreatic cancer progression. *J. Control. Release* **2016**, *232*, 131–142. [CrossRef]
9. Liang, M.; Fan, K.; Zhou, M.; Duan, D.; Zheng, J.; Yang, D.; Feng, J.; Yan, X. H-ferritin–nanocaged doxorubicin nanoparticles specifically target and kill tumors with a single-dose injection. *Proc. Natl. Acad. Sci. USA* **2014**, *111*, 14900–14905. [CrossRef]
10. Kuruppu, A.I.; Zhang, L.; Collins, H.; Turyanska, L.; Thomas, N.R.; Bradshaw, T.D. An apoferritin-based drug delivery system for the tyrosine kinase inhibitor Gefitinib. *Adv. Healthc. Mater.* **2015**, *4*, 2816–2821. [CrossRef]
11. Yang, R.; Tian, J.; Liu, Y.; Yang, Z.; Wu, D.; Zhou, Z. Thermally induced encapsulation of food nutrients into phytoferritin through the flexible channels without additives. *J. Agric. Food Chem.* **2017**, *65*, 9950–9955. [CrossRef] [PubMed]
12. Yang, R.; Liu, Y.; Meng, D.; Chen, Z.; Blanchard, C.L.; Zhou, Z. Urea-driven epigallocatechin gallate (EGCG) permeation into the ferritin cage, an innovative method for fabrication of protein–polyphenol co-assemblies. *J. Agric. Food Chem.* **2017**, *65*, 1410–1419. [CrossRef] [PubMed]
13. Yang, R.; Liu, Y.; Blanchard, C.; Zhou, Z. Channel directed rutin nano-encapsulation in phytoferritin induced by guanidine hydrochloride. *Food Chem.* **2018**, *240*, 935–939. [CrossRef] [PubMed]
14. Wang, Q.; Zhang, C.; Liu, L.; Li, Z.; Guo, F.; Li, X.; Luo, J.; Zhao, D.; Liu, Y.; Su, Z. High hydrostatic pressure encapsulation of doxorubicin in ferritin nanocages with enhanced efficiency. *J. Biotechnol.* **2017**, *254*, 34–42. [CrossRef] [PubMed]
15. Pagadala, N.S.; Syed, K.; Tuszynski, J. Software for molecular docking: A review. *Biophys. Rev.* **2017**, *9*, 91–102. [CrossRef]
16. Subramanian, V.; Evans, D.G. A molecular dynamics and computational study of ligand docking and electron transfer in ferritins. *J. Phys. Chem. B* **2012**, *116*, 9287–9302. [CrossRef]

17. Shahwan, M.; Khan, M.S.; Husain, F.M.; Shamsi, A. Understanding binding between donepezil and human ferritin: Molecular docking and molecular dynamics simulation approach. *J. Biomol. Struct. Dyn.* **2020**, 1–9. [CrossRef]
18. Falvo, E.; Tremante, E.; Arcovito, A.; Papi, M.; Elad, N.; Boffi, A.; Morea, V.; Conti, G.; Toffoli, G.; Fracasso, G.; et al. Improved doxorubicin encapsulation and pharmacokinetics of ferritin-fusion protein nanocarriers bearing proline, serine, and alanine elements. *Biomacromolecules* **2016**, *17*, 514–522. [CrossRef]
19. Sugahara, K.N.; Teesalu, T.; Karmali, P.P.; Kotamraju, V.R.; Agemy, L.; Greenwald, D.R.; Ruoslahti, E. Coadministration of a tumor-penetrating peptide enhances the efficacy of cancer drugs. *Science* **2010**, *328*, 1031–1035. [CrossRef]
20. Yin, S.; Wang, Y.; Zhang, B.; Qu, Y.; Liu, Y.; Dai, S.; Zhang, Y.; Wang, Y.; Bi, J. Engineered Human Heavy-Chain Ferritin with Half-Life Extension and Tumor Targeting by PAS and RGDK Peptide Functionalization. *Pharmaceutics* **2021**, *13*, 521. [CrossRef]
21. Yin, S.; Zhang, B.; Lin, J.; Liu, Y.; Su, Z.; Bi, J. Development of purification process for dual-function recombinant human heavy-chain ferritin by the investigation of genetic modification impact on conformation. *Eng. Life Sci.* **2021**, *21*, 630–642. [CrossRef] [PubMed]
22. Liu, X.; Jiang, J.; Ji, Y.; Lu, J.; Chan, R.; Meng, H. Targeted drug delivery using iRGD peptide for solid cancer treatment. *Mol. Syst. Des. Eng.* **2017**, *2*, 370–379. [CrossRef] [PubMed]
23. Cai, Y.; Cao, C.; He, X.; Yang, C.; Tian, L.; Zhu, R.; Pan, Y. Enhanced magnetic resonance imaging and staining of cancer cells using ferrimagnetic H-ferritin nanoparticles with increasing core size. *Int. J. Nanomed.* **2015**, *10*, 2619–2634.
24. Zhang, C.; Liu, Y.; Feng, C.; Wang, Q.; Shi, H.; Zhao, D.; Yu, R.; Su, Z. Loss of PEG chain in routine SDS-PAGE analysis of PEG-maleimide modified protein. *Electrophoresis* **2015**, *36*, 371–374. [CrossRef] [PubMed]
25. Liu, X.; Jin, W.; Theil, E.C. Opening protein pores with chaotropes enhances Fe reduction and chelation of Fe from the ferritin biomineral. *Proc. Natl. Acad. Sci. USA* **2003**, *100*, 3653–3658. [CrossRef] [PubMed]
26. Liu, L.; Zhang, C.; Li, Z.; Wang, C.; Bi, J.; Yin, S.; Wang, Q.; Yu, R.; Liu, Y.; Su, Z. Albumin binding domain fusing R/K-X-X-R/K sequence for enhancing tumor delivery of doxorubicin. *Mol. Pharm.* **2017**, *14*, 3739–3749. [CrossRef]
27. Blazkova, I.; Nguyen, H.V.; Dostalova, S.; Kopel, P.; Stanisavljevic, M.; Vaculovicova, M.; Stiborova, M.; Eckschlager, T.; Kizek, R.; Adam, V. Apoferritin modified magnetic particles as doxorubicin carriers for anticancer drug delivery. *Int. J. Mol. Sci.* **2013**, *14*, 13391–13402. [CrossRef]
28. Bellini, M.; Mazzucchelli, S.; Galbiati, E.; Sommaruga, S.; Fiandra, L.; Truffi, M.; Rizzuto, M.A.; Colombo, M.; Tortora, P.; Corsi, F.; et al. Protein nanocages for self-triggered nuclear delivery of DNA-targeted chemotherapeutics in cancer cells. *J. Control. Release* **2014**, *196*, 184–196. [CrossRef]
29. Chen, H.; Zhang, S.; Xu, C.; Zhao, G. Engineering protein interfaces yields ferritin disassembly and reassembly under benign experimental conditions. *Chem. Commun.* **2016**, *52*, 7402–7405. [CrossRef]
30. Luzzago, A.; Cesareni, G. Isolation of point mutations that affect the folding of human heavy chain ferritin in *E. Coli*. *EMBO J.* **1989**, *8*, 569–576. [CrossRef]

Review

Living Sample Viability Measurement Methods from Traditional Assays to Nanomotion

Hamzah Al-madani [1,2,†], Hui Du [1,3,†], Junlie Yao [1,3,†], Hao Peng [1,3], Chenyang Yao [1,3], Bo Jiang [1], Aiguo Wu [1,4,*] and Fang Yang [1,4,*]

1. Cixi Institute of Biomedical Engineering, International Cooperation Base of Biomedical Materials Technology and Application, Chinese Academy of Sciences (CAS), Key Laboratory of Magnetic Materials and Devices, Zhejiang Engineering Research Center for Biomedical Materials, Ningbo Institute of Materials Technology and Engineering, CAS, Ningbo 315201, China; hamzah@nimte.ac.cn (H.A.-m.); duhui@nimte.ac.cn (H.D.); yaojunlie@nimte.ac.cn (J.Y.); penghao@nimte.ac.cn (H.P.); yaochenyang@nimte.ac.cn (C.Y.); jiangbo@nimte.ac.cn (B.J.)
2. University of Chinese Academy of Sciences, Beijing 100049, China
3. College of Materials Sciences and Opto-Electronic Technology, University of Chinese Academy of Sciences, Beijing 100049, China
4. Advanced Energy Science and Technology Guangdong Laboratory, Huizhou 516000, China
* Correspondence: aiguo@nimte.ac.cn (A.W.); yangf@nimte.ac.cn (F.Y.); Tel.: +86-574-6387-5030 (F.Y.)
† These authors contributed equally to this work.

Abstract: Living sample viability measurement is an extremely common process in medical, pharmaceutical, and biological fields, especially drug pharmacology and toxicology detection. Nowadays, there are a number of chemical, optical, and mechanical methods that have been developed in response to the growing demand for simple, rapid, accurate, and reliable real-time living sample viability assessment. In parallel, the development trend of viability measurement methods (VMMs) has increasingly shifted from traditional assays towards the innovative atomic force microscope (AFM) oscillating sensor method (referred to as nanomotion), which takes advantage of the adhesion of living samples to an oscillating surface. Herein, we provide a comprehensive review of the common VMMs, laying emphasis on their benefits and drawbacks, as well as evaluating the potential utility of VMMs. In addition, we discuss the nanomotion technique, focusing on its applications, sample attachment protocols, and result display methods. Furthermore, the challenges and future perspectives on nanomotion are commented on, mainly emphasizing scientific restrictions and development orientations.

Keywords: living sample viability measurement; atomic force microscopy; AFM oscillating sensor method; nanomotion

1. Introduction

The development and evaluation of new drugs take several years of investigations on living samples to explore drug pharmacology and toxicology. Compared to in vivo investigations on living sample viability, in vitro investigations are easier to execute and duplicate, the experimental settings are easier to regulate, and they are less morally problematic and costly [1]. In the past few decades, biological, chemical, and physical methods have been used for the rapid and accurate measurement of in vitro living sample behavior [2]. Living sample viability is a measure of the ratio of dead samples to live samples within a sample population. Living sample viability assays are used to assess the general health of samples and to track their survival after treatment with chemical agents or drugs. It is often expressed as a percentage of the control sample [3]. As the central parameter of living samples, viability is mainly measured through single-plate experiments or high-throughput screening, namely, pharmaceutical compound injection and living sample reaction record and assessment [4]. Apparently, living sample viability measurement plays an important role in clarifying the effects of drugs on cell proliferation and cytotoxicity, thus significantly reflecting drug safety and efficacy. For instance, living sample viability measurement provides great opportunities for analyzing the physiological behavior of anticancer drugs, such as selective ingestion and lethality in cancer cells, as well as biosecurity in non-tumor cells [5].

Recently, various chemical, optical, and mechanical methods possessing high accuracy and sensitivity have been developed in response to the demand for living sample viability determination [6]. Obviously, diverse measurement methods have their own superiorities and deficiencies depending on the application. Choosing the appropriate measurement method considers not only the test time, procedures, and the number of samples but also the application, cell line type, and host type. For instance, despite the extreme complexity of interpreting the experimental results of metabolic activity measurements, they have achieved significant progress. Among numerous viability measurement methods (VMMs), the atomic force microscopey (AFM) oscillating sensor method (named nanomotion), taking advantage of the adhesion effect of living samples to an oscillating surface, has emerged as a rapid, quantitative, real-time monitoring technique in the last decade [7]. To some extent, sufficiently detailed understanding of nanomotion strategy, from sampling attachment protocols to results display, will help achieve more reliable and repeatable living sample viability measurement.

This review aims to address the development of VMMs from traditional assays to nanomotion and to shed light on the novelty and practicability of nanomotion VMMs. Above all, various common VMMs are discussed and summarized in order to assess the potential areas of future development by discussing their most significant advantages and drawbacks. The use of nanomotion for monitoring living sample viability is discussed extensively through a comprehensive literature survey which summarizes the applications, the methods of sample adhesion on the microcantilever, and result display methods, and concludes with a consideration of the challenges and deficiencies that need to be addressed in the future. Finally, we hope that the review can promote the development of VMMs and present a promising innovative direction.

2. Living Sample Viability Measurement Methods

Generally, In previous reviews, according to the measurement principle or measurement procedures, VMMs have been classified in previous reviews as direct or indirect, labelled or label-free, and endpoint or real-time [8]. While in this review, VMMs are classified according to the equipment or materials used in the measurement process. VMMs are classified into chemical, optical, and mechanical measurement methods, as shown in Figure 1.

2.1. Chemical Viability Assays

Chemical viability assays work according to a common principle—the injection of living samples with one or more compounds. For instance, an anticancer drug's effectiveness or toxicity is evaluated through living sample interaction with drug's compound(s) [4]. Therefore, chemical assay identification and design depend on the drug's nature and vary according to the biomarkers used. The biomarker can be the outer surface of the sample membrane, nuclear size, or a metabolic process, such as the integrity of the membrane, adenosine triphosphate (ATP), the cellular esterases, enzyme function, and permeability. Chemical viability assays are mainly labelled, endpoint, and multi-sample methods. In general, chemical viability assays have several advantages. They are easy to perform, inexpensive, and rapid. They can be used to measure suspended or adherent samples and do not require complex techniques [1,9,10]. Figure 2 illustrates the wide classification of chemical viability assays and the various techniques they comprise.

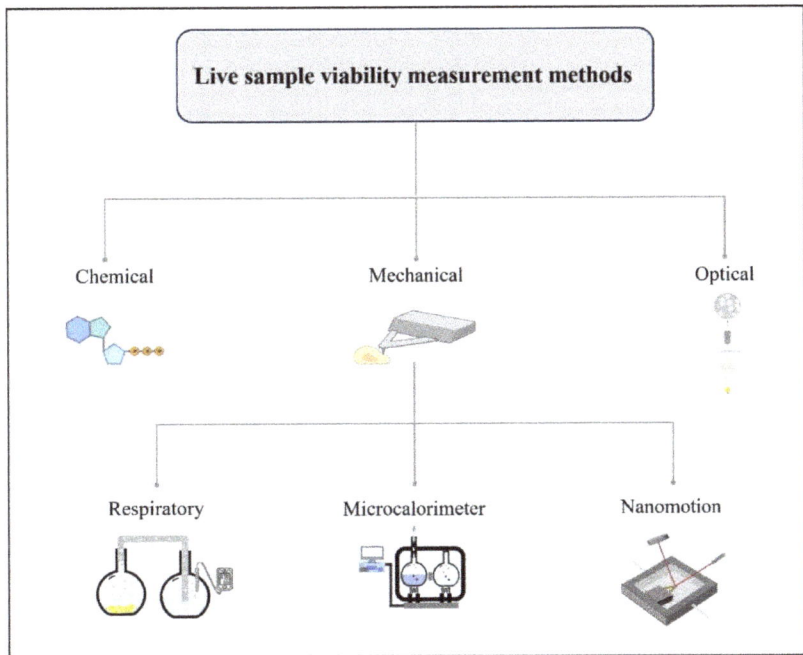

Figure 1. Viability measurement methods are classified according to the equipment or materials used in the measurement process, such as chemical viability assays and optical or mechanical methods.

Chemical viability assays are divided into five main categories: dye exclusion assays, fluorometric assays, luminometric assays, flow cytometry, and colorimetric assays. The principle of dye exclusion assays is based on the determination of membrane integrity. Dye exclusion assays determine the viability of suspension samples, with nonviable samples appearing in blue cytoplasm and living samples appearing in clear cytoplasm. Trypan blue [9–14] is a toxic assay for mammalian cells, and eosin [13,15–17], congo red [18], and erythrosine B stain assays [19,20] are nontoxic assays for mammalian cells. The principle of fluorometric assays is based on cellular esterases' cleavage of a nonfluorescent compound into a fluorescent compound. Fluorometric assays are light-sensitive and are used to determine the viability of both suspensions and adherent samples.

Fluorometric assays include three methods: resazurin (alamarblue) assay, 5-CFDA-AM assay, and fluorescein diacetate-propidium iodide. In the resazurin (alamarblue) assay, a healthy sample undergoes a non-reversible enzymatic reaction that turns the resazurin or alamarblue into a pink color resorufin that spreads in the medium such that, by measuring color change, healthy samples can be calculated [21–23]. In the instance of the 5-CFDA-AM assay, the living sample's enzymatic response transforms the assay into a fluorescent polar and impermeable solution that passes through healthy samples' cellular membranes. [24,25]. Propidium iodide, which interacts with the DNA of a dead sample, is combined with fluorescein diacetate, which is converted to fluorescein by esterase, to indicate apoptosis in the living sample. The combination of the two assays made it possible to measure the viability of living samples more accurately [26,27].

Luminometric assays include three methods: ATP (adenosine triphosphate), luciferase, and bioluminescent-nonlytic methods. The viability of living samples is determined in the ATP assay by using luminometers to assess intracellular ATP levels after the cells have been lysed to release intracellular ATP, which interacts with the luciferase enzyme [28–30]. The luciferase and bioluminescent-nonlytic methods are real-time viability measurement assays and can be used in continuous measurement applications [31–33]. In the luciferase method, cells are not dissolved to release ATP. Still, cells absorb the pro-substrate and turn it into a substrate that spreads in the medium. Bioluminescent-nonlytic assays include fluorescence and luminescent assays [10,34].

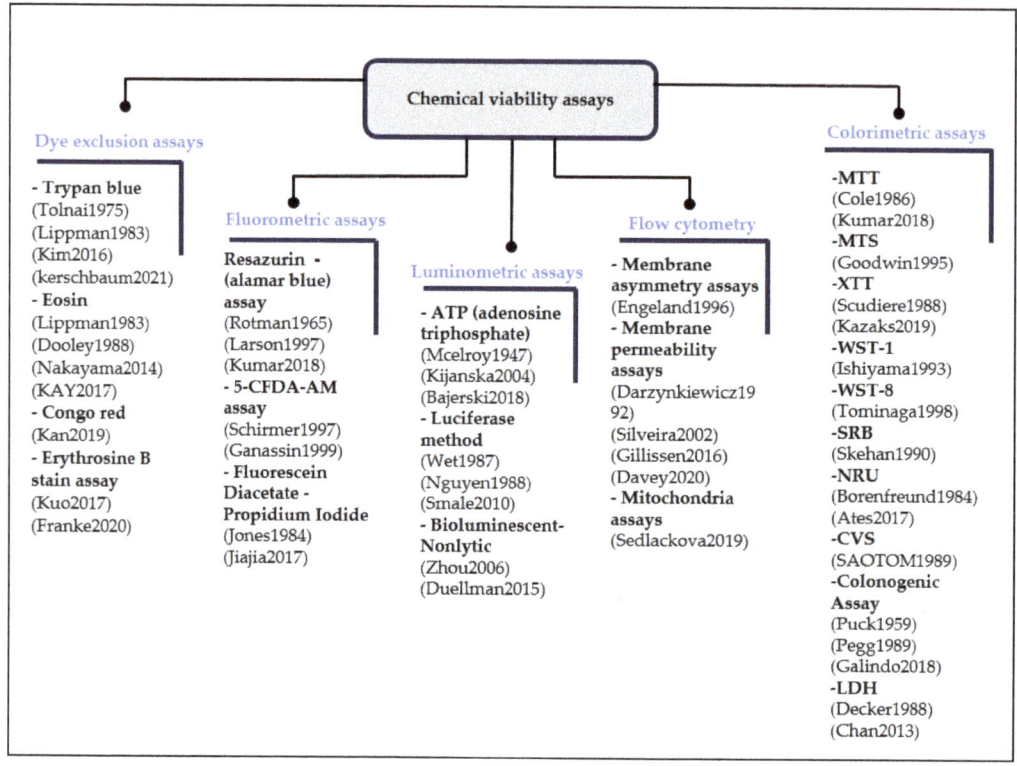

Figure 2. The broad classification of chemical viability assays and the various techniques they involve.

Flow cytometry includes three methods: membrane asymmetry assays, membrane permeability assays, and mitochondria assays. Membrane asymmetry assays are based on detecting changes in a cell membrane's outer surface [35]. However, membrane permeability assays are based on detecting cell membrane integrity and permeability [36–39]. Mitochondria assays involve the detection of the membrane potential, mass, or membrane permeability of mitochondria [40].

Colorimetric assays are based on the determination of metabolic activity and can be applied to both suspensions and adherent living samples. Colorimetric assays include ten methods: MTT, MTS, XTT, WST-1, WST-8, SRB, NRU, CVS, colonogenic assay, and LDH. The MTT assay is converted into a colorful formazan by the active metabolism of viable samples, and the intensity of the colored formazan is proportional to the number of live samples [21,41]. Unlike the MTT assay, the MTS assay is directly soluble in the sample medium [42]. Through online data processing, the XTT assay enables the processing of a large number of samples with high accuracy and speed [43,44]. The WST-1 assay is water-soluble, eliminating the need for a separate formazan dissolving step [45,46]. The WST-8 assay was developed from WST-1 and has the advantage that it is less toxic and more sensitive than other types of colorimetric assays [46,47]. The SRB assay is designed to be more sensitive than the MTT assay based on its ability to bind essential amino acid residues to proteins and does not depend on metabolic activities in measuring the viability of living samples [48,49]. NRU diffuses easily across the plasma membrane and binds to anionic sites in the lysosome. The NRU principle is based on the ability of viable samples to bind the neutral red dye in the lysosomes [50,51]. The CVS assay principle is based on measuring sample adherence by coloring attached samples with CVS, a protein and DNA binding dye [52]. The principle behind the colonogenic assay, also known as the plating assay, is that live samples will generate colonies that are easy to observe for which the number of surviving samples can be easily estimated [53–55]. LDH is released into the extracellular space when the plasma membrane is disrupted, which can be a significant indicator of necrotic cells [56,57].

2.2. Optical Measurement Methods

An optical measurement strategy is a non-invasive way of measuring viability and monitoring the effect of drugs or toxicity [58] which provides an excellent opportunity to observe vital sample processes, including living sample functions and activities. The principle of most optical microscopy and imaging techniques is based on measuring diseased samples by observing their morphology, distribution, or interaction with specific antibodies.

In the case of living cells, the mitochondrial network is dynamic (fuse, divide, and move) [59] but the shape changes yielding vesicular punctiform mitochondria occur at the early stages of cell death [60] and the cell shape is fragmented into small punctuate and round structures that collapse to become isolated, expanded, and more numerous in the case of programmed cell death [61] and elongated or donut-shaped during autophagy [62]. In addition, during necrosis or apoptosis, when cells are under stress, this results in the occurrence of many irregular plasma membrane bulges inside the cells, the formation of many large vacuoles, and the detachment of tissue culture plates. Thus, monitoring the shape and position of cells by optical measurement methods can allow the rapid measurement of cell viability in real-time.

2.2.1. Raman Spectroscopy

Raman spectroscopy (RS) is one of the most popular optical methods used to measure living sample viability [62,63]. Optical spectroscopy detects inelastic photon scattering caused by vibrational bonds in objects [64]. It is non-invasive and can be used to distinguish between healthy and dead samples. In addition, it is a rapid, label-free, real-time method that does not damage samples and works based on the sample's interaction with electromagnetic radiation to provide chemical fingerprints [65]. Analyzing the RS images of living samples has made it possible to calculate samples viability as a percentage of dead or diseased samples compared to healthy samples. The RS images show the morphological changes of living samples, and the multivariate analysis processes these images using a software database. The application of multivariate analysis has enabled the classification of samples according to morphological changes in various subcellular organelles, such as nuclei, mitochondria, and cytoplasm. Therefore, samples can be classified according to their health status [66]. Using RS images and multivariate analysis recognized by custom software, cancer cells were compared with normal cells, providing an apparent discrepancy showing the different shapes of cancer cell components compared with those of normal cells [67]. By applying multivariate analysis to RS data, breast cancer cells were classified into responsive or nonresponsive as a function of drug dosage and type based on the evaluation of metabolic changes [64]. RS enables sample archiving and retesting for more precise therapy response assessment. The advantages of RS maps have been harnessed to quantify dynamic changes at the single-cell level in terms of sensitivity, for spatial and temporal resolution of multiplexed metabolic changes, and for quantitative analysis [64,68].

Recently, an automated platform approach for high-throughput screening RS was created to overcome human factor errors, reduce test time, and increase the number of samples under measurement [62,69]. The automatic development of RS algorithms involves analyzing vast quantities of data and the creation of a reliable and comprehensive database for machine learning [70] so as to increase the speed and reliability of testing. RS was recently combined with deuterium labeling, and the findings indicated that this novel RS detection technology might be used to identify cancer cells at the single-cell level [63]. RS's challenges include the weak Raman signal [71] and light scattering [72], which reduce the method's sensitivity. When interacting with a sample, scattered light can cause frequency deflection due to scattered photons. Spectrum pretreatments and scattered photon filtering can mitigate this effect and increase the quality of the process [72]. As new optical methods have emerged, such as flow imaging microscopy, holography, and on-chip, lensless video microscopy, which will be addressed later, the development of optical measurement methods has helped to overcome the method's drawbacks.

2.2.2. Flow Imaging Microscopy

Flow imaging microscopy (FIM) is a rapid, label-free method used to determine living sample viability [73–75]. It is used to image the flow of fluids that contain vital components, such as human cells or protein particles. FIM captures the morphological changes of living samples and uses multivariate software to analyze FIM images to determine sample viability as a percentage of dead samples compared to healthy samples using the same working principle as Raman spectroscopy, though FIM takes sample images while the sample fluid is in continuous flow [75]. A specialized flow microscope is used for the measurement of living samples. In this system (schematically shown in Figure 3a), sequential bright digital images are captured when the sample passes through the flow cell. Living sample morphological information, number, size, and shape information are collected and then analyzed by software [76]. Flow imaging microscopy is a rapid and straightforward measurement method that reveals very subtle morphological changes in samples related to viability [75], such as mitochondria shape [60] and plasma membrane bulges [58]. The most important thing that distinguishes this technology from other optical microscopy techniques is its high throughput. FIM measures vital components individually one by one and calculates size distributions numerically using deep learning technology and a database generated by custom software [74].

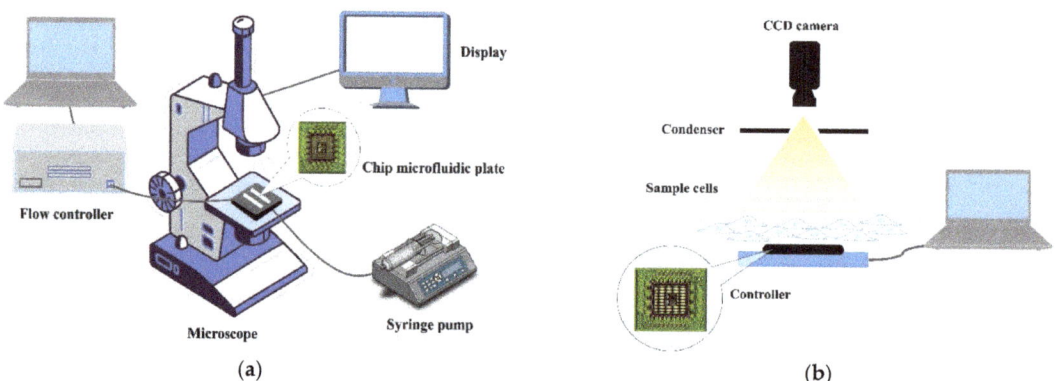

Figure 3. Optical measuring methods: (**a**) schematic of flow imaging microscopy (FIM) techniques; (**b**) digital holographic microscopy.

Currently, the most significant limitation is the speed of data analysis [74]. Microflow imaging provided higher measurement precision, while FlowCAM showed higher-resolution images [75]. Recently, a study using a convolutional neural network for image analysis based on flow imaging microscopy techniques was carried out for a cell-based medicinal products test [73]. However, this required a long time for analysis using algorithms and the application of a machine-learning model to several databases [77]. FIM has proved to be a powerful tool for overcoming vital sample classification difficulties when used in conjunction with image-processing technologies and advanced machine-learning approaches. High classification efficiency improved a dataset by removing nonrepresentative photos logically and methodically. On the other hand, misclassification emphasizes how challenging it is to identify FIM images at a single level [78].

2.2.3. Holography

Holography is a method for detecting living samples by observing morphological changes under stress or vibration resistance. Digital holographic microscopy provides a quantitative, contactless, non-destructive, and marker-free real-time monitoring method of living sample migration, adhesion, and dynamic change. It offers the possibility of measuring the efficacy of drugs in living samples [78,79]. Dead or diseased samples, for instance, will have a different intracellular structure from healthy samples. Changes in live sample structure parameters, such as volume, thickness, and intracellular composition, allow for the classification of living samples based on their health status and the calculation of viability percentages. Holography is a technique for measuring sample structure properties by scattering light after interacting with the sample. The scattering of light is affected by factors such as thickness, roundness, major axis, and intracellular composition [80]. Digital holographic microscopy is an optical microscopy technique that works on the interference between two waves, one from the sample and the other a reference wave from a charge-coupled device (CCD) digital camera, as shown in Figure 3b. In the context of the early diagnosis of cancer, a holographic microscope was used to distinguish between the morphology of cell tissues through a high-magnification optical technique that detects rapid changes resulting from mechanical or morphological changes. The method proved to be effective for cell thickness measurement in a culture medium [81]. It has been used to create high-resolution intensity images of a living sample and provide quantitative light phase and intensity information [80,82].

Due to its advantages, such as high efficiency, low cost, and flexibility to combine with other components, lens-free digital in-line holographic microscopy has become a valuable tool in the characterization and viability analysis of microbiological entities such as cancer cells [83]. Recently, a light-emitting diode has been used with the attachment of a pinhole structure as a practical light source. It enables direct observation of 3D bio-tissue without scanning and in the absence of noise caused by laser light [84,85].

2.2.4. On-Chip, Lensless Video Microscopy Technology

On-chip, lensless video microscopy technology is a label-free, real-time, and non-destructive VMM technology with a field of view twice that of a conventional microscope [86–92]. This technology does not require optical elements, such as lenses, or mechanical elements, such as probes. The areas and dimensions of samples vary according to their health status. By capturing the shadows of living samples and analyzing these images, samples can be divided according to their validity. By analyzing the sample shadows captured in digital images, morphological changes in samples could be monitored in real-time. As a result of the shadow imaging provided by on-chip, lensless microscopy, living sample viability tests could be performed without the need for any labeling or reagents [87]. On-chip, lensless video microscopy technology monitors more than one living sample type simultaneously through the use of microfluidic channels [91]. Large-scale parallel automated imaging can be enabled for large sample populations with a set of microscopes on a chip with low cross-contamination risk [90]. Lens-free imaging allows for a high-throughput screen for living sample viability in situ at the point of use due to its imaging reduced footprint. Data can rarely be collected from such commonly used sites as incubators due to the inhibitory nature of collecting standard microscopic and spectroscopic equipment [93].

The combination of microfluidic microscopy and high pixel resolution eliminates the need for expensive lenses, light sources, and mechanical microscanning [89]. The iterative phase recovery algorithm demonstrated the ability to retrieve and evaluate sample information using image quality algorithms even without references. This was enhanced by using machine-learning techniques [94]. On-chip, lensless video microscopy technology can provide label-free, non-destructive, continuous monitoring in the fields of treatment drug tests and toxicity and proliferation measurements [87]. The on-chip imaging system allows the monitoring of entire populations of living samples while tracking the fate of individual living samples within the population [92]. The main disadvantage of these methods is possible phototoxicity, since the cells and tissues are usually not exposed to direct light during their life cycle. Therefore, the optical microscope process must be designed to minimize phototoxicity. This can be avoided by choosing an efficient microscope and a suitable detector [64].

2.3. Mechanical Measuring Methods

Several mechanical or physical techniques have been developed to quantify living sample viability. These methods are based on the principle of measuring or monitoring one of the vital activities of living samples. For any living organism, adhesion, respiration, proliferation, electrical charge, and thermogenesis process activities are vital signs of life. Several methods have been developed to measure these activities based on monitoring viability.

2.3.1. Respiratory Measuring Methods

Respiratory activity is an essential metabolic activity. The ability to absorb and consume oxygen can be an important factor in indicating the ability of a living sample to survive. Monitoring the harm produced by chemical agents to the breathing activity of a living sample is used to determine the viability of living samples based on respiratory thermodynamic features. The percentage of dead samples relative to healthy samples can be calculated by comparing the oxygen absorbed by live samples with that absorbed by the controls [95]. Measuring the oxygen consumption of living samples requires closed containers isolated from ambient air [96]. Several techniques have been used to measure oxygen consumption, such as the Clark-type oxygen electrode [97] and electron paramagnetic resonance oximetry [98]. However, these methods have some disadvantages, including the difficulty of calibration, the risk of poisoning, the consumption of oxygen, and high costs, especially for large samples [96]. The optical oxygen sensor approach was utilized to avoid the limitations of the Clark-type oxygen electrode and electron paramagnetic resonance oximetry technologies. The optical oxygen sensor approach has been demonstrated to need periodic calibration, consume oxygen, and be sensitive to environmental conditions, such as temperature, pressure, flow, and salinity [99].

The measurement of oxygen consumption by living samples in tissue culture flasks has been carried out using an optical oxygen sensor [100]. The phosphorescence lifetime-based optical oxygen sensor is used to monitor viability response to chemical agents or toxics as a continuous, real-time, rapid, and high-throughput method [96]. While only optical contact between the probe and the fluorescent detector is involved, fluorescence-based oxygen sensors allow non-invasive detection through a clear container. Disposable sensors with fixed calibration are simple, inexpensive, and reliable, making them ideal for contactless microscale measurements. The device uses solid-state oxygen sensor inputs and a phosphorescence phase detector to detect the respiration patterns of living samples in a contactless manner. The sensor changes its phosphorescence lifetime in response to oxygen content, which does not need calibration and is monitored by a phase detector [96].

Oxygen-sensing microplates have been used to measure living sample viability using empirical correlations between fluorescence intensity and viability [101]. Algorithms have been set up to make assessments of the rate of oxygen consumed by living samples and to measure the theoretical correlation between fluorescence and viability [102]. Real-time and non-invasive measurements of oxygen uptake rate [103,104] and oxygen transfer rate [105] were directly correlated with living samples' metabolic activity. Scanning electrochemical microscopy (SECM)-assisted oxygen consumption measurement, which changes with cell nanoscale height, has been used as a real-time method for measuring the viability of a single living sample [106].

All metabolic processes in living organisms are heat-producing reactions [107]. Thus, heat flow indicates the number of metabolic reactions occurring in and the state of living samples [108]. Online oxygen monitoring is carried out by pumping the sample solution into a bioreactor containing a sensor sending signals. A computer translates these signals into data, as shown in Figure 4a. Recently, a method has been developed that combines cellular respiration measurement with measurement of living sample temperature changes and living sample proliferation rates directly by infrared thermal imaging, opening a promising avenue for the future of this technology [109].

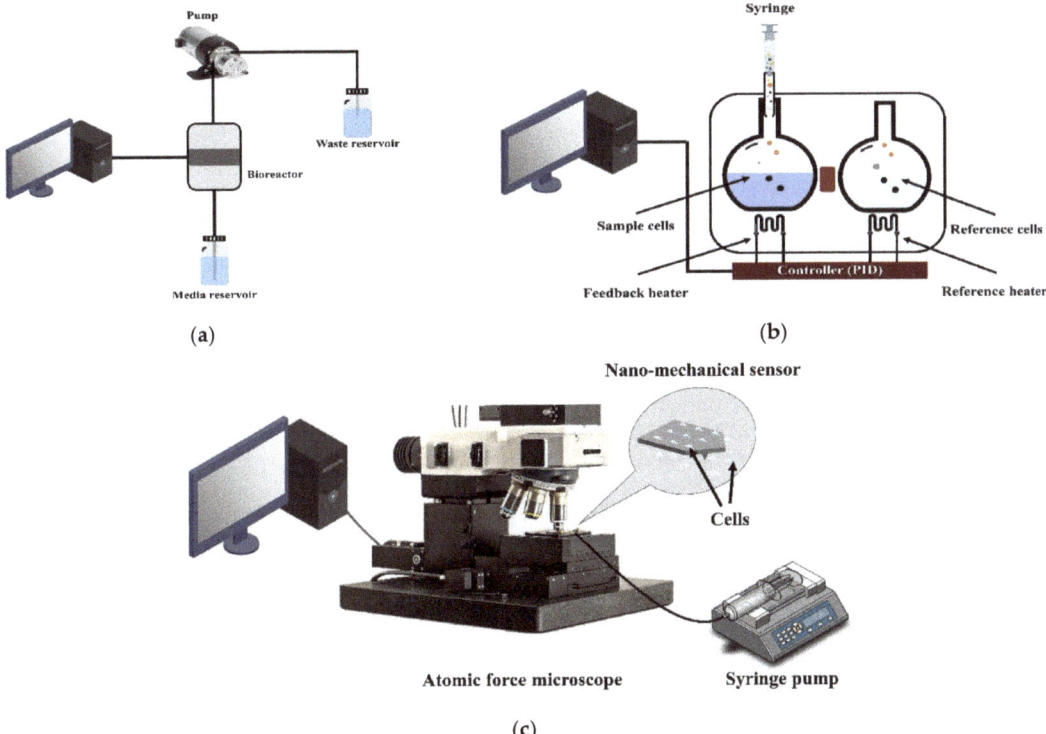

Figure 4. Mechanical measuring methods: (**a**) schematic of an online monitoring system based on respiration activity; (**b**) closed ampoule isothermal microcalorimetry; (**c**) nanomechanical oscillator sensor.

Two central problems with the polydimethylsiloxane (PDMS) materials used today in the manufacture of microfluidic chips show the need to use thermal materials suitable for measuring the oxygen consumed by biological organisms. The high oxygen permeability of PMDS makes respiration and viable oxygen measurements difficult. Owing to its lipophilic nature, it has well-known absorption capacities for biomarkers and medications. Thermoplastic polymer materials with low oxygen permeability, such as polyethylene terephthalate (PET) or cyclic olefin copolymer (COC), are required as chip materials.

On the other hand, the process of manufacturing these thermoplastic polymer materials remains a significant challenge. The use of the manufacturing process for thermoplastic sensor integration is critical; it allows for repeatable measurements across a series of experiments. This technology, which can be easily integrated into existing thermoplastic microfluidic systems and enables living sample respiration monitoring, may pave the way for a more uniform and controlled means of monitoring culture conditions in cell-on-a-chip microfluidic systems [95].

2.3.2. Microcalorimeter Measurement Methods

A microcalorimeter is an instrument designed for measuring the heat produced by microorganisms in closed bioreactors. By measuring the resulting heat employing a microcalorimeter, it is possible to directly monitor living organisms by comparing them with control samples. The percentage of dead samples relative to healthy samples can be calculated by comparing the heat energy produced by live samples to that produced by the controls [110]. Microcalorimetry can provide a continuous, direct real-time measurement of the activities of cellular components [107,110]. The heat energy vs. time curve is a complex construct used to measure a specific metabolic process. Still, it represents an ideal way to indicate total vital activities and living sample fate. Accurate microcalorimetric measurements are made using an isothermal calorimeter, where the measurement is carried out under constant temperature conditions. Many diverse studies in the literature refer to the use of isothermal calorimeters to measure the effect of pharmacokinetics on cells, microorganisms, and tissues, especially anticancer drugs [107,108,111,112]. Three parameters related to a living sample's respiratory system have been used to gauge viability: respiratory intensity, proliferation rate, and normal sample heat. [109].

Closed ampoule isothermal microcalorimetry has been used to evaluate the vital activities of samples in continuous real-time monitoring. In closed ampoule isothermal microcalorimetry, the heat flow between a sample and a heat sink is measured and compared to the heat flow between a reference sample and the heat sink. The measurement is carried out isothermally. The heat flux is recorded as an electrical signal after the calibration [111], as shown in Figure 4b.

The indirect measurements of heat energy produced by living samples include non-contact temperature mapping by temperature-sensing methods. However, besides the complexity of results interpretation, accuracy is affected either by radiation absorption or the limitation of temperature resolutions. In the case of direct methods, the temperature change inside living samples can be measured by nanoscale thermal probing. However, this can cause additional heat production due to the stress response of sample rupture. The limited accuracy of the sensor's energy resolution does not reach the level of single-sample temperature, which is at a level of pW. Therefore, most measurements are made by calculating the average heat generated by colonies of living samples, making determinations based on the calorimetry technique challenging.

The calorimeter principle makes it impossible to set up a sample development environment for an extended time. As a result, calorimeters are unsuitable for research in which living sample viability must be maintained for an extended period of time [113]. The most severe limitations of power resolution are thermal noise and noise created by microcalorimeter sensors. Thermal noise can be decreased by preventing heat exchange with excellent thermal isolation. Differential calorimetry is also an excellent method for removing interference. Furthermore, speedy systems are required for fast biological processes by reducing microcalorimeter chamber size. Moreover, developments in hardware, including data interpretation tools, will make microcalorimetry for living samples a standard tool [114].

Despite the literature indicating the possibility of using this method to measure cellular activities, only a few laboratories use microcalorimeters to measure vital activities. The reason for this may be high costs, which, according to the producers, are due to limited production levels. Still, perhaps by using accurate thermal electronic sensors, the demand for this technology will increase and costs will decrease [108].

2.3.3. Micro-Nanomechanical Oscillator Sensors

The path of research on (bio)sensors has recently turned, with great interest, to micro-nanomechanical systems as precise measurement and monitoring systems. Micro-nanomechanical oscillator sensors appear as experimental real-time measurement techniques. They have enabled the exploration of biological, mechanical, and chemical properties in vital samples and the testing of molecular interactions, biological activities, and dynamic properties at the level of a single molecule and the temporal changes in these properties [115].

Cancer-marker detection and anticancer drug testing are increasingly needed for rapid, real-time, and high-sensitivity techniques. Micro-nanomechanical oscillator sensor techniques may provide the desired alternative [116]. Micro-nanomechanical oscillators are characterized as label-free biosensors. The sample does not require any previous treatment with colorimetric or fluorescent dyes. The other advantage of this technique is that it does not require as much time as colorimetric or fluorescent dye processes. Furthermore, it is possible to measure a single particle or a tiny sensing area of a sample [117]. On the other hand, the accurate measurement of a single living sample in the range of a few nanometers requires direct mechanical contact of the microscale cantilever with a single living sample [118].

Single-living sample measurement is one of the most important advantages of micro-nanomechanical oscillator sensors. Measurement experiments with a sample population can only give an average measurement. This neglects the individual differences between single samples and considers them as homogeneous. A better understanding of biological processes is achieved by considering the heterogeneity of samples, especially in toxicological and anticancer drug tests [119]. Micro-nanomechanical oscillator sensors are susceptible to minimal deflections (at nanometric scale) caused by very small forces (piconewtons). They can be used in low quantitative measurements and in parallel format [120,121].

Micro-nanomechanical oscillator sensors are vibrating mechanical structures that are often cantilevers. This microscale cantilever vibrates as an oscillating mass sensor to which the vital part adheres. The sensitivity of the vibration frequency or the oscillation amplitude depends on the mass of particles adhered to the cantilever. In oscillating mass sensors, either molecular receptors (e.g., protein) stick on the surface of the microscale cantilever or the living sample (e.g., a cell) adhere to the surface of the microscale cantilever.

Antibodies or other molecular receptors stick to the microscale cantilever surface, which is vibration movement-controlled. By moving it towards the sample, molecular recognition occurs between the sample target molecules and the sensor's molecular receptors. This leads to mechanical, optical, or electrical interactions through which vital processes can be monitored [116,119–124].

Micro-nanomechanical oscillator sensors have been used to measure the interactions of surface receptors in vital samples. The chemical reactions of these receptors appear in the form of surface stress that can be traced and measured through micro-nanomechanical oscillator sensors. Surface stress caused by the receptor or the ligand causes micro-nanomechanical sensor deflection and changes in oscillation amplitude [123]. Some studies have used mechanical sensors for the dynamic examination of living cells [117,125].

Other studies have used mechanical sensors to monitor the activities of vital samples, such as cells or bacteria, including viability [120,126–128]. Understanding the mechanisms of the measurement process in the oscillation microscale cantilever is important for obtaining high measurement accuracy. Measurement accuracy depends on considering several factors when designing the measurement process. Factors include the viscosity of the oscillating medium, the adsorbed samples, the cantilever thickness compared to the thickness of the adsorbed samples, the adsorbed samples' locations on the cantilever and their distance from the cantilever clamping region, and the mechanical properties of the cantilever material [117].

Changing the health status of living samples alters some mechanical properties, such as weight and stiffness, and some biological properties, such as adhesion [129–131]. The oscillation amplitude of the micro-nanomechanical sensor varies with the alteration of adhering living sample mechanical properties and this is the principle behind measuring viability [128]. Healthy living samples may undergo death when their adhesion to the surface or the extracellular matrix (ECM) is lost. Undoubtedly, it is notable that cell death is accompanied by the loss of adhesion bonds with the surface or ECM [127]. The significance of this stems from adhesion being essential for viability. As a result, many researchers have developed theories to measure adhesion using nano-micromechanical sensors to monitor viability and toxicity.

One piece of advanced nanomechanical oscillator sensor technology that enables real-time, direct, label-free measurement is the atomic force microscopey (AFM). Living samples are connected to an AFM cantilever and placed in a test chamber. The vibrations of the cantilever are tracked over time. An optical transduction system detects and records cantilever dynamic oscillation deflections by reflecting a laser beam from the oscillation cantilever to a detector, as seen in Figure 4c.

AFM has been used in many measurement techniques and can be described as having three main components: an imaging mode, a force spectroscopy mode, and an oscillating sensor mode. AFM can directly image single membrane proteins and living samples at nanometer resolution in a buffer solution—a crucial advantage over other microscopy techniques. Real-time AFM imaging of single living samples can provide novel insights into dynamic processes [132]. The AFM force spectroscopy mode is, among others, used to measure the interactions of biological systems. The AFM's cantilever tip applies force to the living sample. This force may be a tensile force, a pressure force, or a shear force. AFM can be used to investigate the mechanical properties of microbiological systems ranging from tissues to nucleic acids [133]. Sample shape changes during its life cycle due to mechanical properties, such as internal and external forces [134]. AFM is capable of applying pN to nN forces to microscale indentors, allowing surface tension and tissue stiffness measurements [135]. In the detachment event, this technique directly measures a single sample's detachment force from the surface by applying vertical pressure force [134] or shear force [136–139]. This technique provides a label-free, rapid, and quantitative method to take measurements at the single-cell level [136]. By recording the deflection of the cantilever, the adhesion force can be obtained according to Hooke's law [140].

3. The AFM Oscillating Sensor Mode (Nanomotion)

3.1. Nanomotion Introduction

In measuring adhesion by single-cell force spectroscopy, a living sample is attached by force, which is contrary to the natural adhesion phenomenon. Natural living sample adhesion takes a longer time. It occurs to a lesser degree due to natural factors, such as gravity or the self-propulsion of the living sample [141]. Living sample adhesion is also measured in a detachment event that occurs after adhesion directly. However, adhesion, the number of bonds and the area of contact typically increase with time [142]. This causes the measurement process to be unrealistic and affects the accuracy of the results obtained. This adhesion fact motivated the creation of novel strategies that enable AFM to be used to evaluate adhesion in real-time without putting the living sample under stress by rapidly stretching it or pushing it to detach through cliffs. The AFM oscillating sensor modes are innovative methods that have recently been used to measure the adhesion of living samples, especially in measuring cell viability and the effects of chemical agents, such as drugs and toxic substances [143,144].

On the other hand, the traditional AFM methods provide end-state visualizations of sample fates as effects of chemical agents. They do not show the instantaneous effect of chemical agents on living samples. The real-time quality of nanomotion gives another additional advantage over AFM single-cell force spectroscopy methods.

The operating principle of this method is to take advantage of the high sensitivity of the AFM cantilever. Its high flexibility allows high sensitivity to the nanomotion resulting from the change in mass due to the adhesion of nanoparticles to the cantilever surface. The fundamental principle behind these biological oscillating frequencies is the measurement of the change in frequency response caused by the additional load of biomolecular mass attached to the cantilever surface. In general, AFM oscillating cantilevers measure cantilever deflection or frequency response changes caused by a mass of adhered biomolecules [92]. As a result, AFM oscillating cantilevers are being explored as sensitive mass detectors for biological tracking systems. Figure 5 illustrates a literature survey of AFM nonomotion techniques.

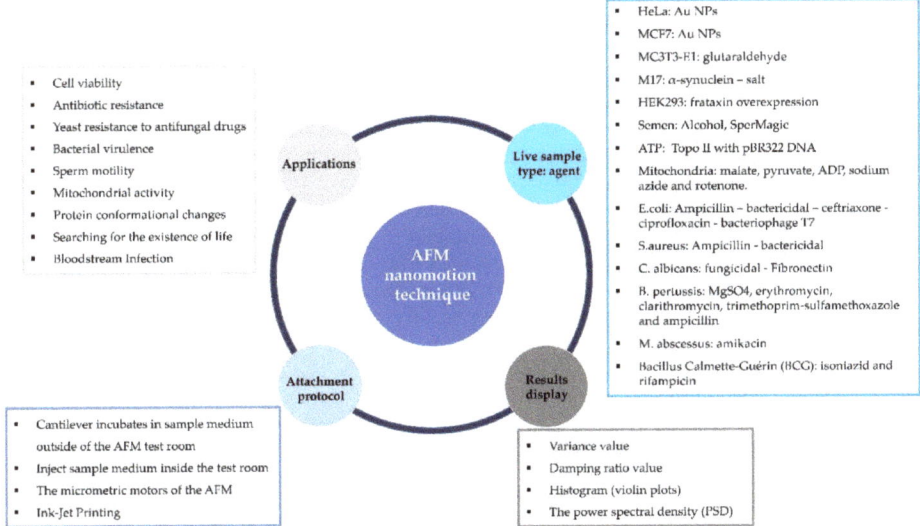

Figure 5. The AFM nanomotion technique. Different applications of nanomotion technology have been used for several types of living samples with different chemical agents. Different protocols were used to adhere the sample to the cantilever surface. Different display methods were used to present the results.

The deflection of a cantilever is proportional to the force. It results from the interaction of the cantilever with the sample according to Hooke's law [127]:

$$\Delta z = k_f^{-1} F \qquad (1)$$

where Δz is the deflection, k_f is the spring constant, and F is the acting force. In case a force is applied on a harmonical oscillation, the amplitude variation depends on the loading force, while the frequency remains constant, as shown in Equation (2) [127]:

$$z = \frac{F \cdot x^2 (3L - x)}{6EI} = \frac{g \cdot x^2 (3L - x)}{6EI} \cdot m_{cell} \qquad (2)$$

where x is the position of cell mass m_{cell} on a cantilever of length L and g, E, and I are the gravity coefficient, Young's modulus of the cantilever, and the moment of inertia of area, respectively.

Biomolecules are attached to the surface of an AFM cantilever that is implanted into a test chamber. The transformation of the cantilever oscillations over time is then tracked. An optical transduction system detects and records the dynamic oscillation frequencies of the cantilever via a laser beam reflected from the oscillating cantilever to a detector. This system's time resolution and sensitivity are ideally suited to studying living organisms at the nanoscale [145,146]. The AFM oscillating cantilever method provides a simple, highly weight-sensitive possibility for direct, real-time and single- or multi-cell measurements with high accuracy. When using the cantilever to evaluate cancer cells, an increase in the cantilever's deflection and a change in oscillation frequency indicates when cancer cells are adhered to the cantilever's surface.

The AFM oscillating sensor mode or AFM nanomotion makes for easier procedures than AFM single-cell force spectroscopy techniques. Still, the method neglects the heterogeneity of cells, though this is an advantage, since we are rarely interested in single-cell behavior but instead in statistical values. The AFM single-cell force spectroscopy methods that measure a single cell's adhesion do not yield high productivity in the measurement of anticancer drugs and cell viability. Furthermore, because of the irregular shape of single cells, theoretical models cannot describe the resulting changes in their shapes due to applying compressive or shear forces to the cells, making methods for calculating cell adhesion strength by surface stress very complex. The shape of a cell strongly influences the applied force, making the calculation of the exact adhesion force difficult.

This novel method serves as a new technique for monitoring cell viability by measuring cell adhesion and thereby as a new technique for testing drug efficacy and toxicity. The changes in the values of the cantilever's deflection and the frequency of vibration appear as a result of changes in cell state and cell detachment from the cantilever's surface as a result of death. As a result, this method could be used to assess the efficacy of drugs and toxins. It has provided many advantages over traditional methods for measuring cell viability, including direct, real-time, and label-free measurement in tens of minutes. In contrast, traditional methods take days or even weeks. This technique complements traditional VMMs. It may be promising for the long-term development of cell viability measurement. Unlike optical and electrical measurement methods, this test can give a quick and reliable direct measurement of the viability of biological organisms, even if they are not characterized [127,147].

3.2. Nanomotion Application

The AFM oscillating cantilevers were initially used as sensors for detecting the presence of bacteria and germs by measuring the effect of bacteria and germs attached to their surfaces in air or liquids. The presence of bacteria or germs was indicated by a change in a cantilever's vibration amplitude and frequency [117,143].

AFM oscillating cantilevers have been used to study the biological activities of bacteria as well as the effects of antibiotics and medicines on them, as shown in more detail in Table 1. The effect of antibiotics on bacteria was detected by changing the vibrations of the AFM oscillating cantilever, as living bacteria produced a larger cantilever deflection compared to antibiotic-treated bacteria [126]. A physical model has been developed to approximate the sum of the spectral frequencies caused by different amplitudes and the frequencies caused by adherent bacteria at various locations on a cantilever [147]. A measuring device based on the AFM oscillating cantilever principle was developed to measure minimal inhibitory and bactericidal concentrations and bacterial metabolic activities [148].

In the case of bloodstream infection, an innovative, rapid early detection of infecting microorganisms was obtained along with an accurate determination of their antibiotic susceptibility using the AFM oscillating cantilever [149]. Nanomotion was employed to detect sperm motility caused by exposed chemical agents. Living sperm produced less deflection with inhibitory chemicals and more deflection after treatment with stimulatory chemicals [150]. One of the recent studies used the AFM oscillating cantilever technique in laboratory experiments that could be used in the search for living organisms in space. The researchers compared the sensitivity of the frequencies and the amplitudes of cantilever vibration resulting from the adhesion of various biological organisms, such as bacteria, yeasts, animal cells, plant cells, and human cells. The study found that when living organisms died, the cantilever's oscillation amplitude decreased [92,120].

The AFM oscillating cantilever technique has been used to monitor cell viability [120,127,144,151,152] by measuring cantilever deflection change due to cell detachment. The effect of anticancer drugs and toxins on a cell's fate and metabolic activities could be successfully derived. The AFM oscillating cantilever technique was also applied to cellular organs, such as mitochondria [145] and ATP [153]. A difference in oscillation was found depending on the state of the mitochondria and their metabolic activity [145]. A group of researchers employed nanomotion to analyze the dynamics of enzymes in response to ligands such as ATP, and this has provided a novel way to investigate protein–ligand interactions [153].

Wu et al. examined the effect of paclitaxel on the MCF-7 breast cancer cell line. The cells settled down onto the microcantilever due to gravity. The attached cells were cultured and incubated for 4 h at 37 °C in a humidified atmosphere of 5% CO_2 until they adhered to the cantilever. The positions of the adhered cells on the cantilever surface were controlled without cells on the free end, which were used to reflect the laser beam [143]. Without controlling the locations of the adherent cells on the surface of the cantilever, but in the same way as adhering cells and under the same conditions applied for incubating the cells, Ruggeri et al. studied the response of a neuron model system to monomeric and toxic amyloid aggregated species of α-syn using an M17 dopaminergic neuroblastoma cell line [152]. To assess dose-dependent toxicity and monitor cell viability by measuring cell adhesion, Yang et al. used an AFM microscope as an early cell-death marker [127]. Three different sizes and surface coatings of Au NPs were added to a HeLa immortal cell line and an MCF-7 breast cancer cell line to measure cell viability. Nanomotion was used to provide information on the cell metabolic changes caused by frataxin deficiency under oxidative stress conditions [151].

3.3. Attachment Protocol

The sample's adhesion to the surface of the microcantilever is crucial to the success of the measurement method. The measurement process principle is based on the adhesion of the cell or the bacteria to the surface of the cantilever as an indication of its viability, with non-adhesion as a sign of death. Choosing the appropriate adhesion protocol depends on the sample's nature, size, and concentration. There are four main factors to consider when choosing a protocol for sample immobilization on the cantilever surface: the process of adhesion should take place in an environment that preserves the live sample; maintenance of the environment under the same conditions during all stages of the process; the possibility of controlling the location and number of cells or bacteria on the cantilever surface; and the risk of contamination, sample death, or cantilever damage. Figure 6 is an illustration of four techniques used in sample immobilization.

Table 1. Literature survey of AFM nonomotion viability measurement method.

Attachment Protocol	Results Display	Application	Cell Type	Time	Agent	Cantilever Type	Cantilever Functionalization	Ref.
Inject sample medium inside AFM test room	Variance value	Antibiotic resistance	E. coli and S. aureus	60–90 min	Ampicillin	DNP-10, Bruker	APTES (0.2%, 1.5 min)	[126]
Cantilever incubates in sample medium outside of the AFM test room	Variance value	Antibiotic resistance	E. Coli	2 h	Ampicillin	DNP-10, Bruker	Glutaraldehyde (0.5%, 7 min)	[154]
Cantilever incubates in sample medium outside of the AFM test room	Variance value; power spectral density	Protein conformational changes	Ligands, such as ATP	<10 min	Topo II enzymes with Pbr322 DNA (200 nm)	DNP-10, Bruker	APTES (0.1%, 1 min)	[153]
Cantilever incubates in sample medium outside of the AFM test room and Micrometric motors of the AFM (AFM single-cell force spectroscopy)	Variance value	Life-searching experiments on Earth and interplanetary missions	E. coli	>190 min	Bactericidal dose (10 μg/mL)	DNP-10, Bruker	Glutaraldehyde (0.5%, 7 min)	[120]
			S. aureus	>190 min	Bactericidal dose (2 μg/mL)		Glutaraldehyde (0.5%, 7 min)	
			C. albicans	>190 min	Fungicidal dose (20 μg/mL)		Glutaraldehyde (0.5%, 7 min)	
			MC3T3-E1	>190 min	5% glutaraldehyde		Fibronection (10 μg/mL, 15 min)	
			M17	>190 min	Salt concentration increasing		Poly-L-lysine (10%, 30 min)	
Cantilever incubates in sample medium outside of the AFM test room	Variance value	Cell viability	MCF7	7 h	Paclitaxel	DNP-10, Bruker	APTES (10%, 30 min)	[144]
Inject sample medium inside AFM test room	Damping value	Cell viability	Hela and MCF7	4–5 h	Au NPs	SNL-10, Bruker	-	[127]
Micrometric motors of the AFM (AFM single-cell force spectroscopy)	Variance value	Single-cell cytotoxicity assays	M17	7 h	Extracellular monomeric and amyloid α-synuclein species	DNP-10, Bruker	Poly-L-lysine (10%, 30 min)	[152]

Table 1. Cont.

Attachment Protocol	Results Display	Application	Cell Type	Time	Agent	Cantilever Type	Cantilever Functionalization	Ref.
Cantilever incubates in sample medium outside of the AFM test room	Variance value	Bloodstream infection	E. coli	90 min	Ceftriaxone, ciprofloxacin and ampicillin	NP-O10, Bruker	Glutaraldehyde (0.5%, 7 min)	[149]
Cantilever incubates in sample medium outside of the AFM test room	Variance value	Mitochondrial activity detected	Mitochondria-embryonic kidney cells	110 min	Malate, pyruvate, ADP, sodium azide, and rotenone	NP-O10, Bruker	Glutaraldehyde (5%, 10 min)	[145]
Inject sample medium inside AFM test room	Variance value	Sperm motility	Semen	-	Alcohol, spermagic	-	APTES (10%, 15 min)	[150]
Cantilever incubates in sample medium outside of the AFM test room	Variance value	Antibiotic resistance	B. pertussis	100 min	Erythromycin (Sigma- E6376); clarithromycin (Sigma - A3487), trimthoprim-sulfamethoxazole	-	Glutaraldehyde (0.5%, 10 min)	[148]
Cantilever incubates in sample medium outside of the AFM test room	Variance value	Antibiotic resistance	Bacillus Calmette-Guérin (BCG) and M. abscessus	200 min	BCG vs. Isoniazid and rifampicin M. abscessus vs. Amikacin	DNP-10, Bruker and SD-qp-CONT, NanoandMore	Glutaraldehyde (0.5%, 15 min)	[155]
The micrometric motors of the AFM (AFM single-cell force spectroscopy)	Variance value	Cell metabolic changes	HEK293	40 min	Frataxin overexpression	DNP-10, Bruker	Poly-D-lysine (20 μg/mL, 15 min)	[151]
Inject sample medium inside AFM test room	Variance value	Antibiotic resistance	E. coli	120 min	Bacteriophage T7	RC800PSA, Olympus	Poly-L-lysine (0.01%, 15 min)	[156]
Cantilever incubates in sample medium outside of the AFM test room	Variance value	Yeast resistance to antifungal drugs	C. albicans	>2 h	Fibronectin	Qp-CONT, nanoandmore	Con A (2 mg/mL, 30 min)	[157]
Cantilever incubates in sample medium outside of the AFM test room	Violin plots	Bacterial virulence	B. pertussis	5 min	Mgso4	SD-qp-CONT, nanoandmore	Poly-L-lysine (0.1%, 5 min)	[158]
Cantilever incubates in sample medium outside of the AFM test room	Variance value	Viability and susceptibility of microorganisms	E. coli and S. aureus	4 h	Ampicillin, glutaraldehyde	SD-qp-CONT, nanoandmore	Glutaraldehyde (0.5%, 10 min)	[159]

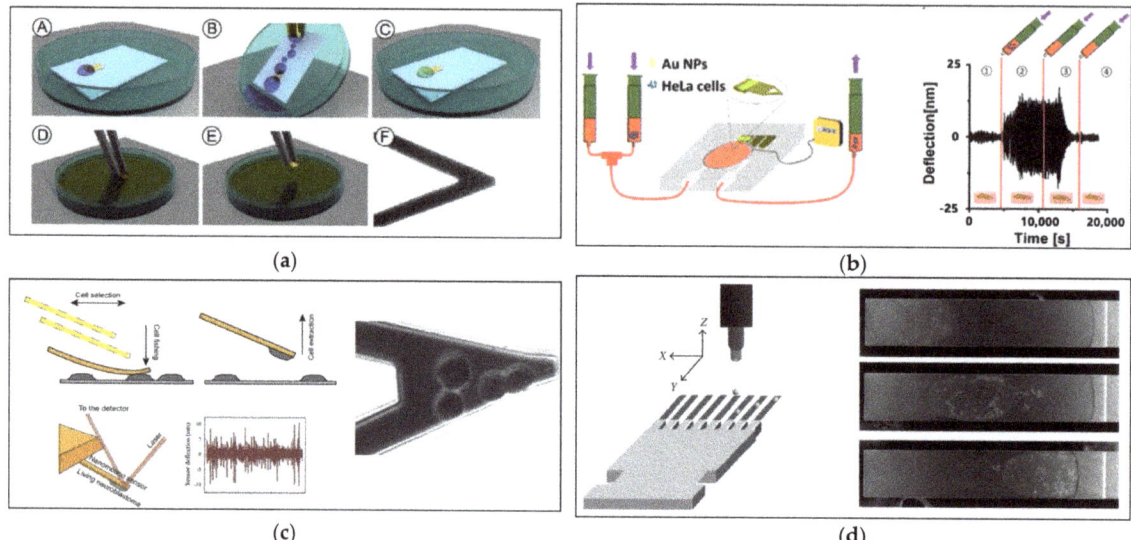

Figure 6. The AFM nanomotion technique attachment protocols: (**a**) cantilever incubated in sample medium outside the AFM test room, (A) functionalizing chemical is placed on the cantilever surface, (B) the remaining chemical is washed using pure water then the cantilever allowed to dry, (C) sample is deposited on the cantilever surface, (D), (E) cantilever is immersed in and out of the culture medium to remove loosing samples, (F) make sure attachment done with sufficient number of sample and no loosely attached samples, "reprinted with permission from Ref. [160]. 2018, École polytechnique fédérale de Lausanne"; (**b**) injection of sample medium inside the test room, (1) the cantilever vibrates at a specific frequency and the deflection is recorded over time, (2) samples are injected and allowed to adhere to the cantilever; as cells attach, deflection increases, (3) chemical agents are injected, and when cells start interacting with the agents, cells start to detach from the cantilever, causing the deflection decrease, (4) the cantilever is washed in preparation for the following measurement cycle, "reprinted with permission from Ref. [127]. 2017, Springer Nature"; (**c**) the micrometric motors of the AFM or the AFM single−cell force spectroscopy, "reprinted with permission from Ref. [152]. 2017, Springer Nature"; (**d**) ink−jet printing method, "reprinted with permission from Ref. [161]. 2012, Hindawi Publishing Corporation".

The direct attachment method is the most commonly used, the easiest, and the least expensive method. The cantilever is incubated in the live sample medium outside the AFM test room; see Figure 6a. This method has been used to measure the viability of bacteria [148,149,151,153–155,157–159] and cells [120,144,145]. The method is carried out by placing a small amount of high-concentration sample medium directly over the cantilever and leaving it for a period of time until the sample settles on the cantilever. After that, the suspended samples are washed by dipping the cantilever in a medium. Then, the cantilever is transferred to the test room and mounted on the AFM. Before placing the sample medium on the cantilever, it is prepared by applying a quantity of certain functionalizing chemicals for a period of time; the cantilever is then washed with water and dried [160]. The drawbacks of this method include the different conditions under which the adhesion process takes place compared to the conditions of the chemical effect process, while the transfer of the cantilever out of the medium between the two stages affects the accuracy and reliability of the test. The adhesion process is carried out under different conditions of the chemical effect process and the inability to control the number of samples and their position above the cantilever's surface, random sedimentation, and the possibility of contamination, sample death, or cantilever damage when handling and installing the cantilever are some of the additional disadvantages of this method.

To overcome the drawbacks of direct immobilization, a high concentration of live sample medium is injected inside the test room [126,127,150,157]. For this purpose, an injection system containing a cantilever was designed inside a fluid chamber inserted into the head of the AFM, as shown in Figure 6b. The AFM cantilever was housed in the thermostatically regulated and sealed test section with in-and-out liquid connections. Samples and chemical agents were collected in syringes. The cantilever oscillated with a constant amplitude at a certain frequency over time. The living samples were then injected and allowed to adhere to the cantilever for a period of time. Thus, the cantilever deflection increased with the increase of adherent living samples. When the cantilever had reached a stable state, the chemicals were injected into the test room. After a while, the effect on the living sample began, and the dead samples detached from the cantilever. In this case, the cantilever's deflection decreased as the number of dead samples increased until it reached its original state after death and all the samples detached. The cantilever's instantaneous deflection was controlled via feedback based on the detection of a camera installed on the microscope [127]. The living samples and chemical agents injected were implemented in the same test room and under the same conditions. Performing all steps under the same conditions enhances the accuracy and reliability of the test. There is no risk of contamination or death of samples due to the handling and installation of the cantilever with attached cells. However, the drawbacks of this method include random sedimentation, the requirement for a high sample concentration, and the inability to control the number of samples and their position above the cantilever's surface.

AFM single-cell force spectroscopy (the micrometric motors of the AFM technique) [151] was used as a sample attachment protocol. The single force technique was originally used to measure the strength of single-cell adhesion. The change in cantilever deflection was measured as the change in the force needed to overcome the adhesion force of a cell (attached to the cantilever) to another cell, to a cell mass, or to a surface [162]. In preparation for the measurement process, a cell is attached to the surface of the cantilever by pressing it against it for a certain period of time, then lifting the cantilever and the cell attached to it; see Figure 6c. Sample immobilization in nanomotion is accomplished by aligning the functionalized cantilever above the single cell and then lowering it until it presses against the cell with a certain force for a certain length of time (5 nN, >3 min). The cantilever is then lifted with the cell attached to its lower surface. The process is repeated once to adhere a new cell in another place on the lower cantilever surface and repeated again depending on the required number of cells. As a result, this method can be used to detect the nanomotion behavior of a small number of cells while maintaining a high level of control over the number and location of samples [120,152]. The AFM single-cell force spectroscopy immobilization protocol is characterized by the possibility of conducting a nanomotion test for a single cell or multiple cells. The location and number of cells or bacteria can be controlled. The adhesion and chemical effect processes are carried out in the same test room and under the same conditions.

However, this technique is more complex and requires expensive equipment. A sample is limited by its size and by cantilever size. There is substantial potential for misinterpretation of data due to cell damage during the adhesion process [163]. Studies have demonstrated the effect of bacterial adhesion position distribution on the cantilever, and the effects increased when the adhesion positions were close to the cantilever's free end [164]. The cantilever vibration due to the attached bacteria was caused not only by the mass effect but also by the bacterial cells' stiffness. This directly affected the sensitivity of nanomotion technology. By comparing theoretical results with measurements in air and deionized water, the viscosity effect of the measuring medium was determined [117].

Ink-jet printing has been employed to immobilize samples on cantilevers, demonstrating its superiority as a method for nanomotion and real-time monitoring measures [161]. The ink-jet technique enables the choice of immobilized samples' positions on cantilevers, which enables the study of the effects of sample location on cantilever nanomotion behavior; see Figure 6d. Controlling the location of samples allows for more flexibility in selecting optimal positions for reflection, resonance, and low noise [143]. The ink-jet printing immobilization process is monitored by a charge-coupled device (CCD) [163]. The ink-jet printing immobilization protocol is characterized by the fact that the location of cells or bacteria can be controlled. The adhesion and chemical effect processes are carried out in the same test room and under the same conditions. However, with this technique, it is not possible to control the number of cells or bacteria; it is also complex and requires expensive equipment. Table 2 provides a summary comparison of AFM nonomotion living sample attachment protocols.

3.4. Results Display

Many sensitive displacement sensors have been developed in order to read out and display the minute deflections of microcantilevers. Optical beam deflection, piezoelectric, and piezoresistive read-out techniques are the ones most commonly used [134]. The most popular read-out technique is optical beam deflection. For a single sensor, it is easy to execute and achieves angstrom resolution. A laser is focused on the cantilever free-end and reflected from it to be detected by a position-sensitive photodiode. A photodetector measures the displacement of the reflected laser beam due to cantilever deflection. The AFM controller receives data from the photodetector and transfers them to a PicoForce spectrometer for deflection recording. A multimeter records the data from the PicoForce spectrometer and then transfers it to the monitoring system [127]. The resulting data do not show static deflection but are presented on a time-dependent deflection chart. The dynamic deflection data collected with the cantilever are usually analyzed using homemade software. The deflection signal is represented as a continuous curve that monitored the cantilever's deflection over time [126]. The amplitude variance of cantilever oscillations that appears in experimental results reflects the metabolic state of the biological samples and the effects of the chemical agents. Several methods were utilized to display deflection data so that the effects of chemical agents on samples could be compared clearly and productively.

Many studies have used the variance of cantilever deflections to compare results from different experiments, as shown in Table 1. When the living samples adhered to the cantilever, the deflection variance increased, but after the injection of chemical agents and their interaction with the samples, the variance value reduced dramatically, indicating that the vital samples had died, as shown in Figure 7a. The following equation was used to determine the variance (Var) values that were utilized to quantify the deflection fluctuations (z_i) [153]:

$$\text{Var} = \frac{1}{N-1} \sum_{i=1}^{N} (z_i - z)^2 \qquad (3)$$

As the number of samples adhering to the cantilever increases, the variance bars used to show the cantilever's deflection increase. The variance decreases as the number of samples adhering to the cantilever decreases. If the number of samples on the cantilever remains constant, the cantilever oscillates with a steady variance [154].

Table 2. AFM nonomotion living sample attachment protocols.

Attachment Protocol	Incubation Condition	Advantages	Drawbacks	Ref.
Cantilever incubated in sample medium outside of the AFM test room	The adhesion process is carried out under different conditions of the chemical effect process	Easy and no need for expensive equipment	The location and number of cells or bacteria cannot be controlled; When handling and installing the cantilever, there is a risk of contamination, sample death, or cantilever damage	[143,148,149,151,153–155,157–159]
Inject sample medium inside the test room	The adhesion and chemical effect processes are carried out in the same test room and under the same conditions	All measurement processes are carried out under the same conditions; There is no risk of contamination or death of cells or bacteria	The location and number of cells or bacteria cannot be controlled; Requires high sample concentration	[126,127,150,156]
The micrometric motors of the AFM—AFM single-cell force spectroscopy	The adhesion and chemical effect processes are carried out in the same test room and under the same conditions	The location and number of cells or bacteria can be controlled; It is a single-cell and multi-cell measurement process	Complex and expensive equipment; There is a risk of cell injury during the adhesion process; A sample is limited by its size and by cantilever size	[120,151,152]
Ink-jet printing	The adhesion and chemical effect processes are carried out in the same test room and under the same conditions	The location of cells or bacteria can be controlled; There is no risk of contamination or death of cells or bacteria	Complex and expensive equipment is needed; The number of cells or bacteria cannot be controlled	[161,165]

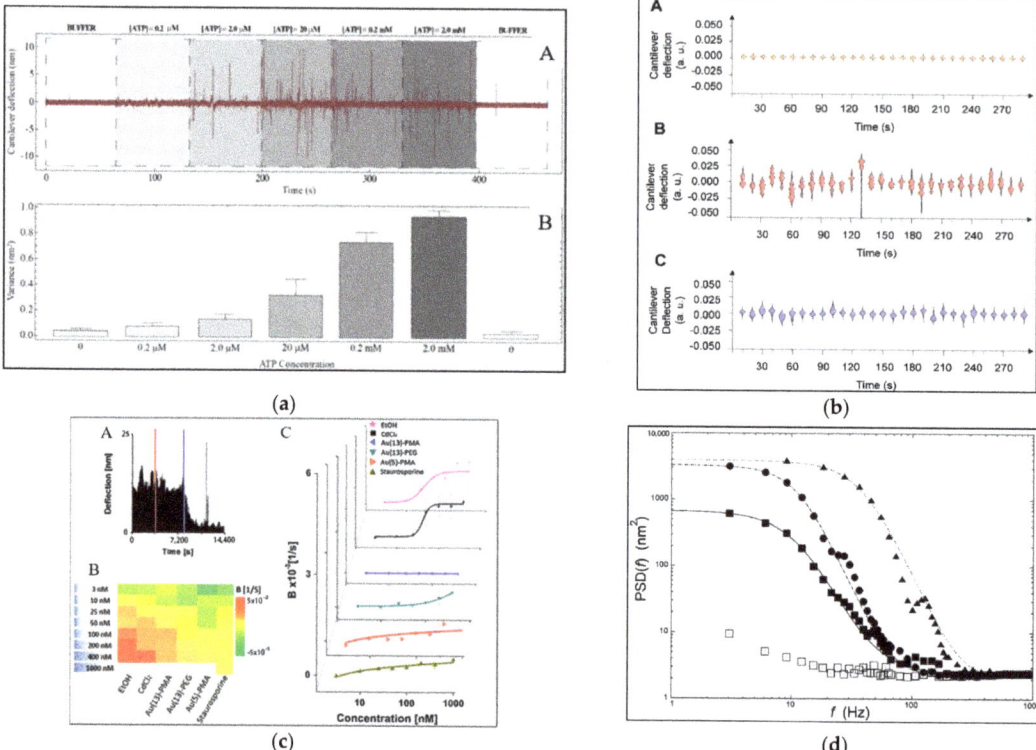

Figure 7. The AFM nanomotion technique results display methods: (**a**) the variance of the cantilever deflection result, (A) the cantilever deflections as a function of ATP concentration, (B) corresponding variance values, "reprinted with permission from Ref. [153]. 2014, Plos One"; (**b**) violin plot for 10 s chunk of the cantilever deflection result, (A) nanomotion cantilever violin plot without samples, (B) with virulent sample and (C) with avirulent sample, "reprinted with permission from Ref. [158]. 2021, MDPI"; (**c**) damping value (B value) of the cantilever deflection result, (A) cantilever oscillation deflection amplitude versus time, (B) heatmap of the damping constants, (C) damping constants B for different agents versus agents' concentration, "reprinted with permission from Ref. [127]. 2017, Springer Nature"; (**d**) the power spectral density (PSD) of the cantilever deflection result (black squares: 2.0 µM ATP concentration, black circles: 0.2 mM, black triangles: 2.0 mM and white squares: baseline), "reprinted with permission from Ref. [153]. 2014, Plos One".

Expressing the results by variance is an easy and practical method that enables clarification of the effects of chemical agents on living samples. Still, the variance shows the end state of the effect and does not show the instantaneous effect during the process. The variance from when the living samples were adhered to the cantilever and before the chemicals were injected is compared with the variance after the living samples died and detached from the cantilever. Referring to the deflection curve in Figure 6b, it can be noticed that there is damping for the oscillation amplitude of the cantilever that decreases from the highest value, when all the living samples have adhered, to the lowest value, which is equal to the deflection value of the free cantilever without any attached samples. This decrease (damping) takes some time and does not suddenly happen, and damping decreases as the number of adhered samples decreases. Variance does not provide a clear picture of the progression of the reaction of the living samples with chemicals from the beginning of the injection until the samples' death. Here, we need to find a result display method that enables us to show the instantaneous effect of chemicals on living samples, especially nanoparticles, whose effect may last for several hours or even days.

To overcome drawbacks in displaying the nanomotion results by variance, several methods have been used to display the instantaneous effect of the nanomotion method, as illustrated in Figure 7b–d. The damping coefficient of exponential attenuation that appeared in the cantilever deflection curve was used to define the deflection of the cantilever [127]. The deflection damping coefficient shows the decreasing deflection period from the start of the chemical agents' effects until the death and detachment of samples from the cantilever surface. Region 3 in Figure 6b shows that the oscillation exhibits exponential reduction, which is defined here by a damping coefficient (B value). As a result, the B value indicates the amplitude of the cantilever, i.e., the damping rate increases as the B value increases. An exponential function was used to represent the oscillation damping, as in the following equation [127]:

$$A(T) = A_0 e^{-Bt} \quad (4)$$

where: A_0 is the amplitude at t = 0. The damping coefficient (B) can then be estimated by solving Equation (4) as:

$$B = -\frac{t}{\tau} \ln\left(\frac{A}{A_0}\right) \quad (5)$$

The B value is equal to the negative logarithmic amplitude ratio at any point t in the chemical agents' effect step and the amplitude at the beginning of the chemical agents' effect step (t = 0). The B value represents the amplitude damping rate; the higher the B value, the higher the amplitude damping rate. Hence, the more significant the chemical agents' effect. As shown in Figure 7c, the value of B increases as the concentration of chemical or toxic (in this case) agents rises, and this is valid for various types of chemical agents.

The deflection values of the cantilever were divided into 10-second chunks. Then, a violin plot was plotted for each chunk [158]. The violin plot is a nanomotion spectrogram reconstructed from a histogram. The vertical axis represents cantilever amplitude and the horizontal axis displays the number of oscillation events symmetrically. Figure 7b shows that violin plots were repeated during the different measurement stages. They were used to compare the deflections of the cantilever. The figure shows a change in violin plot height before and after the samples were attached to the cantilever and after the living samples were affected due to the injection of chemical agents. This "violin plot" may be more complex than the variance method but it may enable real-time monitoring of the measurement process.

Power spectral density (PSD) was used to represent the deflections of the cantilever, as shown in Figure 7d. Fourier analysis was used to calculate the PSD of cantilever deflection. The Fourier analysis or the PSD as a nanomotion autocorrelation function may better describe the dynamic response of nanomotion phenomena; this is due to the measurements' intrinsically stochastic nature. Figure 7d shows the effect of increasing chemical agent concentration on nanomotion deflection compared to the deflection of a cantilever without any chemicals.

3.5. Challenges and Future Perspectives

In the studies on nanomotion, several cantilever types have been used, as shown in Table 1. Most of these cantilevers are made of silicon nitride and coated with a gold layer, some triangular and some rectangular in shape. According to our research, the effects of using cantilevers with different shapes and dimensions or different coating layers on nanomotion deflection when using the same samples and the same experimental conditions have not been studied. The shapes and dimensions of cantilevers have an impact on oscillation specifications and cantilever safety when handling. On the one hand, the gold layer covering the cantilever is crucial to the laser column reflection, which transmits the oscillation deflection of the cantilever to the photodiode. Still, on the other hand, it affects the adhesion of live samples. To ensure the adhesion of living samples on the cantilever surface, cantilevers have been functionalized using different molecules, such as APTES, Glutaraldehyde, Poly-L-lysine, and Poly-D-lysine, as shown in Table 1.

The period in which the measurements were made ranged between ten minutes and seven hours, as shown in Table 1. The time taken for chemical agents to affect living samples varies according to the type of chemical agents and the type of living sample used. Challenge will appear in applications where the chemical agents need a long time to affect the biological samples. Therefore, this will require modifications of the devices used in order to allow real-time monitoring over long periods of up to twenty-four hours or more. Increasing the real-time monitoring period generates other challenges, such as increased noise, increased viscosity of the medium over time, thermal effects, and increasing the time required to process the resulting data.

The AFM oscillating sensor method enables the calculation of the total sample adhesion force by recording the total value of cantilever frequency. The cantilever frequency changes with the number of valid samples still attached to the cantilever surface. The locations of these samples on the surface of the cantilever and away from the free end significantly affect the deflection and frequency of vibrations, but in the AFM oscillating sensor mode, with the final sample, the viability result is not affected by the height of the amplitude but rather by the shape of the drop in amplitude (exponential slope). So, if you assume the localization of samples is neglected in statistical distributions, experiments should be comparable. Thus, we propose that a method be developed in the future which will allow the location of samples on cantilever surfaces to be controlled. This will enable the value of adhesion forces to be calculated. Working on a mathematical model that shows the relationship between adherent samples' masses and locations on the cantilever surface and the cantilever deflection values may develop the measurement of the viability of a single sample as a focal point for increasing accuracy and reliability.

Nanomotion's ability to distinguish between living samples based on their propensity to adhere to a surface opens up a wide field of other applications that enable fingerprinting of different living samples according to their health status. Nanomotion was utilized to collect brain tumor vibration signals from cultured cells based on their vibration, allowing for the differentiation of various brain tumors from the normal brain based on nanomotion characteristics [164].

Nanomotion showed significant applicability in the real-time monitoring of the viability of live samples. However, displaying the results so as to enable the real-time monitoring of the effects of chemical agents on living samples still presents a challenge. The result display methods discussed previously are limited either to the end state of the monitoring process or to a specific point during the measurement. Deflection variance, the most commonly used method, gives a visualization of the final state of chemical agents' effects on living samples but does not show the state of samples instantaneously during the chemical agent step. On the other hand, the deflection damping coefficient value, for example, gives a constant value for the effect on living samples, which may enable a final comparison (depending on the deflection damping curve resulting from the effect of a chemical agent on a living sample) for the state of the living sample at a specific concentration of the chemical agent or comparison between two different chemical agents. Hence, finding a novel display method that enables the display of real-time monitoring of the viability of living samples under the influence of chemical agents may be a promising avenue for future research in this field.

The same living sample and the same chemical agent have not been used in a single study with different result display methods, nor have the different methods been compared. The use of more than one result display method may give a better real-time visualization of chemical agents' effects on viability as measured by nanomotion. A time-dependent study (amplitude modulation (AM) and frequency modulation (FM) over time) was used to show the single-cell force spectroscopy measure of yeast cell metabolism [166]. For all the result display methods, the focus has been on amplitude modulation characters but frequency modulation has not been analyzed for results in order to determine the characteristics of frequencies at the different stages of the measurement process, which information could have revealed a new horizon for the visualization of the results of the viability measurement process using nanomotion. Nanomotion oscillation signals can be converted into sound signals within the human hearing frequency band [164], providing novel concepts that could lead to the development of intelligent detection devices for real-time viability measurement.

4. Conclusions

In this review, the most common methods used for living sample viability measurement have been classified and presented, and their benefits and drawbacks, as well as potential utilities, have been discussed and evaluated. The principles and features of VMMs are summarized in Table 3. Chemical assays are the most widely used method for measuring cell viability. They can be used if the goal is to know the endpoint of a cell's fate. Optical methods measure the viability of living samples by monitoring and imaging morphological changes of the samples, and the efficiency of database processing via deep learning software limits their accuracy and scope. All of the studies in the literature discussed here certify that nanomotion is a promising alternative tool for measuring the viability of living samples which effectively achieves rapid, quantitative, direct, and real-time determination. Nanomotion has been used to study biological activities, toxicity, and drug efficacy in living samples, such as bacteria, cells, sperm, and cellular organs, such as mitochondria and ATP. Sample attachment protocols and result display methods are the critical factors in the progress of VMMs. Further development in these two processes is essential to the enhancement of the efficiency and repeatability of results and the enabling of instantaneous monitoring. To avoid different environmental conditions during the measurement process, sample adhesion to a cantilever surface can be achieved by means of a direct attachment protocol outside the test room or by injecting the sample into the test room to avoid different environmental conditions and reduce the possibility of contamination or cantilever damage. If needed, more complex protocols, such as the micrometric motors of the AFM or ink-jet printing, can be employed to control the numbers and positions of samples on the cantilever surfaces. The present nanomotion result display methods do not show the actions of agents on live samples changing instantly in real-time.

Additionally, cantilever type and functionalization are important factors that affect the success of nanomotion methods, while the numbers and positions of adherent samples on cantilevers are significant factors affecting the reliability and repeatability of the process. We sincerely hope that this review will drive the future development of nanomotion and provide significant thoughts for novel VMMs.

Table 3. Principles and features of VMMs.

Measurement Method	Principle	Features
Chemical viability assays	Injection of chemical compound(s) into living samples and evaluation of sample interaction with these compound(s)	• Labelled and multi-sample methods • Easy, inexpensive, and no need for complex techniques • Suitable for either suspended or adherent samples • Assay identification and design depend on the drug's nature and the type of biomarkers used • Endpoint assays • For a large number of samples, it is time-consuming and labor-intensive
Raman spectroscopy	Detection of morphological changes	• Rapid, label-free, contactless, and multi-sample method • Real-time method, non-invasive and not damaging to samples • RS results are affected by the weak Raman signal and light scattering, which reduce the device's sensitivity • Time-consuming and human factor errors for a large number of samples • Machine-learning algorithms must be used for high-throughput screening
Flow imaging microscopy	Detection of morphological changes of living samples while the sample fluid is in a continuous flow	• Rapid, label-free, contactless, and multi-sample method • Real-time method, non-invasive and not damaging to samples • High throughput • Able to measure samples one by one and numerically calculate size distribution using a convolutional neural network with deep learning technology • Able to solving critical sample classification problems through conjunction with image-processing technology and advanced machine-learning algorithms • The speed of data analysis is the most significant limitation
Holography	Detection of rapid changes in living sample structure parameters resulting from mechanical or morphological changes	• Rapid, label-free, contactless, and multi-sample method • Real-time method, non-invasive and not damaging to samples • Suitable for direct observation of 3D bio-tissue without scanning • Accuracy is affected by light scattering and light source quality

Table 3. Cont.

Measurement Method	Principle	Features
On-chip, lensless video microscopy technology	Detection and evaluation of changes in the shadows of living samples	• Rapid, label-free, contactless, and multi-sample method • Real-time method, non-invasive and not damaging to samples • Has twice the visual field of a conventional microscope • No requirements for optical or mechanical elements, such as lenses or probes • By using microfluidic channels, it is possible to monitor more than one living sample type simultaneously • Machine-learning algorithms must be used for high-throughput screening • Possibility of phototoxicity
Respiratory measuring methods	Detection of the oxygen absorbed and consumed by a living sample	• Rapid, label-free, contactless, and multi-sample method • Real-time method, non-invasive and not damaging to the samples • Continuous high-throughput method • Sensitive to environmental parameters, such as temperature, pressure, flow, and salinity • Calibration difficulty, poisoning risk, oxygen consumption, and high costs, especially for large samples • Sensor materials need to have low oxygen permeability and easy-to-manufacture thermoplastic polymers
Microcalorimeter measuring methods	Detection of the resulting heat from a living sample	• Rapid, label-free, contactless, and multi-sample method • Real-time method, non-invasive and not damaging to samples • Continuous high-throughput method • Sensitive to environmental parameters, such as temperature, pressure, flow, and salinity • The complexity of results interpretation and accuracy affected by radiation absorption • Sensor resolution is not accurate enough to match the single-sample temperature, measured in pW • Calculating the average heat generated by colonies of living samples

Table 3. Cont.

Measurement Method	Principle	Features
Nanomotion	Take advantage of the AFM cantilever's high sensitivity to changes in mass caused by sample adherence to the cantilever surface	• Rapid, label-free method • Real-time method, non-invasive and not damaging to the samples • Applicable to either single or multiple samples • Measurement time is reduced to several hours instead of several days, as with traditional assays • Able to monitor the instantaneous effects of chemical agents on living samples for several hours or even days • Unlike single-cell force spectroscopy, adhesion is evaluated without forcing the living sample to detach through cliffs or stretching • Cantilever surface functionalization is needed for sample attachment • The current nanomotion result display methods do not show the instantaneous effects of chemical agents on living samples

Author Contributions: Conceptualization, H.A.-m. and F.Y.; Validation, H.A.-m. and H.D.; Visualization, H.A.-m., H.D., J.Y., H.P., C.Y. and B.J.; Formal analysis, H.A.-m., J.Y. and H.P.; Investigation, H.A.-m., C.Y. and B.J. Writing - original draft, H.A.-m., C.Y. and B.J., Supervision, A.W. All authors have read and agreed to the published version of the manuscript.

Funding: This research was funded by [the National Natural Science Foundation of China] grant number (51803228), [the Youth Innovation Promotion Association, the Chinese Academy of Sciences] grant number (2022301) and (the Ningbo 3315 Innovative Talent Project) grant number (2018-05-G).

Institutional Review Board Statement: Not applicable.

Informed Consent Statement: Not applicable.

Acknowledgments: Hamzah Al-madani acknowledges financial support from an ANSO Scholarship for Young Talents.

Conflicts of Interest: The authors declare no conflict of interest.

References

1. Mahto, S.K.; Chandra, P.; Rhee, S.W. In vitro models, endpoints and assessment methods for the measurement of cytotoxicity. *Toxicol. Environ. Health Sci.* **2010**, *2*, 87–93. [CrossRef]
2. Hu, C.; He, S.; Lee, Y.J.; He, Y.R.; Anastasio, M.; Popescu, G. Label-free cell viability assay using phase imaging with computational specificity (PICS). *Quant. Phase Imaging VII* **2021**, *11653*, 48. [CrossRef]
3. Kroemer, G.; Galluzzi, L.; Vandenabeele, P.; Abrams, J.; Alnemri, E.S.; Baehrecke, E.H.; Blagosklonny, M.V.; El-Deiry, W.S.; Golstein, P.; Green, D.R.; et al. Classification of cell death: Recommendations of the Nomenclature Committee on Cell Death 2009. *Cell Death Differ.* **2009**, *16*, 3–11. [CrossRef] [PubMed]
4. Single, A.; Beetham, H.; Telford, B.J.; Guilford, P.; Chen, A. A comparison of real-time and endpoint cell viability assays for improved synthetic lethal drug validation. *J. Biomol. Screen.* **2015**, *20*, 1286–1293. [CrossRef] [PubMed]
5. Wei, M.; Zhang, R.; Zhang, F.; Zhang, Y.; Li, G.; Miao, R.; Shao, S. An Evaluation Approach of Cell Viability Based on Cell Detachment Assay in a Single-Channel Integrated Microfluidic Chip. *ACS Sens.* **2019**, *4*, 2654–2661. [CrossRef]
6. Wei, M.; Zhang, R.; Zhang, F.; Zhang, Y. Evaluating cell viability heterogeneity based on information fusion of multiple adhesion strengths. *Biotechnol. Bioeng.* **2021**, *118*, 2360–2367. [CrossRef]
7. Venturelli, L.; Kohler, A.C.; Stupar, P.; Villalba, M.I.; Kalauzi, A.; Radotic, K.; Bertacchi, M.; Dinarelli, S.; Girasole, M.; Pešić, M.; et al. A perspective view on the nanomotion detection of living organisms and its features. *J. Mol. Recognit.* **2020**, *33*, e2849. [CrossRef]
8. Gilbert, D.F. *Cell Viability Assays*; Springer: New York, NY, USA, 2017.
9. Kamiloglu, S.; Sari, G.; Ozdal, T.; Capanoglu, E. Guidelines for cell viability assays. *Food Front.* **2020**, *1*, 332–349. [CrossRef]
10. Duellman, S.J.; Zhou, W.; Meisenheimer, P.; Vidugiris, G.; Cali, J.J.; Gautam, P.; Wennerberg, K.; Vidugiriene, J. Bioluminescent, Nonlytic, Real-Time Cell Viability Assay and Use in Inhibitor Screening. *Assay Drug Dev. Technol.* **2015**, *13*, 456–465. [CrossRef]
11. Kerschbaum, H.H.; Tasa, B.A.; Schürz, M.; Oberascher, K.; Bresgen, N. Trypan blue—Adapting a dye used for labelling dead cells to visualize pinocytosis in viable cells. *Cell. Physiol. Biochem.* **2021**, *55*, 171–184. [CrossRef]
12. Kim, S.I.; Kim, H.J.; Lee, H.J.; Lee, K.; Hong, D.; Lim, H.; Cho, K.; Jung, N.; Yi, Y.W. Application of a non-hazardous vital dye for cell counting with automated cell counters. *Anal. Biochem.* **2016**, *492*, 8–12. [CrossRef] [PubMed]
13. Lippman, M.E. Comparison of dye exclusion assays with a clonogenic assay in the determination of drug-Induced cytotoxicity. *Cancer Res.* **1983**, *43*, 258–264.
14. Tolnai, S. A method for viable cell count. *Tissue Cult. Assoc. Man.* **1975**, *1*, 37–38. [CrossRef]
15. Dooley, M.P. The use of eosin B to assess the viability and developmental potential of rat embryos. *Retrosp. Theses Diss.* **1988**, *8839*, 1–256.
16. Nakayama, Y.; Tsujinaka, T. Acceleration of robust 'biotube' vascular graft fabrication by in-body tissue architecture technology using a novel eosin Y-releasing mold. *J. Biomed. Mater. Res. Part B Appl. Biomater.* **2014**, *102*, 231–238. [CrossRef]
17. Kay, A.B. Paul ehrlich and the early history of granulocytes. *Myeloid Cells Health Dis. A Synth.* **2017**, *4*, 3–15. [CrossRef]
18. Kan, A.; Birnbaum, D.P.; Praveschotinunt, P.; Joshi, N.S. Congo red fluorescence for rapid in situ characterization of synthetic curli systems. *Appl. Environ. Microbiol.* **2019**, *85*, e00434-19. [CrossRef]
19. Kuo, C.T.; Chen, Y.L.; Hsu, W.T.; How, S.C.; Cheng, Y.H.; Hsueh, S.S.; Liu, H.S.; Lin, T.H.; Wu, J.W.; Wang, S.S.S. Investigating the effects of erythrosine B on amyloid fibril formation derived from lysozyme. *Int. J. Biol. Macromol.* **2017**, *98*, 159–168. [CrossRef]
20. Franke, J.D.; Braverman, A.L.; Cunningham, A.M.; Eberhard, E.E.; Perry, G.A. Erythrosin B: A versatile colorimetric and fluorescent vital dye for bacteria. *Biotechnol. J.* **2020**, *68*, 7–13. [CrossRef]
21. Kumar, P.; Nagarajan, A.; Uchil, P.D. Analysis of cell viability by the lactate dehydrogenase assay. *Cold Spring Harb. Protoc.* **2018**, *2018*, 465–468. [CrossRef]
22. Rotman, B.; Papermaster, B.W. Membrane properties of living mammalian cells as studied by enzymatic hydrolysis of fluorogenic esters. *Proc. Natl. Acad. Sci. USA* **1966**, *55*, 134–141. [CrossRef] [PubMed]

23. Larson, E.M.; Doughman, D.J.; Gregerson, D.S.; Obritsch, W.F. A new, simple, nonradioactive, nontoxic in vitro assay to monitor corneal endothelial cell viability. *Investig. Ophthalmol. Vis. Sci.* **1997**, *38*, 1929–1933.
24. Schirmer, K.; Chan, A.G.J.; Greenberg, B.M.; Dixon, D.G.; Bols, N.C. Methodology for demonstrating and measuring the photocytotoxicity of fluoranthene to fish cells in culture. *Toxicol. In Vitro* **1997**, *11*, 107–113. [CrossRef]
25. Ganassin, R.C.; Bols, N.C. Growth of rainbow trout hemopoietic cells in methylcellulose and methods of monitoring their proliferative response in this matrix. *Methods Cell Sci.* **2000**, *22*, 147–152. [CrossRef] [PubMed]
26. Jiajia, L.; Shinghung, M.; Jiacheng, Z.; Jialing, W.; Dilin, X.; Shengquan, H.; Zaijun, Z.; Qinwen, W.; Yifan, H.; Wei, C. Assessment of neuronal viability using fluorescein diacetate-propidium iodide double staining in cerebellar granule neuron culture. *J. Vis. Exp.* **2017**, *2017*, e55442. [CrossRef]
27. Jones, K.H.; Senft, J.A. An improved method to determine cell viability by simultaneous staining with fluorescein diacetate-propidium iodide. *J. Histochem. Cytochem.* **1985**, *33*, 77–79. [CrossRef]
28. Mecelroy, W.D. The energy source for bioluminescence in an isolated system. *Zoology* **1947**, *33*, 342–345. [CrossRef]
29. Bajerski, F.; Stock, J.; Hanf, B.; Darienko, T.; Heine-Dobbernack, E.; Lorenz, M.; Naujox, L.; Keller, E.R.J.; Schumacher, H.M.; Friedl, T.; et al. ATP content and cell viability as indicators for cryostress across the diversity of life. *Front. Physiol.* **2018**, *9*, 921. [CrossRef]
30. Kijanska, M.; Kelm, J. In vitro 3D Spheroids and Microtissues: ATP-based Cell Viability and Toxicity Assays. *Assay Guid. Man.* **2004**, *1*, 1–13.
31. Smale, S.T. Luciferase assay. *Cold Spring Harb. Protoc.* **2010**, *5*, 2008–2011. [CrossRef]
32. Nguyen, V.T.; Morange, M.; Bensaude, O. Firefly luciferase luminescence assays using scintillation counters for quantitation in transfected mammalian cells. *Anal. Biochem.* **1988**, *171*, 404–408. [CrossRef]
33. de Wet, J.R.; Wood, K.V.; DeLuca, M.; Helinski, D.R.; Subramani, S. Firefly luciferase gene: Structure and expression in mammalian cells. *Mol. Cell. Biol.* **1987**, *7*, 725–737. [CrossRef] [PubMed]
34. Zhou, W.; Valley, M.P.; Shultz, J.; Hawkins, E.M.; Bernad, L.; Good, T.; Good, D.; Riss, T.L.; Klaubert, D.H.; Wood, K.V. New bioluminogenic substrates for monoamine oxidase assays. *J. Am. Chem. Soc.* **2006**, *128*, 3122–3123. [CrossRef] [PubMed]
35. van Engeland, M.; Ramaekers, F.C.S.; Schutte, B.; Reutelingsperger, C.P.M. A novel assay to measure loss of plasma membrane asymmetry during apoptosis of adherent cells in culture. *Cytometry* **1996**, *24*, 131–139. [CrossRef]
36. Darzynkiewicz, Z.; Bruno, S.; Del Bino, G.; Gorczyca, W.; Hotz, M.A.; Lassota, P.; Traganos, F. Features of apoptotic cells measured by flow cytometry. *Cytometry* **1992**, *13*, 795–808. [CrossRef] [PubMed]
37. da Silveira, M.G.; Romão, M.V.S.; Loureiro-Dias, M.C.; Rombouts, F.M.; Abee, T. Flow cytometric assessment of membrane integrity of ethanol-stressed Oenococcus oeni cells. *Appl. Environ. Microbiol.* **2002**, *68*, 6087–6093. [CrossRef]
38. Gillissen, M.A.; Yasuda, E.; De Jong, G.; Levie, S.E.; Go, D.; Spits, H.; van Helden, P.M.; Hazenberg, M.D. The modified FACS calcein AM retention assay: A high throughput flow cytometer based method to measure cytotoxicity. *J. Immunol. Methods* **2016**, *434*, 16–23. [CrossRef]
39. Davey, H.; Guyot, S. Estimation of Microbial Viability Using Flow Cytometry. *Curr. Protoc. Cytom.* **2020**, *93*, e72. [CrossRef]
40. Sedlackova, L.; Korolchuk, V.I. Mitochondrial quality control as a key determinant of cell survival. *Biochim. Biophys. Acta Mol. Cell Res.* **2019**, *1866*, 575–587. [CrossRef]
41. Cole, S.P.C. Rapid chemosensitivity testing of human lung tumor cells using the MTT assay. *Cancer Chemother. Pharmacol.* **1986**, *17*, 259–263. [CrossRef]
42. Goodwin, C.J.; Holt, S.J.; Downes, S.; Marshall, N.J. Microculture tetrazolium assays: A comparison between two new tetrazolium salts, XTT and MTS. *J. Immunol. Methods* **1995**, *179*, 95–103. [CrossRef]
43. Kazaks, A.; Collier, M.; Conley, M. Cytotoxicity of Caffeine on MCF-7 Cells Measured by XTT Cell Proliferation Assay (P06-038-19). *Curr. Dev. Nutr.* **2019**, *3*, 548. [CrossRef]
44. Scudiero, D.A.; Shoemaker, R.H.; Paull, K.D.; Monks, A.; Tierney, S.; Nofziger, T.H.; Currens, M.J.; Seniff, D.; Boyd, M.R. Evaluation of a Soluble Tetrazolium/Formazan Assay for Cell Growth and Drug Sensitivity in Culture Using Human and Other Tumor Cell Lines. *Cancer Res.* **1988**, *48*, 4827–4833. [PubMed]
45. Scarcello, E.; Lambremont, A.; Vanbever, R.; Jacques, P.J.; Lison, D. Mind your assays: Misleading cytotoxicity with the WST-1 assay in the presence of manganese. *PLoS ONE* **2020**, *15*, e0231634. [CrossRef] [PubMed]
46. Tominaga, H.; Ishiyama, M.; Ohseto, F.; Sasamoto, K.; Hamamoto, T.; Suzuki, K.; Watanabe, M. A water-soluble tetrazolium salt useful for colorimetric cell viability assay. *Anal. Commun.* **1999**, *36*, 47–50. [CrossRef]
47. Seifabadi, Z.S.; Rezaei-Tazangi, F.; Azarbarz, N.; Nejad, D.B.; Mohammadiasl, J.; Darabi, H.; Pezhmanlarki-Tork, S. Assessment of viability of wharton's jelly mesenchymal stem cells encapsulated in alginate scaffold by WST-8 assay kit. *Med. J. Cell Biol.* **2021**, *9*, 42–47. [CrossRef]
48. Skehan, P.; Storeng, R.; Scudiero, D.; Monks, A.; McMahon, J.; Vistica, D.; Warren, J.T.; Bokesch, H.; Kenney, S.; Boyd, M.R. New colorimetric cytotoxicity assay for anticancer-drug screening. *J. Natl. Cancer Inst.* **1990**, *82*, 1107–1112. [CrossRef]
49. Vajrabhaya, L.o.; Korsuwannawong, S. Cytotoxicity evaluation of a Thai herb using tetrazolium (MTT) and sulforhodamine B (SRB) assays. *J. Anal. Sci. Technol.* **2018**, *9*, 1–6. [CrossRef]
50. Ates, G.; Vanhaecke, T.; Rogiers, V.; Rodrigues, R.M. Assaying cellular viability using the neutral red uptake assay. *Methods Mol. Biol.* **2017**, *1601*, 19–26. [CrossRef]

51. Borenfreund, E.; Puerner, J.A. A simple quantitative procedure using monolayer cultures for cytotoxicity assays (HTD/NR-90). *J. Tissue Cult. Methods* **1985**, *9*, 7–9. [CrossRef]
52. Saotome, K.; Morita, H.; Umeda, M. Cytotoxicity test with simplified crystal violet staining method using microtitre plates and its application to injection drugs. *Toxicol. In Vitro* **1989**, *3*, 317–321. [CrossRef]
53. Puck, T.T. Quantitaive Studies on Mammalian Cells in Vitro. *Rev. Moderen Phys.* **1993**, *46*, 177–188.
54. Pegg, D.E. Viability assays for preserved cells, tissues, and organs. *Cryobiology* **1989**, *26*, 212–231. [CrossRef]
55. Galindo, C.C.; Lozano, D.M.V.; Rodríguez, B.C.; Perdomo-Arciniegas, A.M. Improved cord blood thawing procedure enhances the reproducibility and correlation between flow cytometry CD34+ cell viability and clonogenicity assays. *Cytotherapy* **2018**, *20*, 891–894. [CrossRef]
56. Decker, T.; Lohmann-Matthes, M.L. A quick and simple method for the quantitation of lactate dehydrogenase release in measurements of cellular cytotoxicity and tumor necrosis factor (TNF) activity. *J. Immunol. Methods* **1988**, *115*, 61–69. [CrossRef]
57. Chan, F.K.M.; Moriwaki, K.; de Rosa, M.J. Detection of necrosis by release of lactate dehydrogenase activity. *Methods Mol. Biol.* **2013**, *979*, 65–70. [CrossRef]
58. Ahmad, T.; Aggarwal, K.; Pattnaik, B.; Mukherjee, S.; Sethi, T.; Tiwari, B.K.; Kumar, M.; Micheal, A.; Mabalirajan, U.; Ghosh, B.; et al. Computational classification of mitochondrial shapes reflects stress and redox state. *Cell Death Dis.* **2013**, *4*, e461. [CrossRef]
59. Karbowski, M.; Youle, R.J. Dynamics of mitochondrial morphology in healthy cells and during apoptosis. *Cell Death Differ.* **2003**, *10*, 870–880. [CrossRef]
60. Arnoult, D. Mitochondrial fragmentation in apoptosis. *Trends Cell Biol.* **2006**, *17*, 6–12. [CrossRef]
61. Liu, X.; Hajnoczky, G. Altered fusion dynamics underlie unique morphological changes in mitochondria during hypoxia—Reoxygenation stress. *Cell Death Differ.* **2011**, *18*, 1561–1572. [CrossRef] [PubMed]
62. Mondol, A.S.; Töpfer, N.; Rüger, J.; Neugebauer, U.; Popp, J.; Schie, I.W. New perspectives for viability studies with high-content analysis Raman spectroscopy (HCA-RS). *Sci. Rep.* **2019**, *9*, 1–12. [CrossRef] [PubMed]
63. Wang, J.; Lin, K.; Hu, H.; Qie, X.; Huang, W.E.; Cui, Z.; Gong, Y.; Song, Y. In vitro anticancer drug sensitivity sensing through single-cell raman spectroscopy. *Biosensors* **2021**, *11*, 286. [CrossRef] [PubMed]
64. Wen, X.; Ou, Y.C.; Bogatcheva, G.; Thomas, G.; Mahadevan-Jansen, A.; Singh, B.; Lin, E.C.; Bardhan, R. Probing metabolic alterations in breast cancer in response to molecular inhibitors with Raman spectroscopy and validated with mass spectrometry. *Chem. Sci.* **2020**, *11*, 9863–9874. [CrossRef] [PubMed]
65. Botelho, C.M.; Gonçalves, O.; Marques, R.; Thiagarajan, V.; Vorum, H.; Gomes, A.C.; Neves-Petersen, M.T. Photonic modulation of epidermal growth factor receptor halts receptor activation and cancer cell migration. *J. Biophotonics* **2018**, *11*, e201700323. [CrossRef]
66. Czamara, K.; Petko, F.; Baranska, M.; Kaczor, A. Raman microscopy at the subcellular level: Study on early apoptosis in endothelial cells induced by Fas ligand and cycloheximide. *Analyst* **2016**, *141*, 1390–1397. [CrossRef]
67. Abramczyk, H. Double face of cytochrome c in cancers by Raman imaging. *Sci. Rep.* **2022**, *12*, 1–11. [CrossRef]
68. Pansare, K.; Singh, S.R.; Chakravarthy, V.; Gupta, N.; Hole, A.; Gera, P.; Sarin, R.; Krishna, C.M. Raman Spectroscopy: An Exploratory Study to Identify Post Radiation Cell Survival. *Appl Spectrosc* **2020**, *2*, 553–562. [CrossRef]
69. Schie, I.W.; Rüger, J.; Mondol, A.S.; Ramoji, A.; Neugebauer, U.; Krafft, C.; Popp, J. High-Throughput Screening Raman Spectroscopy Platform for Label-Free Cellomics. *Anal. Chem.* **2018**, *90*, 2023–2030. [CrossRef]
70. Jayan, H.; Pu, H.; Sun, D. Recent developments in Raman spectral analysis of microbial single cells: Techniques and applications. *Crit. Rev. Food Sci. Nutr.* **2021**, *62*, 4294–4308. [CrossRef]
71. Goldrick, S.; Umprecht, A.; Tang, A.; Zakrzewski, R.; Cheeks, M.; Turner, R.; Charles, A.; Les, K.; Hulley, M.; Spencer, C.; et al. High-throughput raman spectroscopy combined with innovate data analysis workflow to enhance biopharmaceutical process development. *Processes* **2020**, *8*, 1179. [CrossRef]
72. Verrier, S.; Zoladek, A.; Notingher, I. Raman Micro-Spectroscopy as a Non-invasive Cell Viability Test. In *Mammalian Cell Viability. Methods in Molecular Biology (Methods and Protocols)*; Stoddart, M., Ed.; Humana Press, Springer: New York, NY, USA, 2011; Volume 740, pp. 179–189. [CrossRef]
73. Grabarek, A.D.; Senel, E.; Menzen, T.; Hoogendoorn, K.H.; Pike-Overzet, K.; Hawe, A.; Jiskoot, W. Particulate impurities in cell-based medicinal products traced by flow imaging microscopy combined with deep learning for image analysis. *Cytotherapy* **2021**, *23*, 339–347. [CrossRef] [PubMed]
74. Farrell, C.J.; Cicalese, S.M.; Davis, H.B.; Dogdas, B.; Shah, T.; Culp, T.; Hoang, V.M. Cell confluency analysis on microcarriers by micro-flow imaging. *Cytotechnology* **2016**, *68*, 2469–2478. [CrossRef]
75. Sediq, A.S.; Klem, R.; Nejadnik, M.R.; Meij, P.; Jiskoot, W. Label-Free, Flow-Imaging Methods for Determination of Cell Concentration and Viability. *Pharm. Res.* **2018**, *35*, 1–10. [CrossRef] [PubMed]
76. Wu, L.; Martin, T.; Li, Y.; Yang, L.; Halpenny, M.; Giulivi, A.; Allan, D.S. Cell aggregation in thawed haematopoietic stem cell products visualised using micro-flow imaging. *Transfus. Med.* **2012**, *22*, 218–220. [CrossRef] [PubMed]
77. Grabarek, A.D.; Jiskoot, W.; Hawe, A.; Pike-overzet, K.; Menzen, T. Forced degradation of cell-based medicinal products guided by flow imaging microscopy: Explorative studies with Jurkat cells. *Eur. J. Pharm. Biopharm.* **2021**, *167*, 38–47. [CrossRef] [PubMed]
78. Gambe-gilbuena, A.; Shibano, Y.; Krayukhina, E.; Torisu, T.; Uchiyama, S. Automatic Identi fi cation of the Stress Sources of Protein Aggregates Using Flow Imaging Microscopy Images. *J. Pharm. Sci.* **2020**, *109*, 614–623. [CrossRef]

79. Kühn, J. Digital holographic microscopy real-time monitoring of cytoarchitectural alterations during simulated microgravity. *J. Biomed. Opt.* **2010**, *15*, 026021. [CrossRef]
80. Pais, D.A.M.; Galrão, P.R.S.; Kryzhanska, A.; Barbau, J.; Isidro, I.A.; Alves, P.M. Holographic imaging of insect cell cultures: Online non-invasive monitoring of adeno-associated virus production and cell concentration. *Processes* **2020**, *8*, 487. [CrossRef]
81. Kemper, B.; Carl, D.D.; Schnekenburger, J.; Bredebusch, I.; Schäfer, M.; Domschke, W.; von Bally, G. Investigation of living pancreas tumor cells by digital holographic microscopy. *J. Biomed. Opt.* **2006**, *11*, 034005. [CrossRef]
82. Odete, M.A.; Philips, L. Label-free Viability Assay using Holographic Video Microscopy Label-free Viability Assay using Holographic Video Microscopy. *Res. Sq.* preprint. **2021**. [CrossRef]
83. Pala, M.A.; Çimen, M.E.; Akgül, A.; Yıldız, M.Z.; Boz, A.F. Fractal dimension-based viability analysis of cancer cell lines in lens-free holographic microscopy via machine. *Eur. Phys. J.* **2021**, *123*, 1–12. [CrossRef]
84. Dubois, F.; Yourassowsky, C.; Monnom, O.; Legros, J.C.; Debeir IV, O.; Van Ham, P.; Kiss, R.; Decaestecker, C. Digital holographic microscopy for the three-dimensional dynamic analysis of in vitro cancer cell migration. *J. Biomed. Opt.* **2006**, *11*, 054032. [CrossRef]
85. Moon, I.; Daneshpanah, M.; Javidi, B.; Stern, A. Automated three-dimensional identification and tracking of micro/nanobiological organisms by computational holographic microscopy. *Proc. IEEE* **2009**, *97*, 990–1010. [CrossRef]
86. Pushkarsky, I.; Liu, Y.; Weaver, W.; Su, T.W.; Mudanyali, O.; Ozcan, A.; Di Carlo, D. Automated single-cell motility analysis on a chip using lensfree microscopy. *Sci. Rep.* **2014**, *4*, 1–9. [CrossRef]
87. Jin, G.; Yoo, I.; Pil, S.; Yang, J.; Ha, U. Biosensors and Bioelectronics Lens-free shadow image based high-throughput continuous cell monitoring technique. *Biosens. Bioelectron.* **2012**, *38*, 126–131. [CrossRef] [PubMed]
88. Kim, S.B.; Bae, H.; Cha, J.M.; Moon, S.J.; Dokmeci, M.R.; Cropek, D.M.; Khademhosseini, A. A cell-based biosensor for real-time detection of cardiotoxicity using lensfree imaging. *Lab Chip* **2011**, *11*, 1801–1807. [CrossRef] [PubMed]
89. Zheng, G.; Lee, S.A.; Yang, S.; Yang, C. Sub-pixel resolving optofluidic microscope for on-chip cell imaging. *Lab Chip* **2010**, *10*, 3125–3129. [CrossRef]
90. Cui, X.; Lee, L.M.; Heng, X.; Zhong, W.; Sternberg, P.W.; Psaltis, D.; Yang, C. Lensless high-resolution on-chip optofluidic microscopes for Caenorhabditis elegans and cell imaging. *Proc. Natl. Acad. Sci. USA* **2008**, *105*, 10670–10675. [CrossRef]
91. Ozcan, A.; Demirci, U. Ultra wide-field lens-free monitoring of cells on-chip. *Lab Chip* **2007**, *8*, 98–106. [CrossRef]
92. Kesavan, S.V.; Momey, F.; Cioni, O.; David-Watine, B.; Dubrulle, N.; Shorte, S.; Sulpice, E.; Freida, D.; Chalmond, B.; Dinten, J.M.; et al. High-throughput monitoring of major cell functions by means of lensfree video microscopy. *Sci. Rep.* **2014**, *4*, 1–11. [CrossRef]
93. Nablo, B.J.; Ahn, J.J.; Bhadriraju, K.; Lee, J.M.; Reyes, D.R. Lens-Free Imaging as a Sensor for Dynamic Cell Viability Detection Using the Neutral Red Uptake Assay. *ACS Appl. Bio Mater.* **2020**, *3*, 6633–6638. [CrossRef] [PubMed]
94. Huang, X.; Li, Y.; Xu, X.; Wang, R.; Yao, J.; Han, W.; Wei, M.; Chen, J.; Xuan, W.; Sun, L. High-precision lensless microscope on a chip based on in-line holographic imaging. *Sensors* **2021**, *21*, 720. [CrossRef] [PubMed]
95. Rothbauer, M.; Ertl, P.; Mayr, T. Measurement of respiration and acidification rates of mammalian cells in thermoplastic microfluidic devices. *Sens. Actuators B Chem.* **2021**, *334*, 129664. [CrossRef]
96. O'Riordan, T.C.; Buckley, D.; Ogurtsov, V.; O'Connor, R.; Papkovsky, D.B. A cell viability assay based on monitoring respiration by optical oxygen sensing. *Anal. Biochem.* **2000**, *278*, 221–227. [CrossRef]
97. Bäckman, P.; Wadsö, I. Cell growth experiments using a microcalorimetric vessel equipped with oxygen and pH electrodes. *J. Biochem. Biophys. Methods* **1991**, *23*, 283–293. [CrossRef]
98. Halpern, H.J.; Yu, C.; Peric, M.; Barth, E.D.; Karczmar, G.S.; River, J.N.; Grdina, D.J.; Teicher, B.A. Measurement of differences in pO2 in response to perfluorocarbon/carbogen in FSa and NFSa murine fibrosarcomas with low-frequency electron paramagnetic resonance oximetry. *Radiat. Res.* **1996**, *145*, 610–618. [CrossRef]
99. Braissant, O.; Astasov-frauenhoffer, M.; Waltimo, T. A Review of Methods to Determine Viability, Vitality, and Metabolic Rates in Microbiology. *Front. Microbiol.* **2020**, *11*, 547458. [CrossRef]
100. Randers-Eichhorn, L.; Bartlett, R.A.; Frey, D.D.; Rao, G. Noninvasive oxygen measurements and mass transfer considerations in tissue culture flasks. *Biotechnol. Bioeng.* **1996**, *51*, 466–478. [CrossRef]
101. Wodnicka, M.; Guarino, R.D.; Hemperly, J.J.; Timmins, M.R.; Stitt, D.; Pitner, J.B. Novel fluorescent technology platform for high throughput cytotoxicity and proliferation assays. *J. Biomol. Screen.* **2000**, *5*, 141–150. [CrossRef]
102. Guarino, R.D.; Dike, L.E.; Haq, T.A.; Rowley, J.A.; Pitner, J.B.; Timmins, M.R. Method for determining oxygen consumption rates of static cultures from microplate measurements of pericellular dissolved oxygen concentration. *Biotechnol. Bioeng.* **2004**, *86*, 775–787. [CrossRef]
103. Mishra, A.; Starly, B. Real time in vitro measurement of oxygen uptake rates for HEPG2 liver cells encapsulated in alginate matrices. *Microfluid. Nanofluidics* **2009**, *6*, 373–381. [CrossRef]
104. Super, A.; Jaccard, N.; Marques, M.P.C.; Macown, R.J.; Griffin, L.D.; Veraitch, F.S.; Szita, N. Real-time monitoring of specific oxygen uptake rates of embryonic stem cells in a microfluidic cell culture device. *Biotechnol. J.* **2016**, *11*, 1179–1189. [CrossRef]
105. Mahfouzi, S.H.; Amoabediny, G.; Doryab, A.; Safiabadi-Tali, S.H.; Ghanei, M. Noninvasive Real-Time Assessment of Cell Viability in a Three-Dimensional Tissue. *Tissue Eng. Part C Methods* **2018**, *24*, 197–204. [CrossRef] [PubMed]
106. Xue, Y.; Lei, J.; Xu, X.; Ding, L.; Zhai, C.; Yan, F.; Ju, H. Real-time monitoring of cell viability by its nanoscale height change with oxygen as endogenous indicator. *Chem. Commun.* **2010**, *46*, 7388–7390. [CrossRef] [PubMed]

107. Wadsö, I. Microcalorimetric techniques for characterization of living cellular systems. Will there be any important practical applications? *Thermochim. Acta* **1995**, *269–270*, 337–350. [CrossRef]
108. Braissant, O.; Wirz, D.; Göpfert, B.; Daniels, A.U. Use of isothermal microcalorimetry to monitor microbial activities. *FEMS Microbiol. Lett.* **2010**, *303*, 1–8. [CrossRef]
109. Yang, N.; Shi, Q.; Zhu, X.; Wei, M.; Ullah, I.; Kwabena, P.O.; Kulik, E.; Mao, H.; Zhang, R. A Cell Viability Evaluation Method Based on Respiratory Thermodynamic Feature Detected by Microscopic Infrared Thermal Imaging Sensor. *IEEE Sens. J.* **2020**, *20*, 637–647. [CrossRef]
110. Tan, A.M.; Lu, J.H. Microcalorimetric study of antiviral effect of drug. *J. Biochem. Biophys. Methods* **1999**, *38*, 225–228. [CrossRef]
111. Spaepen, P.; de Boodt, S.; Aerts, J.; Sloten, J.V. Chapter 21 Digital Image Processing of Live/Dead Staining. *Mamm. Cell Viability Methods Protoc. Methods Mol. Biol.* **2011**, *740*, 209–230. [CrossRef]
112. Lemos, D.; Oliveira, T.; Martins, L.; De Azevedo, V.R.; Rodrigues, M.F.; Ketzer, L.A.; Rumjanek, F.D. Isothermal Microcalorimetry of Tumor Cells: Enhanced Thermogenesis by Metastatic Cells. *Front. Oncol.* **2019**, *9*, 1430. [CrossRef]
113. Wang, F.; Han, Y.; Gu, N. Cell Temperature Measurement for Biometabolism Monitoring. *ACS Sens.* **2021**, *6*, 290–302. [CrossRef] [PubMed]
114. Wang, Y.; Zhu, H.; Feng, J.; Neuzil, P. Recent advances of microcalorimetry for studying cellular metabolic heat. *Trends Anal. Chem.* **2021**, *143*, 116353. [CrossRef]
115. Ilic, B.; Czaplewski, D.; Craighead, H.G.; Neuzil, P.; Campagnolo, C.; Batt, C. Mechanical resonant immunospecific biological detector. *Appl. Phys. Lett.* **2000**, *77*, 450–452. [CrossRef]
116. Zheng, G.; Patolsky, F.; Cui, Y.; Wang, W.U.; Lieber, C.M. Multiplexed electrical detection of cancer markers with nanowire sensor arrays. *Nat. Biotechnol.* **2005**, *23*, 1294–1301. [CrossRef]
117. Ramos, D.; Tamayo, J.; Mertens, J.; Calleja, M.; Villanueva, L.G.; Zaballos, A. Detection of bacteria based on the thermomechanical noise of a nanomechanical resonator: Origin of the response and detection limits. *Nanotechnology* **2008**, *19*, 035503. [CrossRef]
118. Ahmad, M.R.; Nakajima, M.; Kojima, M.; Kojima, S.; Homma, M.; Fukuda, T. Instantaneous and quantitative single cells viability determination using dual nanoprobe inside ESEM. *IEEE Trans. Nanotechnol.* **2012**, *11*, 298–306. [CrossRef]
119. Shen, Y.; Nakajima, M.; Kojima, S.; Homma, M.; Kojima, M.; Fukuda, T. Single cell adhesion force measurement for cell viability identification using an AFM cantilever-based micro putter. *Meas. Sci. Technol.* **2011**, *22*, 944–947. [CrossRef]
120. Kasas, S.; Ruggeri, F.S.; Benadiba, C.; Maillard, C.; Stupar, P.; Tournu, H.; Dietler, G.; Longo, G. Detecting nanoscale vibrations as signature of life. *Proc. Natl. Acad. Sci. USA* **2015**, *112*, 378–381. [CrossRef]
121. Mader, A.; Gruber, K.; Castelli, R.; Hermann, B.A.; Seeberger, P.H.; Rädler, J.O.; Leisner, M. Discrimination of Escherichia coli strains using glycan cantilever array sensors. *Nano Lett.* **2012**, *12*, 420–423. [CrossRef]
122. Sharma, H.; Mutharasan, R. Rapid and sensitive immunodetection of Listeria monocytogenes in milk using a novel piezoelectric cantilever sensor. *Biosens. Bioelectron.* **2013**, *45*, 158–162. [CrossRef]
123. Ndieyira, J.W.; Kappeler, N.; Logan, S.; Cooper, M.A.; Abell, C.; McKendry, R.A.; Aeppli, G. Surface-stress sensors for rapid and ultrasensitive detection of active free drugs in human serum. *Nat. Nanotechnol.* **2014**, *9*, 225–232. [CrossRef] [PubMed]
124. Maciaszek, J.L.; Andemariam, B.; Abiraman, K.; Lykotrafitis, G. AKAP-dependent modulation of BCAM/Lu adhesion on normal and sickle cell disease RBCs revealed by force nanoscopy. *Biophys. J.* **2014**, *106*, 1258–1267. [CrossRef]
125. Liu, Y.; Schweizer, L.M.; Wang, W.; Reuben, R.L.; Schweizer, M.; Shu, W. Chemical Label-free and real-time monitoring of yeast cell growth by the bending of polymer microcantilever biosensors. *Sens. Actuators B. Chem.* **2013**, *178*, 621–626. [CrossRef]
126. Longo, G.; Alonso-Sarduy, L.; Rio, L.M.; Bizzini, A.; Trampuz, A.; Notz, J.; Dietler, G.; Kasas, S. Rapid detection of bacterial resistance to antibiotics using AFM cantilevers as nanomechanical sensors. *Nat. Nanotechnol.* **2013**, *8*, 522–526. [CrossRef] [PubMed]
127. Yang, F.; Riedel, R.; Del Pino, P.; Pelaz, B.; Said, A.H.; Soliman, M.; Pinnapireddy, S.R.; Feliu, N.; Parak, W.J.; Bakowsky, U.; et al. Real-time, label-free monitoring of cell viability based on cell adhesion measurements with an atomic force microscope. *J. Nanobiotechnol.* **2017**, *15*, 1–10. [CrossRef] [PubMed]
128. Bennett, I.; Pyne, A.L.B.; McKendry, R.A. Cantilever Sensors for Rapid Optical Antimicrobial Sensitivity Testing. *ACS Sensors* **2020**, *5*, 3133–3139. [CrossRef]
129. Linna, E.; BinAhmed, S.; Stottrup, B.L.; Castrill, S.R.V. Effect of Graphene Oxide Packing on Bacterial Adhesion using Single Cell Force Spectroscopy. *Biophys. J.* **2018**, *114*, 352a–353a. [CrossRef]
130. Evans, E.A.; Calderwood, D.A. Forces and bond dynamics in cell adhesion. *Science* **2007**, *316*, 1148–1153. [CrossRef]
131. Huang, H.; Dai, C.; Shen, H.; Gu, M.; Wang, Y.; Liu, J.; Chen, L.; Sun, L. Recent advances on the model, measurement technique, and application of single cell mechanics. *Int. J. Mol. Sci.* **2020**, *21*, 6248. [CrossRef]
132. Müller, D.J.; Dufrêne, Y.F. Atomic force microscopy: A nanoscopic window on the cell surface. *Trends Cell Biol.* **2011**, *21*, 461–469. [CrossRef]
133. Ungai-Salánki, R.; Peter, B.; Gerecsei, T.; Orgovan, N.; Horvath, R.; Szabó, B. A practical review on the measurement tools for cellular adhesion force. *Adv. Colloid Interface Sci.* **2019**, *269*, 309–333. [CrossRef] [PubMed]
134. Stewart, M.P.; Hodel, A.W.; Spielhofer, A.; Cattin, C.J.; Müller, D.J.; Helenius, J. Wedged AFM-cantilevers for parallel plate cell mechanics. *Methods* **2013**, *60*, 186–194. [CrossRef] [PubMed]

135. Mathur, A.B.; Collinsworth, A.M.; Reichert, W.M.; Kraus, W.E.; Truskey, G.A. Endothelial, cardiac muscle and skeletal muscle exhibit different viscous and elastic properties as determined by atomic force microscopy. *J. Biomech.* **2001**, *34*, 1545–1553. [CrossRef]
136. Yang, S.P.; Yang, C.Y.; Lee, T.M.; Lui, T.S. Effects of calcium-phosphate topography on osteoblast mechanobiology determined using a cytodetacher. *Mater. Sci. Eng. C* **2012**, *32*, 254–262. [CrossRef]
137. Sagvolden, G.; Giaever, I.; Pettersen, E.O.; Feder, J. Cell adhesion force microscopy. *Proc. Natl. Acad. Sci. USA* **1999**, *96*, 471–476. [CrossRef]
138. Yamamoto, A.; Mishima, S.; Maruyama, N.; Sumita, M. Quantitative evaluation of cell attachment to glass, polystyrene, and fibronectin- or collagen-coated polystyrene by measurement of cell adhesive shear force and cell detachment energy. *J. Biomed. Mater. Res.* **2000**, *50*, 114–124. [CrossRef]
139. Lee, C.C.; Wu, C.C.; Su, F.C. The Technique for Measurement of Cell Adhesion Force. *J. Med. Biol. Eng.* **2004**, *24*, 51–56.
140. Marcotte, L.; Tabrizian, M. Sensing surfaces: Challenges in studying the cell adhesion process and the cell adhesion forces on biomaterials. *Itbm-Rbm* **2008**, *29*, 77–88. [CrossRef]
141. Elbourne, A.; Chapman, J.; Gelmi, A.; Cozzolino, D.; Crawford, R.J.; Truong, V.K. Bacterial-nanostructure interactions: The role of cell elasticity and adhesion forces. *J. Colloid Interface Sci.* **2019**, *546*, 192–210. [CrossRef]
142. Friedrichs, J.; Legate, K.R.; Schubert, R.; Bharadwaj, M.; Werner, C.; Müller, D.J.; Benoit, M. A practical guide to quantify cell adhesion using single-cell force spectroscopy. *Methods* **2013**, *60*, 169–178. [CrossRef]
143. Ramos, D.; Tamayo, J.; Mertens, J.; Calleja, M.; Zaballos, A. Origin of the response of nanomechanical resonators to bacteria adsorption. *J. Appl. Phys.* **2006**, *100*, 106105. [CrossRef]
144. Wu, S.; Liu, X.; Zhou, X.; Liang, X.M.; Gao, D.; Liu, H.; Zhao, G.; Zhang, Q.; Wu, X. Quantification of cell viability and rapid screening anti-cancer drug utilizing nanomechanical fluctuation. *Biosens. Bioelectron.* **2016**, *77*, 164–173. [CrossRef] [PubMed]
145. Stupar, P.; Chomicki, W.; Maillard, C.; Mikeladze, D.; Kalauzi, A.; Radotić, K.; Dietler, G.; Kasas, S. Mitochondrial activity detected by cantilever based sensor. *Mech. Sci.* **2017**, *8*, 23–28. [CrossRef]
146. Kohler, A.C.; Venturelli, L.; Longo, G.; Dietler, G.; Kasas, S. Nanomotion detection based on atomic force microscopy cantilevers. *Cell Surf.* **2019**, *5*, 100021. [CrossRef] [PubMed]
147. Lissandrello, C.; Inci, F.; Francom, M.; Paul, M.R.; Demirci, U.; Ekinci, K.L. Nanomechanical motion of Escherichia coli adhered to a surface. *Appl. Phys. Lett.* **2014**, *105*, 113701. [CrossRef]
148. Villalba, M.I.; Stupar, P.; Chomicki, W.; Bertacchi, M.; Dietler, G.; Arnal, L.; Vela, M.E.; Yantorno, O.; Kasas, S. Nanomotion Detection Method for Testing Antibiotic Resistance and Susceptibility of Slow-Growing Bacteria. *Small* **2018**, *14*, 1702671. [CrossRef]
149. Stupar, P.; Opota, O.; Longo, G.; Prod'Hom, G.; Dietler, G.; Greub, G.; Kasas, S. Nanomechanical sensor applied to blood culture pellets: A fast approach to determine the antibiotic susceptibility against agents of bloodstream infections. *Clin. Microbiol. Infect.* **2017**, *23*, 400–405. [CrossRef]
150. Wu, S.; Zhang, Z.; Zhou, X.; Liu, H.; Xue, C.; Zhao, G.; Cao, Y.; Zhang, Q.; Wu, X. Nanomechanical sensors for direct and rapid characterization of sperm motility based on nanoscale vibrations. *Nanoscale* **2017**, *9*, 18258–18267. [CrossRef]
151. Vannocci, T.; Dinarelli, S.; Girasole, M.; Pastore, A.; Longo, G. A new tool to determine the cellular metabolic landscape: Nanotechnology to the study of Friedreich's ataxia. *Sci. Rep.* **2019**, *9*, 1–9. [CrossRef]
152. Ruggeri, F.S.; Mahul-Mellier, A.L.; Kasas, S.; Lashuel, H.A.; Longo, G.; Dietler, G. Amyloid single-cell cytotoxicity assays by nanomotion detection. *Cell Death Discov.* **2017**, *3*, 1–8. [CrossRef]
153. Alonso-Sarduy, L.; De Los Rios, P.; Benedetti, F.; Vobornik, D.; Dietler, G.; Kasas, S.; Longo, G. Real-time monitoring of protein conformational changes using a nano-mechanical sensor. *PLoS ONE* **2014**, *9*, e103674. [CrossRef] [PubMed]
154. Aghayee, S.; Benadiba, C.; Notz, J.; Kasas, S.; Dietler, G.; Longo, G. Combination of fluorescence microscopy and nanomotion detection to characterize bacteria. *J. Mol. Recognit.* **2013**, *26*, 590–595. [CrossRef] [PubMed]
155. Mustazzolu, A.; Venturelli, L.; Dinarelli, S.; Brown, K.; Floto, R.A.; Dietler, G.; Fattorini, L.; Kasas, S.; Girasole, M.; Longo, G. A rapid unraveling of the activity and antibiotic susceptibility of mycobacteria. *Antimicrob. Agents Chemother.* **2019**, *63*, e02194-18. [CrossRef] [PubMed]
156. Mertens, J.; Cuervo, A.; Carrascosa, J.L. Nanomechanical detection of: Escherichia coli infection by bacteriophage T7 using cantilever sensors. *Nanoscale* **2019**, *11*, 17689–17698. [CrossRef] [PubMed]
157. Kohler, A.C.; Venturelli, L.; Kannan, A.; Sanglard, D.; Dietler, G.; Willaert, R.; Kasas, S. Yeast nanometric scale oscillations highlights fibronectin induced changes in C. Albicans. *Fermentation* **2020**, *6*, 28. [CrossRef]
158. Villalba, M.I.; Venturelli, L.; Willaert, R.; Vela, M.E.; Yantorno, O.; Dietler, G.; Longo, G.; Kasas, S. Nanomotion spectroscopy as a new approach to characterize bacterial virulence. *Microorganisms* **2021**, *9*, 1545. [CrossRef]
159. Venturelli, L.; Harrold, Z.R.; Murray, A.E.; Villalba, M.I.; Lundin, E.M.; Dietler, G.; Kasas, S.; Foschia, R. Nanomechanical bio-sensing for fast and reliable detection of viability and susceptibility of microorganisms. *Sensors Actuators B Chem.* **2021**, *348*, 130650. [CrossRef]
160. Stupar, P. Atomic Force Microscopy of Biological Systems: Quantitative Imaging and Nanomotion Detection. *EPFL* **2018**, *8334*, 1–133.
161. Lukacs, G.; Maloney, N.; Hegner, M. Ink-jet printing: Perfect tool for cantilever array sensor preparation for microbial growth detection. *J. Sens.* **2012**, *3*, 276–283. [CrossRef]

162. Maciaszek, J.L.; Partola, K.; Zhang, J.; Andemariam, B.; Lykotrafitis, G. Single-cell force spectroscopy as a technique to quantify human red blood cell adhesion to subendothelial laminin. *J. Biomech.* **2014**, *47*, 3855–3861. [CrossRef]
163. Zanetti, M.; Chen, S.N.; Conti, M.; Taylor, M.R.; Sbaizero, O.; Mestroni, L.; Lazzarino, M. Microfabricated cantilevers for parallelized cell—cell adhesion measurements. *Eur. Biophys. J.* **2021**, *51*, 147–156. [CrossRef] [PubMed]
164. Nelson, S.L.; Proctor, D.T.; Ghasemloonia, A.; Lama, S.; Zareinia, K.; Ahn, Y.; Al-Saiedy, M.R.; Green, F.H.; Amrein, M.W.; Sutherland, G.R. Vibrational profiling of brain tumors and cells. *Theranostics* **2017**, *7*, 2417–2430. [CrossRef] [PubMed]
165. Braun, T.; Ghatkesar, M.K.; Backmann, N.; Grange, W.; Boulanger, P.; Letellier, L.; Lang, H.P.; Bietsch, A.; Gerber, C.; Hegner, M. Quantitative time-resolved measurement of membrane protein-ligand interactions using microcantilever array sensors. *Nat. Nanotechnol.* **2009**, *4*, 179–185. [CrossRef] [PubMed]
166. Pelling, A.E.; Sehati, S.; Gralla, E.B.; Gimzewski, J.K. Time dependence of the frequency and amplitude of the local nanomechanical motion of yeast. *Nanomed. Nanotechnol. Biol. Med.* **2005**, *1*, 178–183. [CrossRef] [PubMed]

Review

Ion Interference Therapy of Tumors Based on Inorganic Nanoparticles

Yongjie Chi [1,2], Peng Sun [1,3], Yuan Gao [1,4], Jing Zhang [1,2] and Lianyan Wang [1,2,*]

1. Institute of Process Engineering, Chinese Academy of Sciences, Beijing 100190, China; chiyongjie21@ipe.ac.cn (Y.C.); sunpeng23333@163.com (P.S.); m18342201571@163.com (Y.G.); zhangjin@ipe.ac.cn (J.Z.)
2. University of Chinese Academy of Sciences, Beijing 100190, China
3. College of Biological Science and Technology, Shenyang Agricultural University, Shenyang 110866, China
4. Key Laboratory of Forest Plant Ecology, Ministry of Education, College of Chemistry, Chemistry Engineering and Resource Utilization, Northeast Forestry University, Harbin 150040, China
* Correspondence: wanglianyan@ipe.ac.cn

Abstract: As an essential substance for cell life activities, ions play an important role in controlling cell osmotic pressure balance, intracellular acid–base balance, signal transmission, biocatalysis and so on. The imbalance of ion homeostasis in cells will seriously affect the activities of cells, cause irreversible damage to cells or induce cell death. Therefore, artificially interfering with the ion homeostasis in tumor cells has become a new means to inhibit the proliferation of tumor cells. This treatment is called ion interference therapy (IIT). Although some molecular carriers of ions have been developed for intracellular ion delivery, inorganic nanoparticles are widely used in ion interference therapy because of their higher ion delivery ability and higher biocompatibility compared with molecular carriers. This article reviewed the recent development of IIT based on inorganic nanoparticles and summarized the advantages and disadvantages of this treatment and the challenges of future development, hoping to provide a reference for future research.

Keywords: ion interference therapy; cancer; ions; inorganic nanoparticles

1. Introduction

Cancer is still one of the leading causes of death globally [1]. Chemotherapy, as one of the main cancer treatment methods, is widely used especially for advanced cancers in clinic [2]. In order to reduce the side effects of chemotherapy, such as systemic toxicity, multifarious nanoparticles were constructed as drug carriers to achieve targeted and sustained cancer therapy [3]. However, in such a pattern of treatment, some inherent properties of inorganic nanocarriers, such as the properties of absorbing hydrogen ions and releasing metal or nonmetal ions, sometimes seem to be ignored [4]. In the latest research, the released ions in different intracellular environments by inorganic nanoparticles have made a great contribution to inhibiting the activity of cancer cells and enhancing the therapeutic effect of chemotherapy. Therefore, the research based on the ion release of inorganic nanoparticles and its anticancer mechanism has broad prospects. Meanwhile, combining this with immunotherapy to comprehensively improve the anti-cancer ability will be a new option in the future.

Ions exist widely in the human body and have always been the focus of research due to their participation in various life activities, including osmotic pressure, the acid–base balance, catalytic and signal pathway activation, the protein and enzyme composition, targeting biomolecules and so on [5]. The abnormal distribution or specific accumulation of some metal or non-metal ions in cells cause irreversible damage to cells or activate cytotoxicity-related biochemical reactions to induce cell death, which provides a new method for tumor therapy, namely ion interference therapy [6]. However, the application of some small molecule carriers frequently used for ion interference therapy, often encounters obstacles, such as short internal circulation, dose-dependent toxicity, poor specific recognition ability and a low dose of ion release, which limits the effect of cancer treatment [7]. Based on this, inorganic nanoparticles such as Ca_2O, ZnO, $CaCO_3$, BPS (black phosphorus) and NaCl have been constructed for ion interference treatment, with the advantages of long internal circulation, efficient inhibition of tumor development, high biocompatibility and strong specific recognition ability (Table 1) [4,8–11]. Because of these advantages, ion interference therapy based on inorganic nanoparticles, as an emerging cancer treatment, represents strong competition for traditional cancer treatment.

Another major finding about ion interference therapy is that it can enhance the effect of antitumor immunity. Normally, the human immune system can recognize and remove cancer cells. Daniel S. Chen and IRA Mellman divided the fight between the immune system and cancer cells into seven stages in 2013, collectively referred to as the tumor immune cycle. The seven stages are: (1) Tumor cell death releases antigen; (2) Antigen presentation; (3) Start and activate T cells; (4) Cytotoxic T lymphocyte (CTL) is transported to tumor tissue through blood circulation; (5) Cytotoxic T lymphocyte infiltration in tumor tissue; (6) Cytotoxic T lymphocytes recognize tumor cells; (7) Kill cancer cells and finally return to the first step [12]. If the immune process is abnormal at any stage in the cycle, or the cancer cells themselves take some strategy to inhibit the immunity of a link, it can lead to immune escape [13]. In the process of the tumor immune cycle, inducing immunogenic death of tumor cells which can produce a large number of antigens is an important step, so it has been an important focus of immunotherapy for many years. Some traditional methods of cancer treatment have been proved to stimulate the immunogenic death of tumor cells, such as chemotherapy (adriamycin, oxaliplatin, etc.), radiotherapy, etc. Recently, new studies have shown that interfering with the balance of certain ions in the intracellular environment can also lead to the immunogenic death of cells, such as pyroptosis, which was characterized by the continuous swelling of cancer cells until the rupture of the cell membrane and the release of a large number of cell contents, with no obvious side effects compared with chemotherapy and radiotherapy [11]. The immunogenic death of cancer cells finally promotes the antigen release and presentation process, so as to enhance the patient's own ability to clear cancer cells and immunity. Some ions can also accumulate in immunocytes, such as dendritic cells (DCs), so as to promote the activation of immunocytes and the efficiency of antigen presentation, and finally realize the enhancement of immunity [14]. Therefore, using ion interference therapy based on inorganic nanoparticles to activate and promote the patient's immune system to achieve efficient cancer treatment is a reliable means. At the same time, combined it with new immunotherapy methods, such as immune checkpoint inhibitor therapy, will also produce better anti-cancer effects. Consequently, the combination of ion interference therapy and tumor immunotherapy may be a valuable route for humans to fight cancer in the future.

Although chemotherapy and immunotherapy have achieved great success in anti-cancer, there are still some complex troubles in the processes of chemotherapy and tumor immune cycle, which may eventually lead to a low therapeutic effect. In recent years, the research into ion interference therapy based on inorganic nanoparticles has been increasingly explored due to its high efficiency in inhibiting tumor cell proliferation and its ability to promote antitumor immunity. Therefore, making full use of and highlighting the various excellent properties of ion released by nanoparticles to achieve multitherapy is a matter of concern. In this review, we discuss ion interference therapy and ion-related tumor immunotherapy based on inorganic nanoparticles and explore the future perspectives of ion interference therapy in the treatment of cancer.

2. Ion Interference Therapy

Intracellular ions strictly control the life process of cells, such as membrane potential regulation, osmotic balance, protein synthesis, signal transduction etc. [15]. The basic idea of killing cancer cells by regulating the homeostasis of some ions in cells has long been applied in the field of cancer therapy, and has achieved breakthrough results, such as "ferroptosis". In order to pass through the cell membrane barrier favorably, Fe^{2+} ions which are loaded on nanoparticles enter into tumor cells by endocytosis. Then, ferrous ions are converted into Fe^{3+} via the Fenton reaction upon the induction of the intracellular specific microenvironment and finally stimulate cells to produce a large number of reactive oxygen species (ROS) and lipid peroxides, resulting in the death of cancer cells [16,17]. This principle was first classified as "regulated cell death" (RC) because it is highly dependent on Fe^{2+} and regulated by Fe^{2+} [17]. However, with the in-depth study of intracellular ion action and cell signal pathways, Wenbo Bu et al. summarized the research on promoting cancer cell death based on ion induction or regulation as "ion interference therapy" [6]. In recent years, the research into ion interference therapy has attracted much attention. Some metal or non-metal ions (such as H^+, Na^+, K^+, Cl^-, Ca^{2+}, Mn^{2+}, Cu^+) have been applied and have good effects in inhibiting tumors (Figure 1).

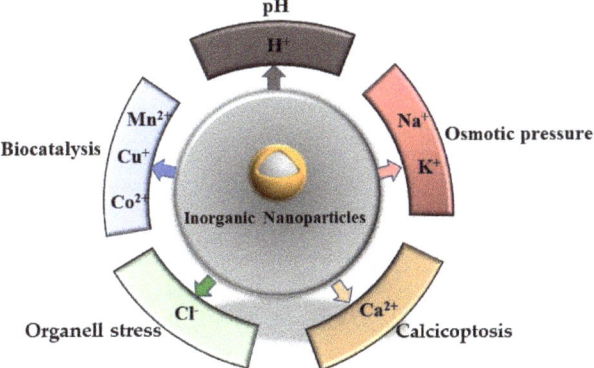

Figure 1. A schematic overview of ion interference antitumor therapy.

2.1. Protons in Tumor Cells

H^+ ions are widely distributed inside and outside cells to regulate the survival microenvironment of cells and the acid–base balance in cells. In particular, the acidic organelle in cells, the lysosome, as a proton store, is the center responsible for degradation, nutrient sensing and immunity [18]. At the same time, lysosomes are also closely related to the growth, proliferation and migration of tumor cells [19]. Targeting lysosomes to achieve H^+-mediated ion interference therapy is also an effective means to treat cancers.

Under normal conditions, to avoid cellular acidosis, cells maintain a balanced internal and external pH, usually 7.2–7.4 inside normal cells and 7.6 inside tumor cells [20]. This is due to the efflux of hydrogen ions in tumor cells, resulting in the pH of cytoplasm inside the tumor cells being higher than that of normal cells, while the pH outside the tumor cells is lower than that of normal cells—that is, the so-called acidic microenvironment. When nanoparticles gain access into the lysosome, the degradation or expansion of nanoparticles can cause lysosome swelling and the extrusion of abundant H^+ into the cytoplasm, which can induce intracellular acidification (Figure 2) [21]. The intracellular abnormal acidic environment will seriously affect the activity of cells. Therefore, we can interfere with the proliferation of tumor cells by regulating the homeostasis of intracellular hydrogen ions.

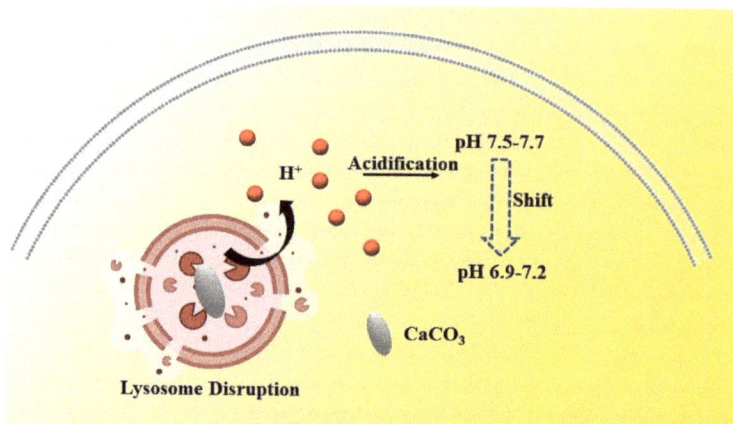

Figure 2. Schematic illustration of H$^+$—induced antitumor therapy. The expansion of CaCO$_3$ nanoparticles will lead to lysosome expansion and the extrusion of a large amount of H$^+$, resulting in intracellular acidification.

Nano-CaCO$_3$ as one kind of ideal vector with high biocompatibility, and can also be used to interfere with the intracellular homeostasis of lysosome hydrogen ions by utilizing the acid response of calcium carbonate, which was so called "lysosome bombs", according to Chenxu Zhang et al. [20]. The nano-vaterite calcium carbonate was coated with disulfide-cross-linked sodium alginate (DSA) and loaded with doxorubicin. When the nanoparticles are swallowed by tumor cells, the DSA on the surface of nano-CaCO$_3$ degrades in response to glutathione. Then, the exposed calcium carbonate particles enter the lysosome inside the tumor cell. Due to the expansion of nano-CaCO$_3$ in the lysosome, lysosome membrane will rupture rapidly and release acidic inclusions. The rapid change in intracellular pH makes tumor cells more sensitive to chemotherapeutic drugs, and finally accelerates the death of cancer cells. It can be seen that it is feasible to realize cancer treatment by interfering with the pH environment inside tumor cells by inorganic nanoparticles. However, there are still many places worth discussing and improving. Through a large number of experimental studies, although nano-calcium carbonate without modification can instantly change the pH environment of cytoplasm in cells, it is difficult to achieve a hydrogen ion interference treatment without relying on chemotherapeutic drugs. This may be due to the reaction between nano-calcium carbonate itself and hydrogen ions, which weakens the decline in pH in the cytoplasm. Secondly, the changes in the pH environment inside tumor cells caused by nanoparticles cannot always be maintained for a long time. When the self-regulatory ability of tumor cells is relatively strong, the effect of this treatment modality is diminished. Therefore, how to design a novel inorganic nanoparticle that can continuously change the intracellular pH and realize a treatment modality based on hydrogen ion interference under conditions independent of other treatments such as chemotherapy will be a new research direction.

At the same time, according to the above research results, the tumor cell death caused by hydrogen ion interference therapy is likely to be a programmed death. Programmed cell death is likely to release more antigens, which can promote the presentation efficiency of antigens to dendritic cells, and finally enhance the anti-cancer immune performance. Therefore, combined with this characteristic, it has potential to explore the application of hydrogen ion interference therapy in immunotherapy.

2.2. Sodium and Potassium Ions in Tumor Cells

The plasma membranes of mammalian cells are not protected by a cell wall, as with plant cells. Therefore, when mammalian cells are under abnormal osmotic pressure, cells swell easily, resulting in cell membrane rupture and cell death [22]. Generally speaking, due to the existence of a sodium and potassium ion exchange pump on the cell membrane, the concentration of potassium ions in cells is higher than that outside cells, so as to maintain their normal structure and physiological activity [23]. Once the concentration gradient of these ions exceeds the limit of cell self-regulation, the osmotic pressure will change sharply, resulting in the destruction of the cytoskeleton, the reduction in cell activity, and even cell lysis [24]. Therefore, interference with ion balance can lead to effective antitumor therapy. Therefore, many Na^+ and K^+ ion carriers have been designed for ion interference therapy of tumors, including molecular carriers and inorganic nanoparticle carriers.

Recently, a molecular carrier, helical polypeptide-based potassium ionophores was designed by Daeyong Lee et al. to achieve transmembrane transport of K^+ [25]. This molecular carrier can easily carry potassium ions through the cell membrane to the outside of tumor cells. When this carrier transports a large amount of K^+ outside the cell, the balance of the K^+ concentration in the cell is broken. The imbalance of intracellular potassium ion levels results in the strong activation of the unfolded protein response (UPR) in the endoplasmic reticulum (ER). The stressed ER leads to oxidative conditions by overproducing ROS, thereby damaging the mitochondria and activating apoptosis signaling. Therefore, K^+ has been proved to have a significant effect in inducing tumor cell death. In addition to the method of inputting relevant ions into cells in the form of carrier, it can also be realized by blocking relevant ion channels on cell membrane. For example, Lin Zhu, Ge Wang et al. used chondroitin sulfate to mineralize a layer of calcium carbonate membrane on the surface of cancer cell membrane, which will seriously affect the function of the Na^+/K^+ ion channel on the surface of the cell membrane and induce the imbalance of ion concentration in cells, so as to promote the death of cancer cells [26]. Although this is a very effective method for ion interference, the number of ions that can be transported by molecular carriers is very low. Therefore, new and efficient ion transport methods have become the focus of attention, and inorganic nanoparticles happen to show more advantages in this regard.

Compared with the molecular ion transporters and ion channel blockers, inorganic nanoparticles that can degrade in cells can achieve larger and more efficient ion transport, so as to interfere with the ion homeostasis in cells. Wen Jiang, Lei Yin et al. synthesized nanoparticles of NaCl (SCNPs), which are stable in the body fluid environment, through a microemulsion reaction. The reaction took place in a hexane/ethanol mixed solvent, with sodium oleate and molybdenum chloride as sodium and chloride precursors, and oleylamine as a surfactant. The as-synthesized SCNPs are hydrophobic because of the oleylamine coating. Additionally, they modified the SCNPs with PEGylated phospholipid. After the tumor cells engulfed the NaCl nanoparticles, the degradation of NaCl nanoparticles caused the explosive increase in the intracellular sodium ion concentration, resulting in the imbalance of intracellular and extracellular sodium ions, the change in cell osmotic pressure, and finally the death of tumor cells (Figure 3) [11]. Similarly, Yang Liu et al. designed a phospholipid-coated $Na_2S_2O_8$ nanoparticle. They also realized the massive accumulation of sodium ions in tumor cells, resulting in the imbalance of osmotic pressure inside and outside the cells, and finally promoted the swelling and death of cells [27]. In their study, they also evaluated the synergistic effect of chemotherapy. They found that the therapeutic effect of chemotherapy could be significantly enhanced by interfering with the homeostasis of intracellular sodium ions [11,27]. Recently, biodegradable K_3ZrF_7:Yb/Er upconversion nanoparticles ZrNPs, which are similar to ion reservoirs, have been developed which can be dissolved inside cancer cells and release high amounts of K^+ and $[ZrF_7]^{3-}$ ions, resulting a surge in intracellular osmolarity and homeostasis imbalance, according to Binbin Ding et al. [28]. Through biodegradable inorganic nanoparticles containing sodium and potassium, this can artificially regulate the internal and external osmotic balance of tumor cells, and finally significantly inhibit the proliferation of tumor cells. In these studies, it is worth noting that, compared with tumor cells, normal cells have higher tolerance to abnormal ion homeostasis. Therefore, these inorganic nanoparticles will not damage normal cells. Therefore, compared with traditional cancer treatment methods, such as chemotherapy, Na$^+$- and K$^+$-based ion interference therapy has higher biocompatibility and is not limited by drug resistance. In general, this treatment is expected to enter clinical trials. However, if better targeting effects can be provided, so that more nanoparticles can enter tumor cells, and a longer in vivo circulation cycle can be developed, as nano-NaCl has only a short cycle of 24 h, it will be more beneficial to the treatment of cancer.

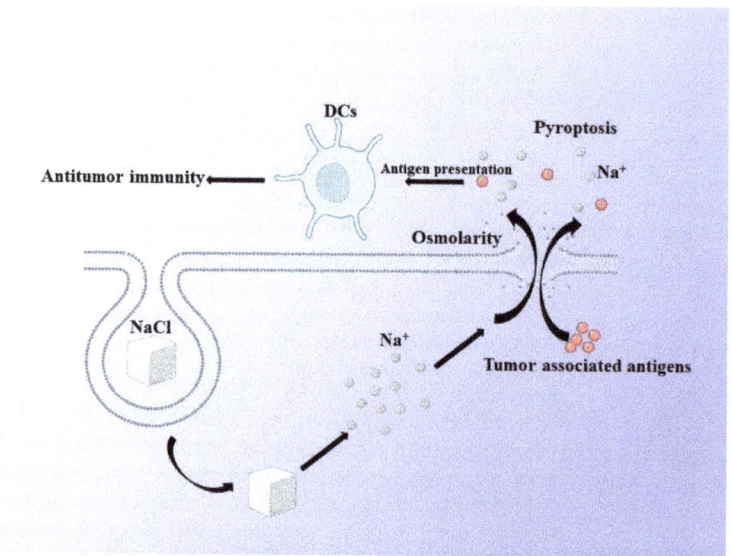

Figure 3. Schematic illustration of Na$^+$—induced antitumor therapy. A large number of Na$^+$ ions being released by NaCl nanoparticles in cancer cells leads to a change in cell osmotic pressure and further induces pyroptosis of cancer cells. Pyroptosis can promote the presentation efficiency of tumor-associated antigens to DCs, thus enhancing the antitumor immunity.

At the same time, the pyroptosis which is induced by the imbalance of intracellular ion homeostasis is more noteworthy. Pyroptosis, also known as inflammatory necrosis, is a kind of programmed cell death which is characterized by the continuous expansion of cells until the rupture of the cell membrane, leading to the release of cell contents and then activating a strong inflammatory response [29]. Regardless of whether NaCl, Na$_2$S$_2$O$_8$, or ZrNPs were used, all could induce pyroptosis, exhibiting superior antitumor immunity activity, as confirmed by enhanced dendritic cell (DC) maturity and the frequency of effector-memory T cells, as well as observably inhibiting tumor growth and pulmonary metastasis. Such research results show that Na$^+$ and K$^+$ ion interference therapy can greatly promote the patient's own immunity, so as to improve the anti-cancer effect. This makes the advantages of ion interference therapy based on inorganic nanoparticles more obvious, but the relevant research cannot be limited to this. For example, whether the balance of sodium and potassium ions can also change the activity of immune cells, such as promoting the activation of dendritic cells, the phenotypic transformation of tumor associated macrophages from M2 to M1, the activation of T cells, etc., needs to be explored. Therefore, research into ion interference therapy in immunity needs to be further conducted, and ion interference therapy combined with immunotherapy may be a promising treatment in the future.

2.3. Calcicoptosis of Tumor Cells

Calcium is an important regulator of many cellular processes, such as muscle contraction, gene transcription, hormone release, neural signal transduction etc. [30]. In 2000,

Michael R. Duchen summarized in detail the complex relationship between calcium ion signals and mitochondria [31]. On the one hand, the accumulation of Ca^{2+} in mitochondria can lead to the transient polarization of mitochondrial membrane potential. On the other hand, under the pathological conditions of intracellular calcium concentration overload, especially related to oxidative stress, the uptake of Ca^{2+} by mitochondria will trigger cell death. Based on this, in 2019, Maike Glitsch claimed that the regulation of calcium by related proteins or ion channels is expected to become another new method for tumor treatment [32]. It can be seen that calcium ions have broad research prospects in the field of tumor therapy [33].

More recently, given the importance of calcium ions in multiple cellular processes, calcium overload, characterized by an abnormal cytoplasm accumulation of free calcium ions (Ca^{2+}), is a widely recognized cause of cell damage and even cell death in numerous cell types. This undesirable destructive process can become a new tool applicable to ion interference therapy. Hence, some calcium-based inorganic nanoparticles were constructed to achieve calcium ion interference therapy, such as CaO_2, CaP and amorphous calcium carbonate.

Calcium-based inorganic NPs, which can mediate intracellular calcium homeostasis interference therapy, namely "calcicoptosis", have attracted extensive attention (Figure 4). On the basis of the unique biological effects of Ca^{2+}, Bu et al. demonstrated a highly efficient strategy for tumor therapy by utilizing pH-sensitive sodium hyaluronate-modified calcium peroxide NPs (SH-CaO_2 NPs) [8]. Effective modification with sodium hyaluronate, which was performed alongside the nucleation and growth of the CaO_2 NPs to confine the grain size, was achieved by the attraction between the negatively charged ions and positively charged nanocrystals. SH-CaO_2 could slowly decompose into free Ca^{2+} and H_2O_2 in the acid tumoral microenvironment, leading to intracellular calcium overload and oxidative stress, resulting in the desensitization of calcium-related channels followed by an uncontrollable cellular accumulation of Ca^{2+} [8]. Overloaded calcium eventually leads to the decrease in mitochondrial membrane potential, mitochondrial damage and cell death. Notably, the killing effect is not limited to tumor types or hypoxic cells, and normal cells are more tolerant of the adverse influence of NPs than tumor cells. In 2021, Shi et al. reported a simple, yet versatile, tumor-targeting "calcium ion nanogenerator" (TCaNG) to reverse drug resistance by inducing intracellular Ca^{2+} bursting [34]. The TCaNG was prepared by loading the antitumor drug doxorubicin (DOX) into calcium phosphate (CaP) nanoparticles and then enveloping them with RGD peptide-decorated DSPE-PEG. Benefiting from the tumor vessel targeting effect of RGD, the TCaNG can be enriched in tumor tissues and internalized by tumor cells. Consequently, the TCaNG could induce Ca^{2+} bursting in acidic lysosomes of tumor cells and then reverse drug resistance to improve cancer treatment. Similarly, the work of Cheng Wang et al. confirmed that the excessive calcium ions produced by amorphous calcium carbonate initiated the apoptosis program and killed cancer cells in cooperation with chemotherapy by using phospholipid-modified amorphous calcium carbonate [7]. Due to the rapid degradation of amorphous calcium carbonate in tumor cells, a large number of calcium ions eventually lead to the death of tumor cells by damaging mitochondria. Yu Bin Dong's team proved that Ca^{2+} overload and photodynamics can produce obvious synergistic killing effect by constructing a nanoscale covalent organic framework (NCOF)-based nanoagent, namely $CaCO_3$@COF-BODIPY-2I@GAG, which is embedded with $CaCO_3$ nanoparticles (NPs) and has a surface decorated with BODIPY-2I as a photosensitizer (PS) and glycosaminoglycan (GAG) targeting agent for CD44 receptors on the digestive tract tumor cells [35]. Xue Feng Yu et al.'s work significantly inhibited the proliferation of cancer cells through calcium overload by utilizing the prepared CaP mineralized black phosphorus material (CaBPs) [4]. CaBPs exhibit enhanced and selective anticancer bioactivity due to the improved pH-responsive degradation behavior and intracellular Ca^{2+} overloading in cancer cells. Furthermore, CaBPs specifically target mitochondria and cause structural damage, thus leading to mitochondria-mediated apoptosis in cancer cells. Jun Lin et al. emphasized the important role of calcium interference therapy of cuprous oxide nanoparticles coated with calcium carbonate [36]. With CaCO$_3$ responsive to pH decomposition and Cu$_2$O responsive to H$_2$S sulfuration, Cu$_2$O@CaCO$_3$ can be turned "on" in the therapeutic mode by the colorectal TME. When the CaCO$_3$ shell decomposes and releases calcium in acidic colorectal TME, excessive calcium accumulates in the cells, and finally cooperates with Cu$_2$O degradation to stimulate the production of a large amount of reactive oxygen species (ROS), which promotes the death of cancer cells. Other calcium-containing nano-formulations were also developed for the purpose of cancer calcicoptosis, and the idea of "calcium-interference therapy" would potentially open up new opportunities for the development of antitumor strategies with high safety and precision.

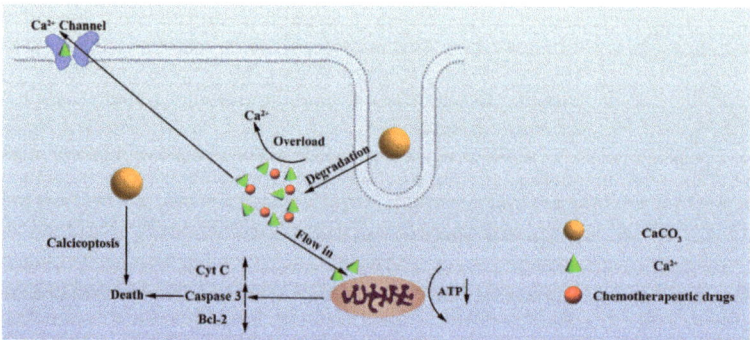

Figure 4. Schematic illustration of calcium overload-induced calcicoptosis. The overload of calcium ions in cancer cells promotes the damage of mitochondria, which induces specific calcium overload-induced cell death, which is called "calcicoptosis".

Although the ion therapy based on calcicoptosis has achieved great success both in vitro anti-tumor experiments and in vivo anti-tumor experiments in mice, the anti-tumor effect of single intracellular calcium interference therapy does not seem to be so obvious. Therefore, calcicoptosis combined with other cancer treatments, such as photothermal therapy, chemokinetic therapy and immunotherapy, is a more effective means. At present, research on the effect of calcium on immunotherapy has also made some progress.

The acid-sensitive PEG-decorated calcium carbonate (CaCO$_3$) nanoparticle incorporating curcumin, which was designed by Pan Zheng et al., can successfully induce calcium overload in the mitochondria in cells under ultrasonic environment, and cause immunogenic death of cancer cells, a manner of tumor cell death that can trigger antitumor immune responses [37]. The overload of intracellular calcium under ultrasonic conditions leads to cell immunogenic death, which greatly promotes the process of antigen delivery to mature dendritic cells, and finally produces strong antitumor immunity.

At the same time, as verified by Jingyi an and Kaixiang Zhang et al., calcium can also enhance antitumor immunity in other ways. They prepared honeycomb calcium carbonate nanoparticles (OVA@CaCO$_3$, denoted as HOCN) with ovalbumin (OVA) as the skeleton [14]. Firstly, their research found that the degradation behavior of HOCN in response to the microenvironment at the tumor site improved the low-pH environment, which inhibited the maturation of dendritic cells. Secondly, the calcium ions released after the nanoparticles were ingested by dendritic cells destroyed the autophagy inhibition conditions of dendritic cells (DCs). Autophagy of DCs can further activate T cells and eventually enhance antitumor immunity. Finally, after the tumor cells swallowed the nanoparticles, overloaded calcium ions in tumor cells promote the release of damage-associated molecular patterns (DAMPs) and the maturation of dendritic cells. This study provides us with a new research idea. Calcium-based inorganic nanoparticles can also regulate the calcium concentration in immune cells. At the same time, the calcium concentration in immune cells can also affect the related immune process, so as to promote antitumor immunity.

2.4. Intracellular Chloridion

In addition to cations, anion homeostasis in cells also has a great impact on cell life activities. The balance of normal anion concentration in cells provides a basis for maintaining cell morphology and function [38]. Compared with normal cells, cancer cells are more sensitive to anion homeostasis [38]. Therefore, interfering with anion homeostasis in cancer cells is also an effective means to achieve cancer treatment. Chloridion, as the most common anion in cells, has been widely studied. Synthetic anion transporters, a type of small molecular organic compound with transmembrane anion transport activity, can interfere with the homeostasis of cell anions, especially chloride ions, and trigger cell death [38]. Adil S. Aslam et al. proved that disturbed chloride ion concentration induced by an ion transporter in a cellular level affects endoplasmic reticulum (ER) stress by increasing Ca^{2+} concentration, and leads to apoptosis [39]. Similarly, Dong Un Lee et al. utilized cetrimonium bromide (CTAB: cationic quaternary amino group-based) gold nanorods to prove that the burst release of Cl^-, as a result of lysosomal swelling by gold nanorods, induced a massive Ca^{2+} influx, which eventually promotes the apoptosis of cancer cells [40]. Meanwhile, Yuping Jiang et al. prepared a ClO_2-loaded $CaSiO_3$ nanoparticle which can produce large numbers of Cl^- by capturing methionine to disturb the balance of homeostasis in cancer cells (Figure 5). The released Cl^- by ClO_2, which can enter the mitochondria through the voltage-dependent anion channel (VDAC), leads to mitochondrial damage and membrane potential decline, which further induce cell apoptosis [41]. In general, chloride ions directly or indirectly affect the ion homeostasis at different organelle levels in cells, including the endoplasmic reticulum, lysosome and mitochondria, thus affecting cell activity.

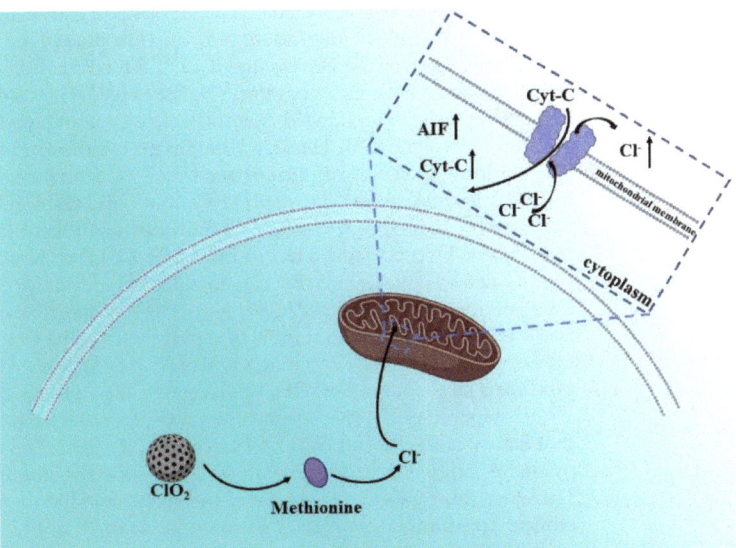

Figure 5. Schematic illustration of Cl^-—induced antitumor therapy. The released Cl^- by ClO_2, which can enter mitochondria through the voltage-dependent anion channel (VDAC), leads to mitochondrial damage and membrane potential decline, which further induce cell apoptosis.

Similar to other ion interference treatment methods, chloride-based treatment often has better antitumor effects in combination with other treatment methods, such as starvation therapy. Therefore, this combination strategy is expected to be adopted to build more multifunctional platforms for in vivo treatment. Through the previous research, it can be found that ion interference therapy always needs to be combined with other treatment methods in order to achieve a better therapeutic effect.

Although there are few studies on the antitumor immunity of chloride ions, we can still expect that chloride-mediated apoptosis will have a positive effect on cancer immunotherapy.

2.5. Manganese, Copper, and Cobalt That Activate Biocatalysis

Enzymes are powerful catalysts to complete various chemical reactions related to life activities, such as metabolism, detoxification and biosynthesis [42]. Due to the high catalytic efficiency, some biological enzymes are used to treat various diseases, including inhibiting tumor growth. Glucose oxidase (GO_x), which has the function of catalyzing the decomposition of glucose into H_2O_2, has been recognized as a "star" enzyme catalyst involved in cancer treatment [43]. Hanchun Yao et al. used collagenase, which can etch the dense extracellular matrix around tumor cells to remodel tumor microenvironment, so as to treat cancer more effectively [44]. However, due to the high synthesis cost and specific constraints of microenvironment where cells survive, biological protease used to treat diseases is difficult to popularize in the clinic.

With the development of nanotechnology, some ions with a catalytic function were developed to replace enzymes, and inorganic nanoparticles were used as carriers to inhibit tumor proliferation. Iron-based nanozymes (INs) are one of the earliest inorganic nanomaterials with exploitable catalytic behaviors [45]. Fe^{2+} can catalyze the reaction of H_2O_2 into the highly toxic $\cdot OH$, which can inhibit tumor proliferation [46]. Mn^{2+}, Cu^+ and Co^{2+}, which are delivered by inorganic nanoparticles into tumor cells, can also produce similar catalytic reactions as Fe^{2+}, so they are also used for ion interference therapy.

Lian-Hua Fu et al. constructed a Cu^{2+}-doped calcium phosphate nanoparticles. The released Cu^{2+} with the degradation of calcium phosphate in tumor cells can react with glutathione to form the Fenton agent Cu^+, which further triggers the H_2O_2 to generate $\cdot OH$ to enhance the antitumor effect (Figure 6) [47]. They also proved that Mn^{2+}-mediated Fenton-like reaction enhanced chemotherapy by using biodegradable manganese-doped calcium phosphate [43]. Hanjing Kong et al. synthesized fine CaO_2 nanoparticles with Cu–ferrocene molecules at the surface (CaO_2/Cu–ferrocene). Under an acidic condition, the particles release Ca^{2+} ions and H_2O_2 in a rapid fashion. The Fenton reaction between Cu^+ (from Cu–ferrocene) and H_2O_2 induced significant in vitro and in vivo antitumor phenomena by producing a large amount of $\cdot OH$ [48]. Shutao Gao et al. also proved that Co^{2+} is also a Fenton agent that can enhance the antitumor effect by using a ZIF-67 coated CaO_2 nanoparticle (CaO_2@ZIF-67) [49]. The nanoparticle is broken down in the weakly acidic environment within tumors to rapidly release the Fenton-like catalyst Co^{2+}. As with other Fenton reagents, Co^{2+} reacts with H_2O_2, which was released by degraded CaO_2 to produce a large number of $\cdot OH$, which eventually leads to the death of tumor cells.

Compared with other forms of ion interference therapy, this ion interference therapy that activates the biocatalysis process does not depend on other traditional cancer treatments. A single ion treatment can produce good therapeutic effects in vitro and in vivo. Therefore, this treatment method is more reliable and has more potential to achieve practical clinical applications. For this treatment, it is worth noting that they all rely on H_2O_2 to start the tumor killing program, and Ca_2O is both a good donor of H_2O_2 and Ca^{2+}. Therefore, the combination of calcicoptosis and manganese, copper, and cobalt ion interference therapy using Ca_2O may be a more effective anti-tumor plan. In addition, there are also some new developments in the field of immunotherapy. Mengyu Chang et al. conducted further research by using Cu_2O@$CaCO_3$ modified with hyaluronic acid. They found that the oxidative stress caused by Cu^+ from Cu_2O@$CaCO_3$ nanocomposites can efficiently reprogram the macrophages from the M2 phenotype to the M1 phenotype and initiate a vaccine-like immune effect after primary tumor removal, which further induces an immune-favorable tumor microenvironment and intense immune responses for anti-CD47 antibody to simultaneously inhibit distant metastasis and recurrence by immunotherapy [36]. It can be predicted that the combination of ion interference therapy and immunotherapy will have a more significant antitumor effect.

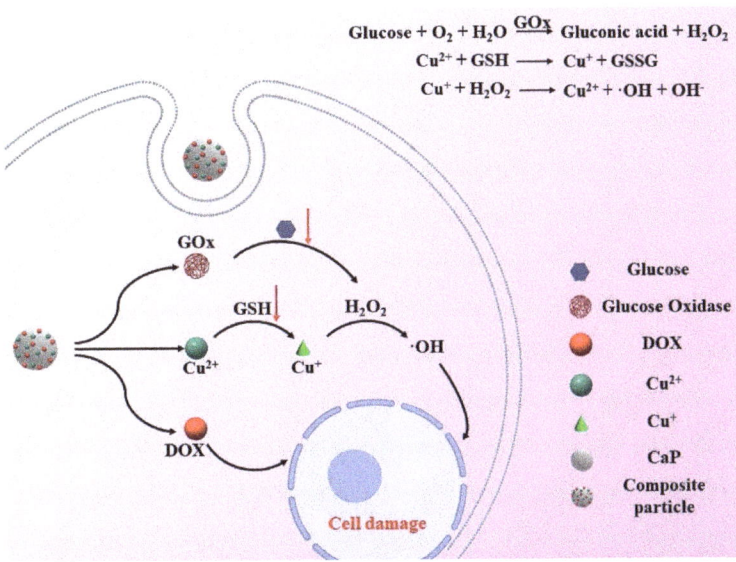

Figure 6. Schematic illustration of Cu$^+$—induced antitumor therapy. The released Cu^{2+} with the degradation of calcium phosphate in tumor cells can react with glutathione to form the Fenton agent Cu$^+$, which further triggers the H$_2$O$_2$ to generate ·OH to enhance the antitumor effect.

3. Conclusions and Perspectives

In summary, once the homeostasis of some common ions in cells is abnormal, such as specific accumulation or reduction, it will affect the normal physiological activities of cells. The treatment of inhibiting cancer cell proliferation by artificially interfering with the balance of ion homeostasis in tumor cells is called ion interference therapy. Compared with small molecular ion transport carriers with limited transport efficiency, inorganic nanoparticles that can release or absorb a large number of ions more quickly have become the best medium for ion interference therapy (Table 1). The homeostasis imbalance of different ions mediated by different inorganic nanoparticles induces the death of cancer cells by different principles. Several different principles of different ions on interference therapy are summarized as follows: (a) a great quantity of H$^+$ released from damaged lysosomes, caused by the rapid degradation or expansion of nanoparticles acidify the cytoplasmic environment and induce cancer cell death; (b) the imbalance of a large amount of sodium or potassium concentration gradient, caused by the transportation of sodium or potassium nano-inorganic salt into cancer cells, leads to the change in cell osmotic pressure, makes the cells swell, and finally induces pyroptosis; (c) calcium overload in tumor cells mediated by calcium-based inorganic nanoparticles damages the normal function of mitochondria and further induces apoptosis, which is called calciocoptosis; (d) The specific accumulation of chloride ions in the mitochondria of tumor cells leads to the decrease in mitochondrial membrane potential, the release of mitochondrial contents and the induction of apoptosis; (e) some Fenton agents produced by the degradation of inorganic nanoparticles, including Cu$^+$, Mn^{2+} and Mn^{2+}, can trigger the H$_2$O$_2$ to generate ·OH to enhance the anti-tumor effect. On the other hand, apoptosis and pyroptosis are both forms of programmed cell death, mediated by ion interference therapy, which enhances the effects of antitumor immunity. Among these nanoparticles, Ca^{2+} promotes the effects of antitumor immunity by destroying the autophagy inhibition conditions of dendritic cells, a special mechanism compared with others.

Table 1. The classification of inorganic nanomaterials for IIT.

Inorganic Nanoparticles	Interfering Ions	Mechanism	Reference
$CaCO_3$ NPs	H^+	Intracellular pH	[20]
NaCl NPs	Na^+	Osmotic pressure	[11]
$Na_2S_2O_8$ NPs	Na^+	Osmotic pressure	[27]
K_3ZrF_7:Yb/Er	K^+, $[ZrF_7]^{3-}$	Osmotic pressure	[28]
SH-CaO_2 NPs	Ca^{2+}	Calcicoptosis	[8]
CaP NPs	Ca^{2+}	Calcicoptosis	[34]
Amorphous calcium carbonate (ACC NPs)	Ca^{2+}	Calcicoptosis	[7]
$CaCO_3$@COF-BODIPY-2I@GAG	Ca^{2+}	Calcicoptosis	[35]
CaBPs	Ca^{2+}	Calcicoptosis	[4]
Cu_2O@$CaCO_3$	Cu^+, Ca^{2+}	Biocatalysis, calcicoptosis	[36]
$CaCO_3$@PEG	Ca^{2+}	Calcicoptosis	[37]
OVA@$CaCO_3$	Ca^{2+}	Calcicoptosis	[14]
ClO_2	Cl^-	Organelle stress	[41]
CaO_2/Cu–ferrocene	Ca^{2+}, Cu^+	Calcicoptosis, biocatalysis	[48]
Cu-CaP NPs	Cu^{2+}	Biocatalysis	[47]
Mn-CaP NPs	Mn^{2+}	Biocatalysis	[43]
CaO_2@ZIF-67	Co^{2+}	Biocatalysis	[49]

Compared with traditional tumor treatment, such as chemotherapy and radiotherapy, ion interference treatment based on inorganic nanoparticles shows higher biocompatibility and further therapeutic effects in vivo due to the lower tolerance of tumor cells to abnormal ion concentration compared to normal cells. Although ion interference therapy is a practical method, and different ions have different action mechanisms, it still has many aspects worthy of in-depth exploration.

Through previous research, it can be found that ion interference therapy always needs to be combined with other treatment methods in order to achieve better therapeutic effects. Ion interference therapy alone does not seem to produce obvious antitumor effects, such as for Ca^{2+}, which may be related to the degradation efficiency of inorganic nanoparticles, the in vivo circulation time, the ability to release or adsorb ions, and the metabolic rate of tumor cells. Therefore, finding ways to improve the performance of inorganic nanoparticles based on the above aspects is the next research direction that needs to be focused on. At the same time, it should also be considered whether there is synergy between different ion therapy methods to promote antitumor effects, such as Cu^+ and Ca^{2+}. The combination of various types of ion interference therapy also has broad development prospects. In addition, the combination of ion interference therapy and immunotherapy cannot be ignored. The effects of different ions on immune cells and the process of antitumor immunity need to be further studied. Additionally, the use of new composite inorganic nanoparticles combined with ion interference therapy and immunotherapy will provide a promising strategy for enhancing antitumor immunity.

Author Contributions: The manuscript was written with contributions from all authors. All authors have read and agreed to the published version of the manuscript.

Funding: This research was funded by National Natural Science Foundation of China, grant number 81973262, 82011530138, 81970660.

Institutional Review Board Statement: Not applicable.

Informed Consent Statement: Not applicable.

Data Availability Statement: Not applicable.

Conflicts of Interest: The authors declare no conflict of interest.

References

1. Cao, W.; Chen, H.-D.; Yu, Y.-W.; Li, N.; Chen, W.-Q. Changing profiles of cancer burden worldwide and in China: A secondary analysis of the global cancer statistics 2020. *Chin. Med. J.* **2021**, *134*, 783–791. [CrossRef]
2. Claessens, A.K.M.; Ibragimova, K.I.E.; Geurts, S.M.E.; Bos, M.; Erdkamp, F.L.G.; Tjan-Heijnen, V.C.G. The role of chemotherapy in treatment of advanced breast cancer: An overview for clinical practice. *Crit. Rev. Oncol. Hematol.* **2020**, *153*, 102988. [CrossRef]
3. Lee, J.C.; Lamanna, N. Is there a role for chemotherapy in the era of targeted therapies? *Curr. Hematol. Malig. Rep.* **2020**, *15*, 72–82. [CrossRef]
4. Pan, T.; Fu, W.; Xin, H.; Geng, S.; Li, Z.; Cui, H.; Zhang, Y.; Chu, P.K.; Zhou, W.; Yu, X.F. Calcium phosphate mineralized black phosphorous with enhanced functionality and anticancer bioactivity. *Adv. Funct. Mater.* **2020**, *30*, 2003069. [CrossRef]
5. Chu, X.; Jiang, X.; Liu, Y.; Zhai, S.; Jiang, Y.; Chen, Y.; Wu, J.; Wang, Y.; Wu, Y.; Tao, X.; et al. Nitric oxide modulating calcium store for Ca 2+ -initiated cancer therapy. *Adv. Funct. Mater.* **2021**, *31*, 2008507. [CrossRef]
6. Liu, Y.; Zhang, M.; Bu, W. Bioactive nanomaterials for ion-interference therapy. *View* **2020**, *1*, e18. [CrossRef]
7. Wang, C.; Yu, F.; Liu, X.; Chen, S.; Wu, R.; Zhao, R.; Hu, F.; Yuan, H. Cancer-specific therapy by artificial modulation of intracellular calcium concentration. *Adv. Healthc. Mater.* **2019**, *8*, e1900501. [CrossRef] [PubMed]
8. Zhang, M.; Song, R.; Liu, Y.; Yi, Z.; Meng, X.; Zhang, J.; Tang, Z.; Yao, Z.; Liu, Y.; Liu, X.; et al. Calcium-overload-mediated tumor therapy by calcium peroxide nanoparticles. *Chem* **2019**, *5*, 2171–2182. [CrossRef]
9. Zhao, Y.-Z.; Lin, M.-T.; Lan, Q.-H.; Zhai, Y.-Y.; Xu, H.-L.; Xiao, J.; Kou, L.; Yao, Q. Silk fibroin-modified disulfiram/zinc oxide nanocomposites for pH triggered release of Zn2+ and synergistic antitumor efficacy. *Mol. Pharm.* **2020**, *17*, 3857–3869. [CrossRef]
10. Li, Y.; Zhou, S.; Song, H.; Yu, T.; Zheng, X.; Chu, Q. CaCO3 nanoparticles incorporated with KAE to enable amplified calcium overload cancer therapy. *Biomaterials* **2021**, *277*, 121080. [CrossRef]
11. Jiang, W.; Yin, L.; Chen, H.; Paschall, A.V.; Zhang, L.; Fu, W.; Zhang, W.; Todd, T.; Yu, K.S.; Zhou, S.; et al. NaCl nanoparticles as a cancer therapeutic. *Adv. Mater.* **2019**, *31*, e1904058. [CrossRef]
12. Chen, D.S.; Mellman, I. Oncology meets immunology: The cancer-immunity cycle. *Immunity* **2013**, *39*, 1–10. [CrossRef]
13. Wang, Z.; Geng, Z.; Shao, W.; Liu, E.; Zhang, J.; Tang, J.; Wang, P.; Sun, X.; Xiao, L.; Xu, W.; et al. Cancer-derived sialylated IgG promotes tumor immune escape by binding to Siglecs on effector T cells. *Cell. Mol. Immunol.* **2020**, *17*, 1148–1162. [CrossRef]
14. An, J.; Zhang, K.; Wang, B.; Wu, S.; Wang, Y.; Zhang, H.; Zhang, Z.; Liu, J.; Shi, J. Nanoenabled disruption of multiple barriers in antigen cross-presentation of dendritic cells via calcium interference for enhanced chemo-immunotherapy. *ACS Nano* **2020**, *14*, 7639–7650. [CrossRef] [PubMed]
15. Rao, S.G.; Patel, N.J.; Singh, H. Intracellular chloride channels: Novel biomarkers in diseases. *Front. Physiol.* **2020**, *11*, 96. [CrossRef]
16. Song, R.; Li, T.; Ye, J.; Sun, F.; Hou, B.; Saeed, M.; Gao, J.; Wang, Y.; Zhu, Q.; Xu, Z.; et al. Acidity-activatable dynamic nanoparticles boosting ferroptotic cell death for immunotherapy of cancer. *Adv. Mater.* **2021**, *33*, e2101155. [CrossRef]
17. Zuo, S.; Yu, J.; Pan, H.; Lu, L. Novel insights on targeting ferroptosis in cancer therapy. *Biomark. Res.* **2020**, *8*, 50. [CrossRef]
18. Zhang, Z.; Yue, P.; Lu, T.; Wang, Y.; Wei, Y.; Wei, X. Role of lysosomes in physiological activities, diseases, and therapy. *J. Hematol. Oncol.* **2021**, *14*, 79. [CrossRef] [PubMed]
19. Machado, E.R.; Annunziata, I.; van de Vlekkert, D.; Grosveld, G.C.; D'Azzo, A. Lysosomes and cancer progression: A malignant liaison. *Front. Cell Dev. Biol.* **2021**, *9*, 373. [CrossRef] [PubMed]
20. Zhang, C.; Li, S.; Yu, A.; Wang, Y. Nano CaCO3 "Lysosomal bombs" enhance chemotherapy drug efficacy via rebalancing tumor intracellular pH. *ACS Biomater. Sci. Eng.* **2019**, *5*, 3398–3408. [CrossRef]
21. Liu, C.-G.; Han, Y.-H.; Kankala, R.K.; Wang, S.-B.; Chen, A.-Z. Subcellular performance of nanoparticles in cancer therapy. *Int. J. Nanomed.* **2020**, *15*, 675–704. [CrossRef]
22. Li, Y.; Konstantopoulos, K.; Zhao, R.; Mori, Y.; Sun, S.X. The importance of water and hydraulic pressure in cell dynamics. *J. Cell Sci.* **2020**, *133*, jcs240341. [CrossRef] [PubMed]
23. Chen, D.; Song, M.; Mohamad, O.; Yu, S.P. Inhibition of Na+/K+-ATPase induces hybrid cell death and enhanced sensitivity to chemotherapy in human glioblastoma cells. *BMC Cancer* **2014**, *14*, 1–15. [CrossRef]
24. Grigorov, E.; Kirov, B.; Marinov, M.B.; Galabov, V. Review of microfluidic methods for cellular lysis. *Micromachines* **2021**, *12*, 498. [CrossRef] [PubMed]
25. Lee, D.; Lee, S.H.; Noh, I.; Oh, E.; Ryu, H.; Ha, J.; Jeong, S.; Yoo, J.; Jeon, T.J.; Yun, C.O.; et al. A helical polypeptide-based potassium ionophore induces endoplasmic reticulum stress-mediated apoptosis by perturbing ion homeostasis. *Adv. Sci.* **2019**, *6*, 1801995. [CrossRef]
26. Zhu, L.; Wang, G.; Shi, W.; Ma, X.; Yang, X.; Yang, H.; Guo, Y.; Yang, L. In situ generation of biocompatible amorphous calcium carbonate onto cell membrane to block membrane transport protein—A new strategy for cancer therapy via mimicking abnormal mineralization. *J. Colloid Interface Sci.* **2019**, *541*, 339–347. [CrossRef]

27. Liu, Y.; Zhen, W.; Wang, Y.; Song, S.; Zhang, H. Na$_2$S$_2$O$_8$ nanoparticles trigger antitumor immunotherapy through reactive oxygen species storm and surge of tumor osmolarity. *J. Am. Chem. Soc.* **2020**, *142*, 21751–21757. [CrossRef]
28. Ding, B.; Sheng, J.; Zheng, P.; Li, C.; Li, D.; Cheng, Z.; Ma, P.; Lin, J. Biodegradable upconversion nanoparticles induce pyroptosis for cancer immunotherapy. *Nano Lett.* **2021**, *21*, 8281–8289. [CrossRef] [PubMed]
29. Wang, L.; Qin, X.; Liang, J.; Ge, P. Induction of pyroptosis: A promising strategy for cancer treatment. *Front. Oncol.* **2021**, *11*, 635774. [CrossRef]
30. Ma, Z.; Zhang, J.; Zhang, W.; Foda, M.F.; Zhang, Y.; Ge, L.; Han, H. Intracellular Ca^{2+} cascade guided by NIR-II photothermal switch for specific tumor therapy. *iScience* **2020**, *23*, 101049. [CrossRef]
31. Duchen, M.R. Mitochondria and calcium: From cell signalling to cell death. *J. Physiol.* **2000**, *529*, 57–68. [CrossRef] [PubMed]
32. Glitsch, M. Mechano- and pH-sensing convergence on Ca(2+)-mobilising proteins—A recipe for cancer? *Cell Calcium* **2019**, *80*, 38–45. [CrossRef] [PubMed]
33. Delierneux, C.; Kouba, S.; Shanmughapriya, S.; Potier-Cartereau, M.; Trebak, M.; Hempel, N. Mitochondrial calcium regulation of redox signaling in cancer. *Cells* **2020**, *9*, 432. [CrossRef] [PubMed]
34. Liu, J.; Zhu, C.; Xu, L.; Wang, D.; Liu, W.; Zhang, K.; Zhang, Z.; Shi, J. Nanoenabled intracellular calcium bursting for safe and efficient reversal of drug resistance in tumor cells. *Nano Lett.* **2020**, *20*, 8102–8111. [CrossRef]
35. Guan, Q.; Zhou, L.L.; Lv, F.H.; Li, W.Y.; Li, Y.A.; Dong, Y.B. A glycosylated covalent organic framework equipped with BODIPY and CaCO$_3$ for synergistic tumor therapy. *Angew. Chem. Int. Ed.* **2020**, *59*, 18042–18047. [CrossRef]
36. Chang, M.; Hou, Z.; Jin, D.; Zhou, J.; Wang, M.; Wang, M.; Shu, M.; Ding, B.; Li, C.; Lin, J. Colorectal tumor microenvironment-activated bio-decomposable and metabolizable Cu$_2$O@CaCO$_3$ nanocomposites for synergistic oncotherapy. *Adv. Mater.* **2020**, *32*, e2004647. [CrossRef] [PubMed]
37. Zheng, P.; Ding, B.; Jiang, Z.; Xu, W.; Li, G.; Ding, J.; Chen, X. Ultrasound-augmented mitochondrial calcium ion overload by calcium nanomodulator to induce immunogenic cell death. *Nano Lett.* **2021**, *21*, 2088–2093. [CrossRef]
38. Yu, X.-H.; Hong, X.-Q.; Mao, Q.-C.; Chen, W.-H. Biological effects and activity optimization of small-molecule, drug-like synthetic anion transporters. *Eur. J. Med. Chem.* **2019**, *184*, 111782. [CrossRef] [PubMed]
39. Aslam, A.S.; Fuwad, A.; Ryu, H.; Selvaraj, B.; Song, J.-W.; Kim, D.W.; Kim, S.M.; Lee, J.W.; Jeon, T.-J.; Cho, D.-G. Synthetic anion transporters as endoplasmic reticulum (ER) stress inducers. *Org. Lett.* **2019**, *21*, 7828–7832. [CrossRef]
40. Lee, D.U.; Park, J.-Y.; Kwon, S.; Park, J.Y.; Kim, Y.H.; Khang, D.; Hong, J.H. Apoptotic lysosomal proton sponge effect in tumor tissue by cationic gold nanorods. *Nanoscale* **2019**, *11*, 19980–19993. [CrossRef]
41. Jiang, Y.; Tan, Y.; Xiao, K.; Li, X.; Shao, K.; Song, J.; Kong, X.; Shi, J. pH-regulating nanoplatform for the "double channel chase" of tumor cells by the synergistic cascade between chlorine treatment and methionine-depletion starvation therapy. *ACS Appl. Mater. Interfaces* **2021**, *13*, 54690–54705. [CrossRef]
42. Sutrisno, L.; Hu, Y.; Hou, Y.; Cai, K.; Li, M.; Luo, Z. Progress of iron-based nanozymes for antitumor therapy. *Front. Chem.* **2020**, *8*, 680. [CrossRef] [PubMed]
43. Fu, L.-H.; Hu, Y.-R.; Qi, C.; He, T.; Jiang, S.; Jiang, C.; He, J.; Qu, J.; Lin, J.; Huang, P. Biodegradable manganese-doped calcium phosphate nanotheranostics for traceable cascade reaction-enhanced anti-tumor therapy. *ACS Nano* **2019**, *13*, 13985–13994. [CrossRef]
44. Yao, H.; Guo, X.; Zhou, H.; Ren, J.; Li, Y.; Duan, S.; Gong, X.; Du, B. Mild acid-responsive "nanoenzyme capsule" remodeling of the tumor microenvironment to increase tumor penetration. *ACS Appl. Mater. Interfaces* **2020**, *12*, 20214–20227. [CrossRef] [PubMed]
45. Cramer, F.; Kampe, W. Inclusion compounds. XVII.1 catalysis of decarboxylation by cyclodextrins. A model reaction for the mechanism of enzymes. *J. Am. Chem. Soc.* **1965**, *87*, 1115–1120. [CrossRef] [PubMed]
46. Ploetz, E.; Zimpel, A.; Cauda, V.; Bauer, D.; Lamb, D.C.; Haisch, C.; Zahler, S.; Vollmar, A.M.; Wuttke, S.; Engelke, H. Metal–organic framework nanoparticles induce pyroptosis in cells controlled by the extracellular pH. *Adv. Mater.* **2020**, *32*, e1907267. [CrossRef]
47. Fu, L.; Wan, Y.; Qi, C.; He, J.; Li, C.; Yang, C.; Xu, H.; Lin, J.; Huang, P. Nanocatalytic theranostics with glutathione depletion and enhanced reactive oxygen species generation for efficient cancer therapy. *Adv. Mater.* **2021**, *33*, e2006892. [CrossRef]
48. Kong, H.; Chu, Q.; Fang, C.; Cao, G.; Han, G.; Li, X. Cu–ferrocene-functionalized CaO$_2$ nanoparticles to enable tumor-specific synergistic therapy with GSH depletion and calcium overload. *Adv. Sci.* **2021**, *8*, e2100241. [CrossRef]
49. Gao, S.; Jin, Y.; Ge, K.; Li, Z.; Liu, H.; Dai, X.; Zhang, Y.; Chen, S.; Liang, X.; Zhang, J. Self-supply of O$_2$ and H$_2$O$_2$ by a nanocatalytic medicine to enhance combined chemo/chemodynamic therapy. *Adv. Sci.* **2019**, *6*, 1902137. [CrossRef]

Review

NIR-II Aggregation-Induced Emission Luminogens for Tumor Phototheranostics

Yonghong Tan [1,†], Peiying Liu [1,†], Danxia Li [1], Dong Wang [1,*] and Ben Zhong Tang [2]

[1] Center for AIE Research, Shenzhen Key Laboratory of Polymer Science and Technology, Guangdong Research Center for Interfacial Engineering of Functional Materials, College of Materials Science and Engineering, Shenzhen University, Shenzhen 518060, China; 2110343013@email.szu.edu.cn (Y.T.); 2110343005@email.szu.edu.cn (P.L.); 2110343102@email.szu.edu.cn (D.L.)

[2] Shenzhen Institute of Aggregate Science and Technology, School of Science and Engineering, The Chinese University of Hong Kong, Shenzhen 518172, China; tangbenz@cuhk.edu.cn

* Correspondence: wangd@szu.edu.cn
† These authors contributed equally to this work.

Abstract: As an emerging and powerful material, aggregation-induced emission luminogens (AIEgens), which could simultaneously provide a precise diagnosis and efficient therapeutics, have exhibited significant superiorities in the field of phototheranostics. Of particular interest is phototheranostics based on AIEgens with the emission in the range of second near-infrared (NIR-II) range (1000–1700 nm), which has promoted the feasibility of their clinical applications by virtue of numerous preponderances benefiting from the extremely long wavelength. In this minireview, we summarize the latest advances in the field of phototheranostics based on NIR-II AIEgens during the past 3 years, including the strategies of constructing NIR-II AIEgens and their applications in different theranostic modalities (FLI-guided PTT, PAI-guided PTT, and multimodal imaging-guided PDT–PTT synergistic therapy); in addition, a brief conclusion of perspectives and challenges in the field of phototheranostics is given at the end.

Keywords: aggregation-induced emission; NIR-II emission; phototheranostics; cancer treatment

1. Introduction

Cancer, one of the deadliest diseases in recent decades, has remained a global health concern due to its growing morbidity rate, developing relapse rate, and low survival rate [1–3]. Traditional cancer diagnostic methods, including magnetic resonance imaging (MRI), positron emission tomography (PET), and computed tomography (CT), exhibit some respective and collective drawbacks such as insufficient sensitivity and specificity, high cost, and cumbersome instrumentation [4,5]. Those conventional therapeutic methods toward cancers (such as surgical removal, chemotherapy, and radiotherapy) commonly cause side effects, systematic toxicity, unavoidable invasion, and high relapse rate [6]. In general, conventional protocols for cancer diagnostics and therapeutics are individually conducted, which could result in the inefficiency of the curing process and the inaccuracy of treatments. Given the circumstances, tremendous efforts have been made to explore more effective approaches for cancers treatment, among which phototheranostics is a significant advancement that enables the ingenious integration of precise photodiagnostic imaging with phototherapeutic intervention in a single system within spatial colocalization [7–9]. This inspiration stirs researchers' increasing interest in both fundamental studies and clinical trials, mainly on account of its intrinsic advantages, such as simultaneously accurate diagnosis with in situ effective therapy, improved pharmacokinetics, maximized efficacy, optimized drug safety, elevated sensitivity, and specificity in comparison with traditional cancer treatments.

Various modalities are involved in phototheranostic systems, including therapeutic methods such as photodynamic therapy (PDT) and photothermal therapy (PTT), and diagnostic technologies such as photothermal imaging (PTI), photoacoustic imaging (PAI), and fluorescence imaging (FLI). As an emerging strategy for cancer treatments via generating reactive oxygen species (ROS) with the assistance of light, tissue oxygen, and photosensitizer (PS), PDT has a remarkable light-controllable ability, specific spatiotemporal selectivity, and minimized invasiveness [10]. PTT is another effective light-triggered cancer therapy modality, which affords excellent tumor suppression by sufficient thermal production upon photoirradiation [11]. Moreover, the thermal signal generated during PTT can be detected by thermal imaging systems for PTI, providing images with great temperature sensitivity for tumor detection. Apart from that, the generated thermal signal gives rise to the rapid thermoelastic expansion of tissue, based on which the light-triggered diagnostic protocol, PAI, can be established, sufficing to provide imaging with high penetration depth and portray clear tumor profiles [12]. Among all photodiagnostic modalities, FLI has aroused intense interest on account of its simple operation, high sensitivity, noninvasive features, and preferable biosafety especially organic fluorophores [13–15]. However, FLI generally suffers from some drawbacks in terms of tissue penetration and spatial resolution, which hinders its practical utilization. Moreover, conventional organic fluorophores are ordinarily hydrophobic, which inherently form aggregates in a physiological environment that is generally composed of water, causing local concentration increasing and fluorescence quenching, which is the notorious aggregation-caused quenching (ACQ) effect, consequentially leading to unsatisfactory imaging outcomes.

Fortunately, aggregation-induced emission (AIE), a unique phenomenon discovered in 2001 by Tang, has solved this predicament, which is shown in some twisted-structure molecules with propeller-shaped conformation, tetraphenylethene (TPE) derivatives, for instance. The emissions of AIEgens demonstrate a low intensity in a single molecular state but are enhanced in aggregated state, exhibiting completely contrary features to ACQ [16–18]. Numerous endeavors have been made to explore the mechanism of AIE phenomenon, and the restriction of intramolecular motion (RIM) that includes restriction of intramolecular rotation (RIR) and restriction of intramolecular vibration (RIV) has been widely approved, according to which the twisted structure and the sufficient structural rotors and/or vibrators jointly endow AIE luminogens (AIEgens) with the distinct characteristics [7,19]. Due to the structural superiorities, most of the excited-state energy of AIEgens can be consumed through the nonradiative decay pathway, resulting from vigorous intramolecular motions in the single molecular state, consequently promoting photothermal conversion. On the contrary, the intramolecular motions can be suppressed in an aggregated state; thus, the radiative decay pathway is in the dominant position, consequently boosting fluorescent emission. In addition, AIEgens have been recognized to possess many intrinsic advantages including good biocompatibility, large Stokes shift, excellent tolerance for high concentration, turn-on feature, high photobleaching threshold and outperformed photosensitizing features, which all allow the great potential for efficient phototheranostics.

On the other hand, enthused by the remained shortcomings of fluorescence imaging with visible (400–680 nm) and first near-infrared region (NIR-I, 700–900 nm), including low tissue penetration, unsatisfactory spatial resolution, etc., researchers pay attention to develop fluorescent materials with emission in the range of second near-infrared (NIR-II) window to overcome these drawbacks [20–25]. NIR-II fluorophores possess the capability of surmounting the inherent deficiencies of conventional FLI, by virtue of its remarkable features including deep penetration, reduced tissue scattering, minimal damage, and high spatial resolution endowed by the extremely long wavelength [26–28]. The combination of the advantages of both NIR-II fluorophores and AIEgens unprecedentedly complemented each other with excellent imaging and extraordinary therapy, thus allowing a better application in the clinical field and accelerating the progression of contemporary precision medicine [29,30]. Witnessing the rapid development and great significance of theranostic researches based on NIR-II AIEgens, it is crucial to publish a comprehensive review article to systematically generalize the merits of NIR-II AIEgens in cancer theranostics and to provide an integrated picture of this area through the introduction of basic concepts and recent trends as well as novel perspectives.

In this minireview, we summarize the recent advances of NIR-II AIEgens (Table 1) in the cancer theranostic field during the past three years. In the first section, the breakthroughs in photothermal therapy under the guidance of FLI generated by NIR-II AIEgens are primarily elaborated. In the second section, up-to-date signs of progress observed in photothermal therapy under the guidance of PAI are presented subsequently, as well as the design strategies and mechanistic insights of theranostics. In the third section, boosted multimodality theranostics systems based on NIR-II AIEgens are listed, and the complex construction and the favorable superiorities of the systems are discussed in detail as well. The existing limitations and novel perspectives in this field conclude the study. We expect that this review will provide valuable insights into NIR-II AIEgens-based theranostics and serve as an inspiration for developing integrated systems of diagnostics and therapeutics, thereby stimulating more studies at this research frontier.

Table 1. The chemical structures and essential photophysical properties of NIR-II AIEgens.

Name	Chemical Structure	$\lambda_{Abs}/\lambda_{em}$ (nm)	Extinction Coefficient	QY	Photothermal Conversion
BPBBT		≈705/ ≈1020 (in water, NPs)	0.9×10^4 $M^{-1} \cdot cm^{-1}$ at 808 nm	0.145% [a] (NPs)	27.5% (NPs, 808 nm laser)
PBPTV		≈700/≈960 (in water, NPs)	N.A.	8.6% [b] (in DCM)	45.3% (NPs, 808 nm laser)
DTPA–BBTD		752/975 (in THF)	7.09 $L \cdot g^{-1} \cdot cm^{-1}$ at 753 nm	0.151% [a] (NPs)	13.2% (NPs, 660 nm laser)
NIRb14		822/1090 (in THF)	N.A	N.A.	31.2% (NPs, 808 nm laser)
BITT		595/741 (in ethanol)	3.9×10^4 $M^{-1} \cdot cm^{-1}$	5.8% [b] (aggregate)	35.76% (660 nm laser)
TSSI		636/992 (in water, NPs)	N.A.	N.A.	46.0% (NPs, 660 nm laser)
TSSAM		595/1022 (in DMSO)	N.A.	0.034% [a] (NPs)	40.1% (NPs, 660 nm laser)
TTT-4		568/830 (in DMSO)	N.A.	0.8% [b] (aggregate, DMSO:toluene = 19:1)	39.9% (NPs, 660 nm laser)

[a] Relative PL quantum yields of these molecules recalculated based on IR-26 = 0.05%. [b] Absolute PL quantum yields.

2. NIR-II FLI-Guided PTT

Nowadays, FLI in NIR-II has become a momentous facility for cancer diagnosis owing to its prominent merits for in vivo monitoring and visualizing of lesions [31,32]. Additionally, the combination of NIR-II FLI and PTT could provide unlimited prospects to construct outstanding theranostic systems [33,34]. As illustrated in Figure 1a [35], when a fluorophore absorbs photons or other energy, it can be promoted to the excited states (S_n) from the ground state (S_0) and transfers back to the ground state via either radiative or nonradiative decay. Nevertheless, it is not difficult to find that these two modes of energy dissipation are in competition with each other since energy is conserved. As a result, the strategy to keep the equilibrium between fluorescence (radiative decay) and photothermal effect (nonradiative decay) is the focus of FLI-guided PTT.

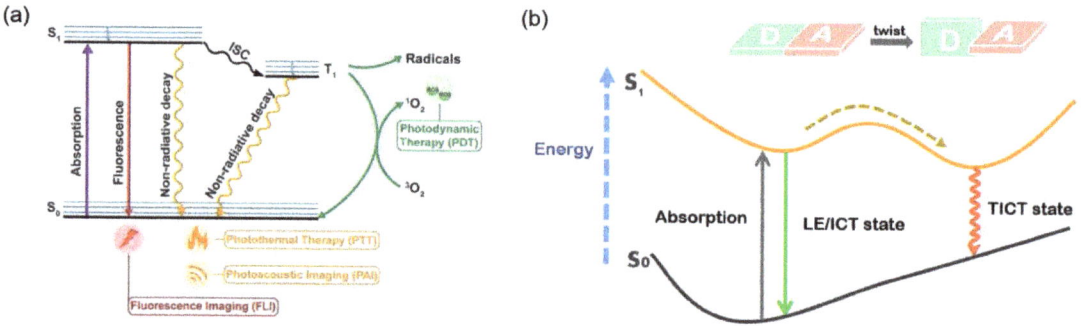

Figure 1. (a) Schematic illustration of Jablonski diagram. Reprinted with permission from Ref. [35]. Copyright 2021, Light Publishing Group; (b) the illustration of TICT mechanism.

Fortunately, AIEgens exhibit free-moving molecular rotators or vibrators in their structure, which are ideal agents to keep the equilibrium between fluorescence and photothermal effect [36]. Using reverse thinking of the AIE process, researchers devised numerous strategies to maximize molecular motion in the aggregated state of AIEgens to exhibit superior heat transitions without compromising FLI [37]. In addition, it was found that twisted intramolecular charge transfer (TICT) states in AIEgens typically abate the fluorescence signals but enhance their photothermal capability, which quickly sparked strong interest among researchers [38–40].

AIEgens with long emission wavelengths generally have powerful electron donor (D)–acceptor (A) strength and, therefore, are candidates to modulate TICT formations, since increasing the D–A effect can achieve red-shifted emission and stabilize the TICT state by facilitating charge separation (Figure 1b) [41,42]. When AIEgen is under unbound and free rotating conditions, nonradiative decay would dominate the excited-state energy consumption. In contrast, upon reaching an aggregated state, the physical constraints disable TICT formations; thus, the equilibrium moves to the radiative decay pathway accompanied with bright fluorescence [43,44]. Therefore, tailoring intramolecular motion is a feasible strategy to realize a subtle balance between fluorescence and photothermal effect [45,46].

Lu et al. [47] reported a strategy inspired by the theory of RIR to tailor the equilibrium of fluorescence and photothermal efficiency. They combined NIR-II AIEgen (BPBBT, Figure 2a) with human serum albumin (HSA), in order to restrict the intramolecular rotation of BPBBT. Fluorescence emission spectra demonstrated that the fluorescence intensity of BPBBT decreased at a fraction value of water (f_w) below 30% but increased when further raising f_w. This phenomenon could be explained by the fact that BPBBT transitioned from LE state to TICT state when the polarity of solvent elevated, then the increase in poor solvent contributed to forming the aggregated state of BPBBT, which prevented TICT formations and enhanced the fluorescence emission (Figure 2b). It was found that with the enhancement of the HSA ratio in BPBBT/HSA complexes (BPBBT NPs), the photothermal effect was further increased. The energy difference between S$_1$ and S$_0$ of BPBBT at NPs state was determined to be narrower than at the AIE state but broader than at the TICT state (Figure 2c), which evinced that the addition of HSA successfully altered LE and TICT state by raising the dihedral angles to provide a chance for the equilibrium to move to the TICT state. In vivo biological imaging showed that fluorescence signal was detected in orthotopic and metastatic tumors accurately and reached a maximum at 30 h postinjection. Notably, NIR-II imaging-guided PTT based on BPBBT NPs could precisely distinguish lesions with dimensions as small as 0.5 mm × 0.3 mm and completely cure tumor-bearing mice with the optimized laser doses (5 out of 5) without recurrence in 30 days (Figure 2d). Additionally, compared with HSA/indocyanine green (ICG) complexes that were applied to NIR-I imaging-guided PTT, BPBBT NPs provided more accurate and sensitive imaging and exhibited a higher photothermal conversion effect and better photostability, which dramatically enhanced the efficiency of PTT and prevented from omitting small lesions. Above all, the BPBBT NPs displayed great potential in NIR-II FLI-guided PTT, particularly for colon cancer theranostics.

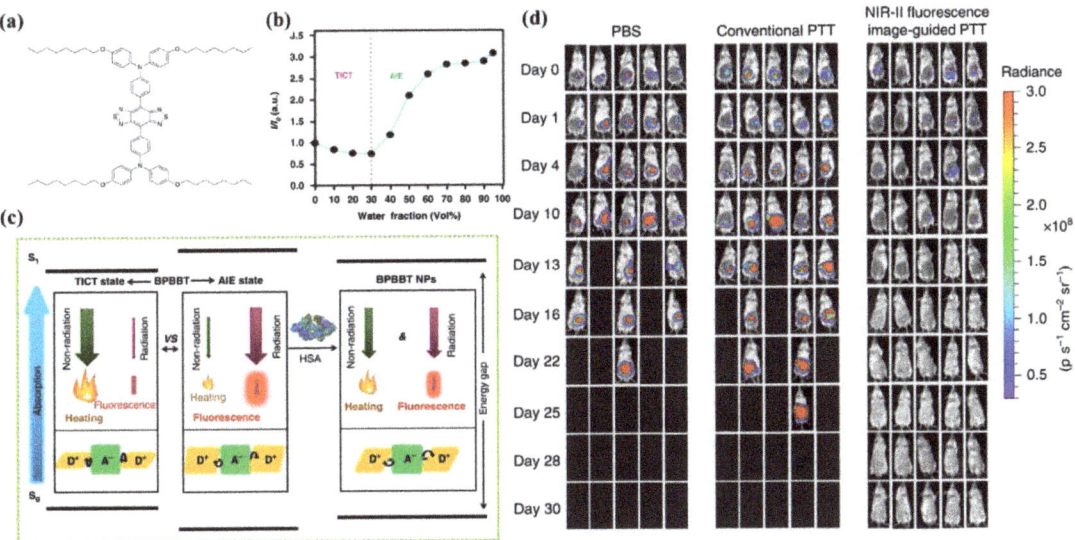

Figure 2. (**a**) Chemical structure of BPBBT; (**b**) plot of fluorescence intensity ratio of BPBBT (10 μM) in water/THF mixture; (**c**) the illustration of has-altering radiative decay and nonradiative decay of BPBBT; (**d**) the in vivo fluorescence imaging of BALB/c mice bearing orthotopic CT26 colon cancer before or after different treatment (n = 5). Reprinted with permission from Ref. [47]. Copyright 2019, Nature Publishing Group.

Thus far, there are few reliable strategies available for through-skull imaging and therapy, because blood–brain barrier (BBB) is an intractable obstruction for various nanoparticles/macromolecule into the brain [36,48]. Tang group [49] developed the natural killer (NK) cell-mimic nanorobots with highly bright NIR-II fluorescence, named NK@AIEdots, to construct smart and safe multifunctional nanoplatforms for BBB-crossing and brain-tumor-targeting through-skull imaging and therapy (Figure 3a). NK@AIEdots wrap a natural kill cell membrane on a reported highly bright NIR-II AIE-active conjugated polymer nanoendoskeleton, PBPTV. The inspiration for this strategy came from the remarkable properties of NK cells whose membrane can form a "green channel" to help NK@AIEdots realize the BBB crossing [50]. PBPTV is the low-bandgap-conjugated polymer with a high quantum yield (QY) up to 8.6%, which is constructed by using a strong and twisted dual-electron acceptor

(BPT). BPT results in long-wavelength absorption and also promotes the intramolecular motion, thus tailoring the equilibrium of the TICT and AIE states (Figure 3b). It was observed that NK@AIEdots displayed bright and long emissions at the NIR-II region, as well as having outstanding photothermal effects (Figure 3c,d). Meanwhile, NK@AIEdots could successfully pass through the BBB and spontaneously accumulate in glioma cells in vivo owing to tumor-targeting proteins of the NK cell membrane, as well as lit up the glioma as intense NIR-II fluorescence even at 48 h postinjection. Moreover, upon NIR light irradiation, NK@AIEdots could effectively inhibit the growth of brain tumor cells with less weight loss in mice, compared with the two control groups. In brief, the NK@AIEdots-based theranostics platform successfully applied to the BBB-crossing and brain-tumor-targeting through-skull FLI-guided PTT.

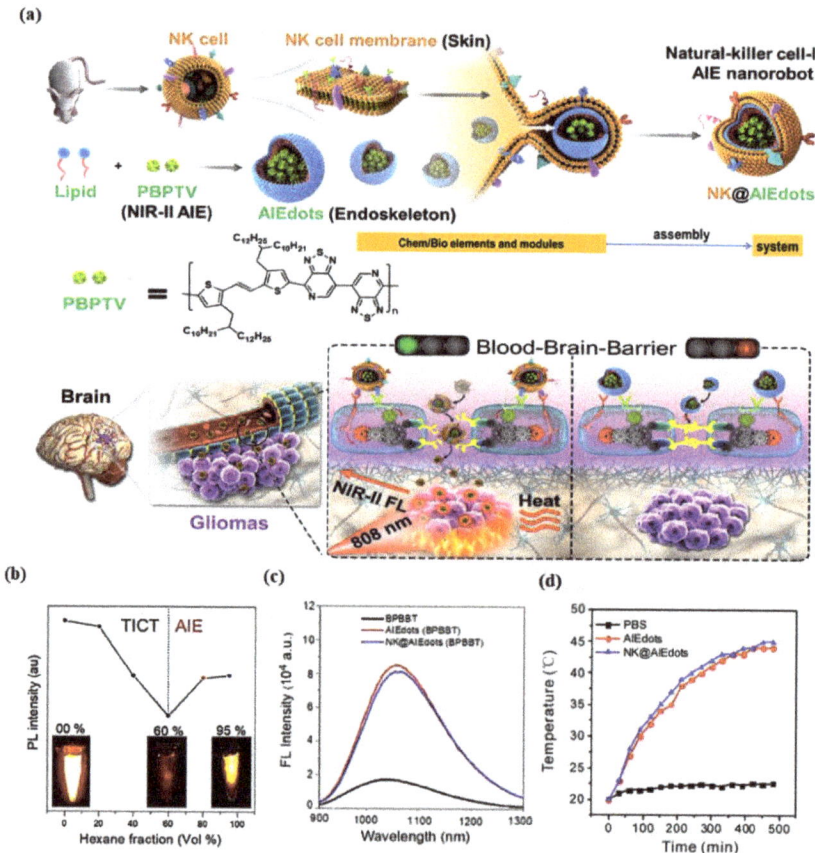

Figure 3. (**a**) Schematic illustration of the preparation and assembly process of NK-cell-mimic AIE nanoparticles (NK@ AIEdots); (**b**) plot of fluorescence intensity of PBPTV in dichloromethane/hexane mixtures; (**c**) fluorescence spectra of BPBBT, AIEdots (BPBBT), NK@AIEdots (BPBBT) in water; (**d**) photothermal effect of PBS, AIEdots, and NK@AIEdots; Reprinted with permission from Ref. [49]. Copyright 2020, American Chemical Society.

Xu et al. [51] developed a phototheranostics platform based on a single molecule with intense fluorescence in the NIR-II region. In this study, they prepared DTPA–BBTD by means of changing the electron-withdrawing substituent groups in the molecular backbone of a D–A–D-type AIEgen, which was previously designed by the same group. Compared with the original molecule, DTPA–BBTD exhibited more twisted conformation and smaller bandgap, both of which enhanced TICT and photothermal effect. As-prepared DTPA–BBTD-based AIE dots could produce bright and well-distributed fluorescence signals through the liver/spleen/head regions even in the blood vessels. Researchers found that the presented AIE dots could efficiently suppress tumor growth upon laser irradiation. Meanwhile, pathological examinations also suggested the AIE dots exhibited excellent biocompatibility with negligible cytotoxicity to normal tissues. In summary, a feasible tactic was conducted to design an AIE agent with PPT effect, which is a potential candidate for NIR-II FLI and NIR-I PAI-guided high-efficiency PTT.

3. PAI-Guided PTT Based on NIR-II Fluorophores

With the explorations of the potential of NIR-II phototheranostics, the phenomenon that the brightness of organic fluorophores including AIEgens generally decreases with the bathochromic shift of emission wavelength has become significant in the NIR-II region. Tang et al. [52] have synthesized a series of NIR-II emissive fluorophores, whose emissions are all located in the NIR-II region with the presence of common solvents (including PhMe, DCM, CHCl$_3$, THF, and DMF), while the emission intensities are relatively inferior for FLI. On the basis of the "energy gap law", the situation above can be attributed to the ascendancy of nonradiative decay pathways when the electronic bandgap decreases, and these inherent features endow the NIR-II fluorophores with the intrinsic superiority in PAI-guided PTT, because both PAI and PTT are closely associated with the nonradiative decay [53,54].

In terms of PAI, it relies on the signal of phonons generated by the light irradiation, exceeding the traditional optical diffusion limit caused by photons after light excitation, which endows it the ability to provide higher spatial resolution [55–58] and penetrate deeper depths as high as 11 cm in NIR-II region [59,60]. More specifically, photons are converted into localized heat that induces transient thermoelastic expansion and wideband acoustic waves in the process of PAI, according to which the process of photo-to-thermo transitions is involved [61–63]. Therefore, the nonradiative decay pathway is closely related to the photothermal conversion property, and NIR-II fluorophores exhibit good potential for PAI.

As for the PTT process, the NIR-II fluorophores demonstrate relatively better photothermal conversion efficiency, compared with those emissions within visible spectroscopy, whose temperature variation generated by photothermal effect reaches merely 13 °C [64]. Thus, the strategy to enhance nonradiative decay, which significantly improves photothermal conversion efficiency to achieve PAI-diagnosis-guided PTT, is another appealing approach of the utilization of NIR-II fluorophores in the photothernostic field.

Inspired by the inherent superiorities mentioned above, Tang et al. put forward a strategy to boost nonradiative decay so as to elevate the photothermal conversion efficiency using reverse thinking of the AIE process, aiming to maximize molecular motions in the aggregated state to enhance heat transitions through extending the side chain length or adding twisted groups, among which the studies of Liu et al. [52] have unprecedentedly integrated the superiorities of reversed AIE and dark TICT to achieve improved photothermal conversion, which can be described as "adjusting TICT in aggregates for boosting photothermal properties".

In this case, bulky alkyl chains were introduced into the planar D–A–D skeleton with molecular rotors as the branches (Figure 4a), from which a series of NIR-II fluorophores with different substituent groups were synthesized including NIRb14, NIRb10, NIRb6, and NIR6, where typical TICT and AIE properties were manifested. Structurally, only when the branched alkyl chains' structure is placed in the second carbon of thiophene can the suitable steric hindrance be provided, avoiding the overwhelming hindrance that leads to intense fluorescence. NIRb14 exhibited the most outstanding photothermal conversion efficiency than the other branched molecules and the widely used golden nanorods, due to the larger bulky chains serving as shielding units that limited intermolecular interactions and conserved intramolecular motions in the aggregated state. PAE-b-PCL and PEG-b-PCL were employed to integrate NIRb14 into mixed-shell NPs (named NIRb14-PAE/PEG NPs) via nanoprecipitation method with the intention of prolonging in vivo blood circulation time and enhancing accumulation in the tumor region, since poly(β-amino esters) (PAE) is able to respond to the cancer tissue [65], and polyethylene glycol (PEG) has excellent biocompatibility. Subsequently, in thermal imaging and in vitro experiment, NIRb14-PAE/PEG NPs exhibited desirable PAI abilities and noteworthy charge conversion features in response to the acidic tumor microenvironment, as well as superior PTT performance, which motivated further investigations of in vivo PAI-guided PTT capability by using the xenograft 4T1 tumor mouse model. As indicated in Figure 4b, the thermal imaging of injected mice exhibited significant temperature variation (Δ_T = 29 °C) at the tumor region after irradiation, in contrast to the negligible results observed from other control groups. Furthermore, by monitoring the tumor volumes during 16 days, the in vivo antitumor efficacies were examined, the results of which are illustrated in Figure 4c. Compared with those of control groups that failed in suppressing the tumor growth, in the presence of both NIRb14-PAE/PEG NPs and laser, the average tumor volume after treatment was much smaller than the initial size, demonstrating the most preeminent antitumor efficacy and great potential of PAI-guided PTT for clinical applications. This study introduces an ingenious molecular design tactic for constructing a photothermal conversion-boosted NIR-II phototheranostics by means of the stabilization of the dark TICT state or the restriction of radiative decay, in addition to demonstrating its potential in cancer diagnosis and therapy by PAI-guided PTT model.

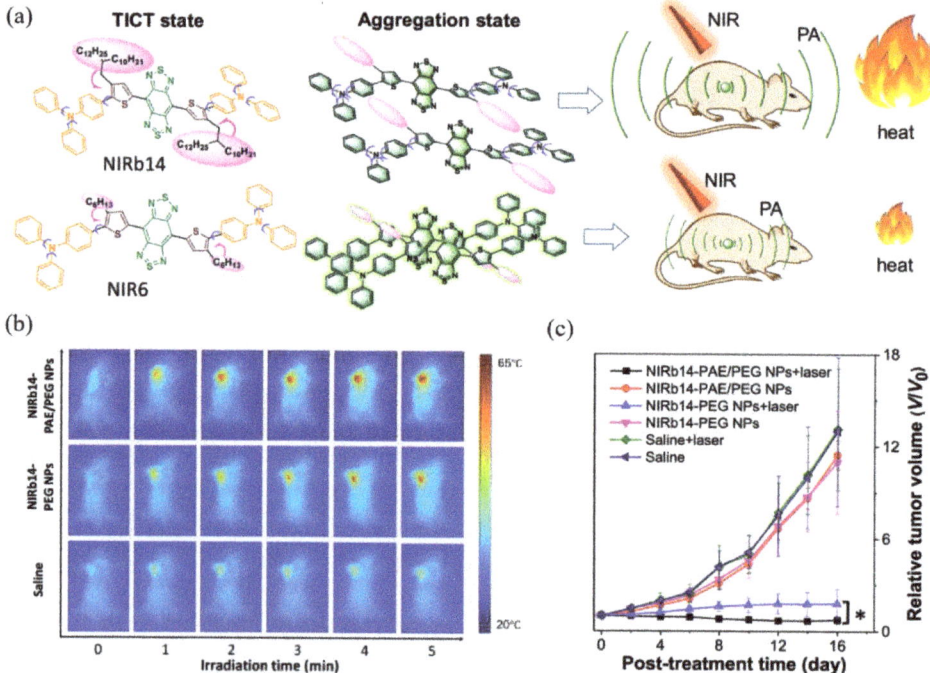

Figure 4. (**a**) Molecular structure and the schematic illustration of the NIR-II fluorophores TICT state in solution, in aggregation state, and the scheme for PAI-guided PTT; (**b**) infrared radiation thermal images of mice with 4T1 tumor under 808 nm laser irradiation; (**c**) tumor growth curves of different treatment groups after 16 days. *: V/V_0 < 0.05 comparing "NIRb14-PAE/PEG NPs + laser" and "NIRb14-PEG NPs + laser" groups. Reprinted with permission from Ref. [52]. Copyright 2019, American Chemical Society.

4. Multimodal Imaging-Guided Synergistic Therapy

Although AIEgens displayed great potential in FLI/PAI-guided PTT, difficulties are still remained in realizing the optimal treatments via one-to-one modality. For instance, the imaging information with both favorable sensitivity and penetration depth is not able to be obtained by a single imaging modality [58,66–69], and it is also burdensome to achieve satisfactory treatment via PDT or PTT alone, which is attributed to hypoxic tumor microenvironment for PDT and heat shock effect in PTT [70,71]. Constructing multimodality phototheranostic platforms is a smart strategy that is able to achieve "1 + 1 > 2" to solve these problems, which can afford precise diagnosis and efficacious therapy via combining different kinds of imaging technologies with therapy methods, and has induced great interest recently. However, it is a challenging task because keeping the equilibrium between radiative and nonradiative decays is intractable for conventional materials, which is crucial to building up a versatile phototheranostic system with favorable fluorescent and photothermal properties concurrently. By virtue of affluent free-motioned molecular rotators or vibrators in structure, the photophysical properties of AIEgens could be manipulated easily by boosting or inhibiting intramolecular motions [16,20,72]. Furthermore, endowing AIEgens with twisted conformations could lead to relatively loose packing in the aggregated state through tactful molecular regulation, which is beneficial to balance radiative and nonradiative decays. In addition, considering the superior features of NIR-II imaging mentioned at the beginning of this review, we believe that constructing multimodality theranostic systems based on NIR-II AIEgens is a win–win integration [73].

4.1. NIR-II FLI-Guided PDT–PTT Synergistic Therapy

It is hard to achieve satisfactory treatment via PDT and PTT alone, as described in the previous section, and the combination of them is regarded as a groundbreaking strategy to overcome respective shortcomings and realize boosted synergistic therapeutic outcomes [8,11,74] because PTT could heighten the oxygen concentration by increasing the flow rate of blood to strengthen the PDT effect, thus promoting the elimination of heat-resistant tumor in PTT reversely. Recently, a novel zwitterion-type AIEgen was facilely synthesized, which could afford effective NIR-II FLI-guided synergistic PDT-PTT [75]. In this study, the author designed a series of zwitterionic compounds (BITT, BITB, ITT, and ITB in Figure 5b). The extra benzene ring in BITT would elongate π conjugation and increase D–A strength, resulting in a longer absorption wavelength than ITT in ethanol solution and admirable ROS generation capacity [76]. Meanwhile, compared with BITB, the AIE performance of BITT would be dramatically strengthened due to the introduction of triphenylamine moiety. The alkyl chain with the sulfonic acid group and twisted TPA moiety would enlarge the intermolecular distance, which is beneficial to restrain fluorescence quenching by intermolecular π–π stacking in aggregated state and lead to relatively loose intermolecular packing (reserve partial intramolecular motion), endowing BITT with both intense NIR-II fluorescence and high photothermal conversion capability through keeping the balance between radiative and nonradiative decays. Moreover, BITT would also exhibit superior biocompatibility owing to the zwitterionic structure [77]. Photophysical experiments demonstrated that BITT NPs in aqueous solution possessed a long emission wavelength peak at 810 nm, with the tail partially located in the NIR-II region, high photothermal conversion efficiency (35.76%), excellent ROS generation capability, as well as high photostability and photothermal stability.

Motivated by prominent photophysical properties within BITT NPs, in vivo imaging and therapy performance of BITT NPs for 4T1 tumor-bearing nude mice were estimated subsequently. As shown in Figure 5c,d, compared with NIR-I FLI, an intense and durable NIR-II fluorescence signal was observed from 0.5 to 24 h postinjection. After being exposed to laser irradiation for only 2 min, the temperature of the tumor region raised from 34.7 to 52.4 °C (Figure 5c), which was competent to suppress tumor tissue efficaciously. As indicated in Figure 5f,g, the tumors were completely eliminated without any recidivation via synergistic PDT–PTT therapy in the presence of BITT NPs and laser irradiation, and no obvious body weight loss was observed in four groups, demonstrating excellent tumoricidal capability and biocompatibility. This study thus provided a new strategy to construct a NIR-II AIEgen with both FLI imaging and PTT–PDT therapy.

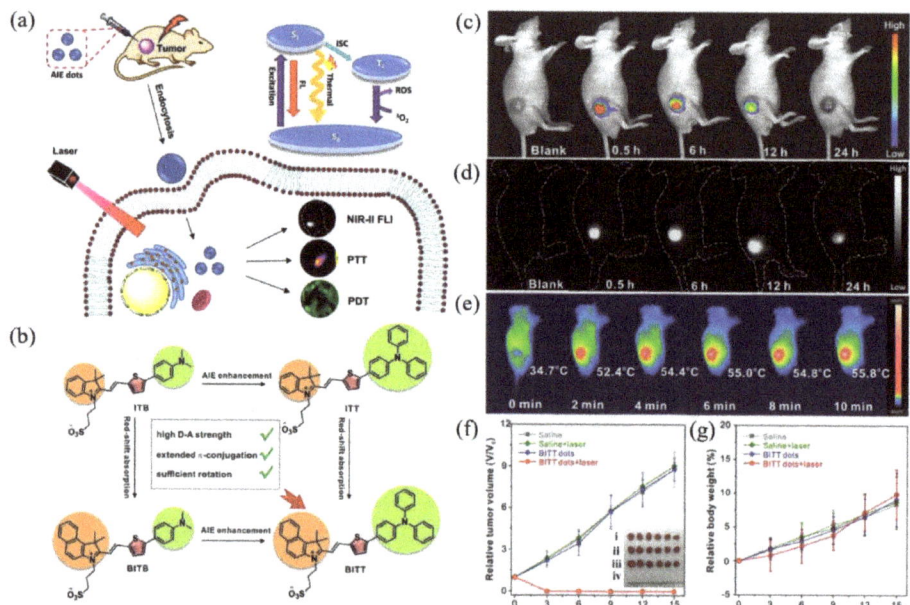

Figure 5. (a) Diagram of NIR-II FLI-guided PDT–PTT synergistic therapy for tumor-bearing mice with BITT NPs; (b) principle for designing BITT; (c) NIR-I FLI and (d) NIR-II FLI of tumor-bearing mice after administration of BITT NPs at different monitoring times; (e) IR imaging of tumor-bearing mice under laser irradiation at 12 h postinjection of BITT NPs; (f) growth curves of tumors with various treatments. Inset: photograph of tumors harvested from the mice at day 15 after different treatment; (g) body weight curves of tumor-bearing mice after different treatments at day 15. Reprinted with permission from Ref. [75]. Copyright 2021, Wiley-VCH.

4.2. NIR-II FLI–PAI–PTI Trimodal-Guided Synergistic PDT–PTT

Considering the complementary advantages of FLI, PAI, and PTI, as well as boosted therapeutic effect of PDT and PTT, constructing single-component theranostic platforms that can afford all these phototheranostic modalities simultaneously would be extremely important [78–82]. Motivated by the superiorities of NIR-II AIEgens mentioned before, a number of remarkable multifunctional phototheranostic platforms have been developed upon NIR-II AIEgens. A simple and one-for-all phototheranostic platform with NIR-II AIE characteristics was reported [83], which could afford NIR-II FLI–PAI–PTI trimodal-imaging-guided synergistic PDT–PTT. As illustrated in Figure 6a, three novel AIE compounds composed of 1,3-bis (dicyanomethylidene) indane, thiophene and triphenylamine were prepared. The twisted conformation of TPA would extend the intermolecular distance and lead to loosened packing in an aggregated state, which is helpful to retain intramolecular motions partially, and the stretching vibrations of carbon–nitrogen bonds would maintain considerable frequency even in the aggregated state, thus strengthening the heat generation efficiency of AIEgens. Benefiting from the increasing number of thiophene units, compared with TI and TSI, TSSI, as well as TSSI NPs fabricated with DSPE-mPEG2000, would exhibit more admirable properties for phototheranostics since the addition of thiophene moiety could enhance D–A strength and the capacity of ROS generation and also enlarge the distance of donor and acceptor within the AIEgens, further boosting intramolecular motions. As expected, TSSI NPs exhibited a maximum emission near 1000 nm, indicating remarkable properties for NIR-II imaging, and they also possessed superior ROS and heat generation capabilities, both of which confirmed the perfect equilibrium between radiative and nonradiative decays. Furthermore, TSSI NPs were utilized to implement in vivo experiments for tumor-bearing mice. Obvious fluorescence signals (Figure 6b, upper) were observed from 6 to 36 h and reached a maximum at 12 h after injection in NIR-II FLI. Surprisingly, the time-dependent tendency of PA intensity (Figure 6b, lower) was nearly unanimous with NIR-II FLI outcomes for the same tumor model. In in vivo experiments of PTI and PTT (Figure 6c), upon NIR irradiation for only 10 min, the temperature of the tumor region raised from 37.3 °C to 54.8 °C, and insignificant temperature variation was observed in the normal tissue. As shown in Figure 6d, e, the tumor tissues of mice were obliterated completely upon both TSSI NPs and NIR irradiation through only one injection and irradiation. This study offered a smart tactic to construct one-for-all AIEgen, but it also demonstrated its great potential in multimodality theranostics.

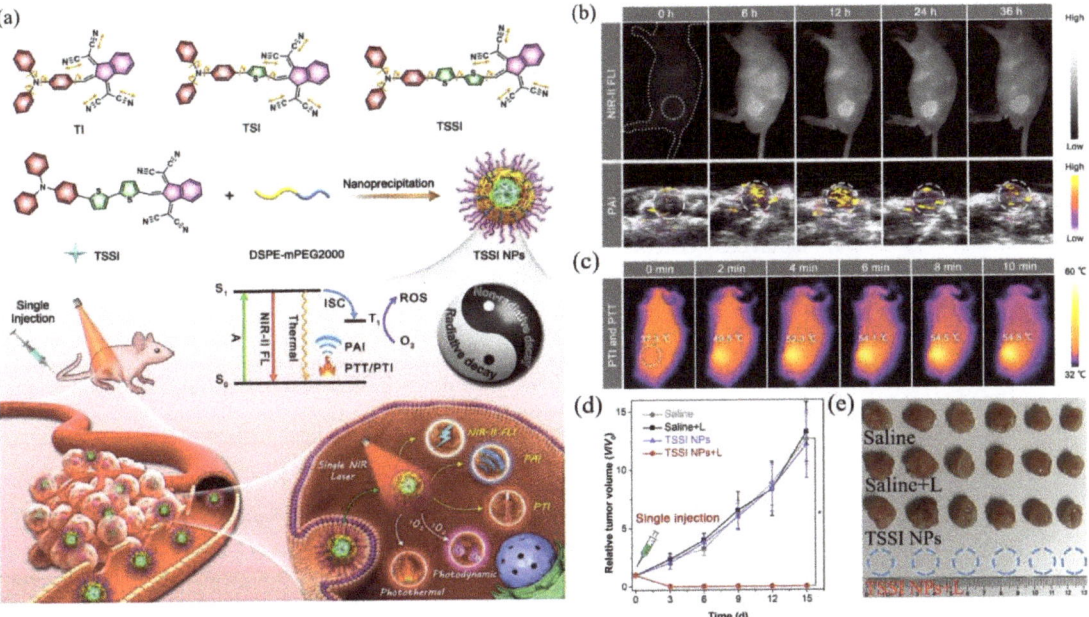

Figure 6. (**a**) Diagram of all-around AIE molecular structures, nanoparticles fabricating, and multimodal phototheranostic system; (**b**) NIR-II FLI (upper) and PAI (lower) for tumor-bearing mice from 0 h to 36 h postinjection of TSSI NPs; (**c**) PTI and PTT for tumor-bearing mice at 12 h postinjection of TSSI NPs upon laser irradiation; (**d**) tumor growth curves of mice with various treatments; (**e**) photograph of 4T1 tumors harvested from the mice at day 15 after different treatment. Reprinted with permission from Ref. [83]. Copyright 2020, Wiley-VCH.

Shortly thereafter, several outstanding single-component multimodal theranostic platforms were constructed tactfully. A novel type of theranostic AIEgen was designed under the guidance of a similar principle as the above-mentioned [84]. In this study, three AIE compounds (TAM, TSAM, TSSAM) with none, one, two thiophenes were prepared, respectively. As expected, TSSAM possessed NIR-II emission, high photothermal conversion efficiency, and ROS generation capability, as well as outstanding photostability and photothermal stability. TSSAM NPs exhibited favorable accordance and intratumor retention capability for in vivo NIR-II FLI and PAI, where the signal intensities of NIRI-I FLI and PAI reached a plateau at 12 h postinjection and remained durable at 24 h postinjection. Upon laser irradiation for only 2 min, the temperature of the tumor region increased from 37.1 to 57.6 °C. In the presence of TSSAM NPs and laser irradiation, all tumors were eradicated on day 3. Later, Wen et al. [85] also designed four novel NIR-emissive AIEgens. Benefiting from the highly bright emission of the tail located in the NIR-II region in the aggregated state, photothermal conversion efficiency, and high ROS generation capability of TTT-4, in vivo multimodal imaging and therapy performance of TTT-4 dots was evaluated. Significant fluorescence signals were detected in both NIR-I and NIR-II FLI from 1 to 12 h postinjection at the tumor sites, as well as intense PA signals captured by the PAI system. The temperature of the tumor region raised from 37.3 to 55 °C upon laser irradiation for only 3 min, and the results of in vivo synergistic phototherapeutic experiments illustrated that tumor tissue was completely eradicated after 15-day treatment. In general, TSSAM and TTT-4 enrich the types of versatile phototheranostic AIEgens and display great potential in NIR-II FLI–PAI–PTI-guided synergistic PDT–PTT.

5. Conclusions

In this minireview, we summarized the existing strategies for constructing efficacious theranostic platforms based on NIR-II AIEgens and their great potential in basic studies and practical applications. Through subtle regulation of hierarchical structures at molecular and aggregated levels, respectively, the AIE and TICT properties could be manipulated, thus achieving admirable equilibrium between fluorescence and photothermal effect, as well as superior performance in NIR-II FLI-guided PTT against cancer. To further boost photothermal properties for PAI and PTT, the strategy named "reverse thinking of AIE" is proposed. With the existence of long-branched alkyl chains and molecular motors in the molecule, it displayed excellent photothermal conversion efficiency and photoacoustic effect, thus achieving remarkable antitumor efficacy in PAI-guided PTT. Furthermore, in order to overcome the inherent drawbacks in one-to-one modality, a series of versatile theranostic molecules with AIE characteristics were prepared. On account of tactful molecule design, these AIE molecules can make the best utilization of excited-state energy and keep a remarkable balance between radiative and nonradiative decays, thus exhibiting highly bright emission, prominent photothermal conversion efficiency, and efficacious ROS generation to achieve NIR-II FLI-guided PTT–PDT. Some of them even suffice to NIR-II FLI–PAI–PTI trimodal-guided synergistic PDT–PTT, and in vivo experiments revealed that these AIE NPs could afford precise tumor imaging and thorough tumor elimination outcomes.

Although reported phototheranostic systems based on NIR-II AIEgens validated great potential in both basic studies and clinical practices, there are still challenges that should be addressed. Firstly, the development of new AIEgens with a longer emission wavelength for the theranostic study is urgent, since the maximum emission wavelength of all reported theranostic AIEgens is below 1100 nm. Secondly, as a vital guideline for designing theranostic molecules, the structure–property relationship of AIEgens for theranostic applications is still obscure [7]. Thirdly, the current achievements of phototheranostics are still far from clinical applications. Although the biocompatibility of AIEgens has been demonstrated by substantial experimental data, more comprehensive research studies are in urgent need for further clinical trials, as well as investigation of their long-term safety using in vivo experiments with diverse animal models. The results of this review also call for constructing functional materials based on NIR-II AIEgens to target specific organs or in response to external stimuli.

Overall, NIR-II AIEgen is a predominant candidate to construct a phototheranostic system with superior performance in basic research and practical applications, which opens up new avenues for advanced theranostics and will realize clinical trials in the near future.

Author Contributions: Y.T. and P.L. contributed equally to this work. Conceptualization, Y.T., P.L. and D.L.; writing—original draft preparation, Y.T., P.L. and D.L.; writing—review and editing, Y.T., P.L. and D.L.; supervision, D.W.; project administration, D.W. and B.Z.T.; funding acquisition, D.W. All authors have read and agreed to the published version of the manuscript.

Funding: This research was funded by the National Natural Science Foundation of China (52122317, 22175120), the Developmental Fund for Science and Technology of Shenzhen government (RCYX20200714114525101, JCYJ20190808153415062), and the Natural Science Foundation for Distinguished Young Scholars of Guangdong Province (2020B1515020011).

Institutional Review Board Statement: Not applicable.

Informed Consent Statement: Not applicable.

Data Availability Statement: Not applicable.

Conflicts of Interest: The authors declare no conflict of interest.

References

1. Sung, H.; Ferlay, J.; Siegel, R.L.; Laversanne, M.; Soerjomataram, I.; Jemal, A.; Bray, F. Global cancer statistics 2020: GLOBOCAN estimates of incidence and mortality worldwide for 36 cancers in 185 countries. *CA Cancer J. Clin.* **2021**, *71*, 209–249. [CrossRef]
2. Ferlay, J.; Colombet, M.; Soerjomataram, I.; Parkin, D.M.; Pineros, M.; Znaor, A.; Bray, F. Cancer statistics for the year 2020: An overview. *Int. J. Cancer.* **2021**, *149*, 778–789. [CrossRef] [PubMed]
3. Shi, J.J.; Kantoff, P.W.; Wooster, R.; Farokhzad, O.C. Cancer nanomedicine: Progress, challenges and opportunities. *Nat. Rev. Cancer* **2017**, *17*, 20–37. [CrossRef] [PubMed]
4. Dierolf, M.; Menzel, A.; Thibault, P.; Schneider, P.; Kewish, C.M.; Wepf, R.; Bunk, O.; Pfeiffer, F. Ptychographic X-ray computed tomography at the nanoscale. *Nature* **2010**, *467*, 436–439. [CrossRef]
5. Gambhir, S.S. Molecular imaging of cancer with positron emission tomography. *Nat. Rev. Cancer* **2002**, *2*, 683–693. [CrossRef] [PubMed]
6. Yang, P.; Gai, S.; Lin, J. Functionalized mesoporous silica materials for controlled drug delivery. *Chem. Soc. Rev.* **2012**, *41*, 3679–3698. [CrossRef] [PubMed]
7. Wang, D.; Lee, M.M.S.; Xu, W.; Kwok, R.T.K.; Lam, J.W.Y.; Tang, B.Z. Theranostics based on AIEgens. *Theranostics* **2018**, *8*, 4925–4956. [CrossRef]
8. Wang, D.; Lee, M.M.S.; Xu, W.; Shan, G.; Zheng, X.; Kwok, R.T.K.; Lam, J.W.Y.; Hu, X.; Tang, B.Z. Boosting Non-Radiative Decay to Do Useful Work: Development of a Multi-Modality Theranostic System from an AIEgen. *Angew. Chem. Int. Ed.* **2019**, *58*, 5628–5632. [CrossRef] [PubMed]
9. Muthu, M.S.; Leong, D.T.; Mei, L.; Feng, S.S. Nanotheranostics—Application and Further Development of Nanomedicine Strategies for Advanced Theranostics. *Theranostics* **2014**, *4*, 660–677. [CrossRef]
10. Fan, W.; Huang, P.; Chen, X. Overcoming the Achilles' heel of photodynamic therapy. *Chem. Soc. Rev.* **2016**, *45*, 6488–6519. [CrossRef]
11. Chen, C.; Ni, X.; Jia, S.; Liang, Y.; Wu, X.; Kong, D.; Ding, D. Massively Evoking Immunogenic Cell Death by Focused Mitochondrial Oxidative Stress using an AIE Luminogen with a Twisted Molecular Structure. *Adv. Mater.* **2019**, *31*, 1904914. [CrossRef]
12. Attia, A.B.E.; Balasundaram, G.; Moothanchery, M.; Dinish, U.S.; Bi, R.Z.; Ntziachristos, V.; Olivo, M. A review of clinical photoacoustic imaging: Current and future trends. *Photoacoustics* **2019**, *16*, 100144. [CrossRef]
13. Zhang, R.R.; Schroeder, A.B.; Grudzinski, J.J.; Rosenthal, E.L.; Warram, J.M.; Pinchuk, A.N.; Eliceiri, K.W.; Kuo, J.S.; Weichert, J.P. Beyond the margins: Real-time detection of cancer using targeted fluorophores. *Nat. Rev. Clin. Oncol.* **2017**, *14*, 347–364. [CrossRef] [PubMed]
14. Day, R.N.; Davidson, M.W. The fluorescent protein palette: Tools for cellular imaging. *Chem. Soc. Rev.* **2009**, *38*, 2887–2921. [CrossRef] [PubMed]
15. Weissleder, R.; Pittet, M.J. Imaging in the era of molecular oncology. *Nature* **2008**, *452*, 580–589. [CrossRef] [PubMed]
16. Luo, J.; Xie, Z.; Lam, J.W.Y.; Cheng, L.; Chen, H.; Qiu, C.; Kwok, H.S.; Zhan, X.; Liu, Y.; Zhu, D.; et al. Aggregation-induced emission of 1-methyl-1,2,3,4,5-pentaphenylsilole. *Chem. Commun.* **2001**, *18*, 1740–1741. [CrossRef] [PubMed]
17. Song, N.; Zhang, Z.; Liu, P.; Yang, Y.; Wang, L.; Wang, D.; Tang, B.Z. Nanomaterials with Supramolecular Assembly Based on AIE Luminogens for Theranostic Applications. *Adv. Mater.* **2020**, *32*, 2004208. [CrossRef]
18. Kang, M.; Zhang, Z.; Song, N.; Li, M.; Sun, P.; Chen, X.; Wang, D.; Tang, B.Z. Aggregation-enhanced theranostics: AIE sparkles in biomedical field. *Aggregate* **2020**, *1*, 80–106. [CrossRef]
19. Feng, G.; Liu, B. Multifunctional AIEgens for Future Theranostics. *Small* **2016**, *12*, 6528–6535. [CrossRef]
20. Wang, D.; Tang, B.Z. Aggregation-Induced Emission Luminogens for Activity-Based Sensing. *Acc. Chem. Res.* **2019**, *52*, 2559–2570. [CrossRef]
21. Mei, J.; Hong, Y.; Lam, J.W.Y.; Qin, A.; Tang, Y.; Tang, B.Z. Aggregation-Induced Emission: The Whole Is More Brilliant than the Parts. *Adv. Mater.* **2014**, *26*, 5429–5479. [CrossRef]
22. Mei, J.; Leung, N.L.C.; Kwok, R.T.K.; Lam, J.W.Y.; Tang, B.Z. Aggregation-Induced Emission: Together We Shine, United We Soar! *Chem. Rev.* **2015**, *115*, 11718–11940. [CrossRef]
23. Smith, A.M.; Mancini, M.C.; Nie, S.M. Bioimaging Second window for in vivo imaging. *Nat. Nanotechnol.* **2009**, *4*, 710–711. [CrossRef] [PubMed]

24. Hong, G.; Lee, J.C.; Robinson, J.T.; Raaz, U.; Xie, L.; Huang, N.; Cooke, J.P.; Dai, H. Multifunctional in vivo vascular imaging using near-infrared II fluorescence. *Nat. Med.* **2012**, *18*, 1841–1846. [CrossRef] [PubMed]
25. Luo, S.; Zhang, E.; Su, Y.; Cheng, T.; Shi, C. A review of NIR dyes in cancer targeting and imaging. *Biomaterials* **2011**, *32*, 7127–7138. [CrossRef] [PubMed]
26. Qian, G.; Dai, B.; Luo, M.; Yu, D.; Zhan, J.; Zhang, Z.; Ma, D.; Wang, Z. Band Gap Tunable, Donor-Acceptor-Donor Charge-Transfer Heteroquinoid-Based Chromophores: Near Infrared Photoluminescence and Electroluminescence. *Chem. Mater.* **2008**, *20*, 6208–6216. [CrossRef]
27. Xie, C.; Zhou, W.; Zeng, Z.; Fan, Q.; Pu, K. Grafted semiconducting polymer amphiphiles for multimodal optical imaging and combination phototherapy. *Chem. Sci.* **2020**, *11*, 10553–10570. [CrossRef] [PubMed]
28. Wang, W.; He, X.; Du, M.; Xie, C.; Zhou, W.; Huang, W.; Fan, Q. Organic Fluorophores for 1064 nm Excited NIR-II Fluorescence Imaging. *Front. Chem.* **2021**, *9*, 769655. [CrossRef]
29. Shao, A.; Xie, Y.; Zhu, S.; Guo, Z.; Zhu, S.; Guo, J.; Shi, P.; James, T.; Tian, H.; Zhu, W. Far-Red and Near-IR AIE-Active Fluorescent Organic Nanoprobes with Enhanced Tumor-Targeting Efficacy: Shape-Specific Effects. *Angew. Chem. Int. Ed.* **2015**, *54*, 7275–7280. [CrossRef] [PubMed]
30. Liu, J.; Chen, C.; Ji, S.; Liu, Q.; Ding, D.; Zhao, D.; Liu, B. Long wavelength excitable near-infrared fluorescent nanoparticles with aggregation-induced emission characteristics for image-guided tumor resection. *Chem. Sci.* **2017**, *8*, 2782–2789. [CrossRef]
31. Li, Y.; Hu, D.; Sheng, Z.; Min, T.; Zha, M.; Ni, J.; Zheng, H.; Li, K. Self-assembled AIEgen nanoparticles for multiscale NIR-II vascular imaging. *Biomaterials* **2021**, *264*, 120365. [CrossRef] [PubMed]
32. Niu, G.; Zhang, R.; Shi, X.; Park, H.; Xie, S.; Kwok, R.T.K.; Lam, J.W.Y.; Tang, B.Z. AIE luminogens as fluorescent bioprobes. *Trends Anal. Chem.* **2020**, *123*, 115769. [CrossRef]
33. Jiang, Y.; Li, J.; Zhen, X.; Xie, C.; Pu, K. Dual-Peak Absorbing Semiconducting Copolymer Nanoparticles for First and Second Near-Infrared Window Photothermal Therapy: A Comparative Study. *Adv. Mater.* **2018**, *30*, 1705980. [CrossRef] [PubMed]
34. Wang, K.; Zhuang, J.; Chen, L.; Xu, D.; Zhang, X.; Chen, Z.; Wei, Y.; Zhang, Y. One-pot synthesis of AIE based bismuth sulfide nanotheranostics for fluorescence imaging and photothermal therapy. *Colloids Surf. B Biointerfaces* **2017**, *160*, 297–304. [CrossRef]
35. Zhang, Z.; Kang, M.; Wang, Y.; Song, G.; Wen, H.; Wang, D.; Tang, B.Z. Recent Advances of Aggregation-induced Emission Materials in Phototheranostics. *Chin. J. Lumin.* **2021**, *42*, 361–378. [CrossRef]
36. Zhang, M.; Wang, W.; Mohammadniaei, M.; Zheng, T.; Zhang, Q.; Ashley, J.; Liu, S.; Sun, Y.; Tang, B.Z. Upregulating Aggregation-Induced-Emission Nanoparticles with Blood-Tumor-Barrier Permeability for Precise Photothermal Eradication of Brain Tumors and Induction of Local Immune Responses. *Adv. Mater.* **2021**, *33*, 2008802. [CrossRef] [PubMed]
37. Li, H.; Wen, H.; Zhang, Z.; Song, N.; Kwok, R.T.K.; Lam, J.W.Y.; Wang, L.; Wang, D.; Tang, B.Z. Reverse Thinking of the Aggregation-Induced Emission Principle: Amplifying Molecular Motions to Boost Photothermal Efficiency of Nanofibers. *Angew. Chem. Int. Ed.* **2020**, *59*, 20371–20375. [CrossRef] [PubMed]
38. Zhang, M.; Li, J.; Yu, L.; Wang, X.; Bai, M. Tuning the fluorescence based on the combination of TICT and AIE emission of a tetraphenylethylene with D-pi-A structure. *Rsc. Adv.* **2020**, *10*, 14520–14524. [CrossRef]
39. Wang, D.; Chen, L.; Zhao, X.; Yan, X. Enhancing near-infrared AIE of photosensitizer with twisted intramolecular charge transfer characteristics via rotor effect for AIE imaging-guided photodynamic ablation of cancer cells. *Talanta* **2021**, *225*, 122046. [CrossRef]
40. Wang, C.; Chi, W.; Qiao, Q.; Tan, D.; Xu, Z.; Liu, X. Twisted intramolecular charge transfer (TICT) and twists beyond TICT: From mechanisms to rational designs of bright and sensitive fluorophores. *Chem. Soc. Rev.* **2021**, *50*, 12656–12678. [CrossRef] [PubMed]
41. Hu, R.; Lager, E.; Aguilar-Aguilar, A.; Liu, J.; Lam, J.W.Y.; Sung, H.H.Y.; Williams, I.D.; Zhong, Y.; Wong, K.; Pena-Cabrera, E.; et al. Twisted Intramolecular Charge Transfer and Aggregation-Induced Emission of BODIPY Derivatives. *J. Phys. Chem. C* **2009**, *113*, 15845–15853. [CrossRef]
42. Yang, Q.; Ma, Z.; Wang, H.; Zhou, B.; Zhu, S.J.; Zhong, Y.; Wang, J.Y.; Wan, H.; Antaris, A.; Ma, R.; et al. Rational Design of Molecular Fluorophores for Biological Imaging in the NIR-II Window. *Adv. Mater.* **2017**, *29*, 1605497. [CrossRef]
43. Morris, W.A.; Kolpaczynska, M.; Fraser, C.L. Effects of alpha-Substitution on Mechanochromic Luminescence and Aggregation-Induced Emission of Difluoroboron beta-Diketonate Dyes. *J. Phys. Chem. C* **2016**, *120*, 22539–22548. [CrossRef]
44. Ge, S.; Li, B.; Meng, X.; Yan, H.; Yang, M.; Dong, B.; Lu, Y. Aggregation-induced emission, multiple chromisms and self organization of N-substituted-1,8-naphthalimides. *Dyes Pigm.* **2018**, *148*, 147–153. [CrossRef]
45. Cai, Y.; Liang, P.; Tang, Q.; Yang, X.; Si, W.; Huang, W.; Zhang, Q.; Dong, X. Diketopyrrolopyrrole-Triphenylamine Organic Nanoparticles as Multifunctional Reagents for Photoacoustic Imaging-Guided Photodynamic/Photothermal Synergistic Tumor Therapy. *ACS Nano* **2017**, *11*, 1054–1063. [CrossRef] [PubMed]
46. Wang, Y.; Zhu, W.; Du, W.; Liu, X.; Zhang, X.; Dong, H.; Hu, W. Cocrystals Strategy towards Materials for Near-Infrared Photothermal Conversion and Imaging. *Angew. Chem. Int. Ed.* **2018**, *57*, 3963–3967. [CrossRef] [PubMed]
47. Gao, S.; Wei, G.; Zhang, S.; Zheng, B.; Xu, J.; Chen, G.; Li, M.; Song, S.; Fu, W.; Xiao, Z.; et al. Albumin tailoring fluorescence and photothermal conversion effect of near-infrared-II fluorophore with aggregation-induced emission characteristics. *Nat. Commun.* **2019**, *10*, 2206. [CrossRef]
48. Wang, H.; Mu, X.; Yang, J.; Liang, Y.; Zhang, X.; Ming, D. Brain imaging with near-infrared fluorophores. *Coord. Chem. Rev.* **2019**, *380*, 550–571. [CrossRef]

49. Deng, G.; Peng, X.; Sun, Z.; Zheng, W.; Yu, J.; Du, L.; Chen, H.; Gong, P.; Zhang, P.; Cai, L.; et al. Natural-Killer-Cell-Inspired Nanorobots with Aggregation-Induced Emission Characteristics for Near-Infrared-II Fluorescence-Guided Glioma Theranostics. *ACS Nano* **2020**, *14*, 11452–11462. [CrossRef]
50. Chiossone, L.; Dumas, P.Y.; Vienne, M.; Vivier, E. Natural killer cells and other innate lymphoid cells in cancer. *Nat. Rev. Immunol.* **2018**, *18*, 671–688. [CrossRef] [PubMed]
51. Xu, Y.; Dang, D.; Zhu, H.; Jing, X.; Zhu, X.; Zhang, N.; Li, C.; Zhao, Y.; Zhang, P.; Yang, Z.; et al. Boosting the AIEgen-based photo-theranostic platform by balancing radiative decay and non-radiative decay. *Mater. Chem. Front.* **2021**, *5*, 4182–4192. [CrossRef]
52. Liu, S.; Zhou, X.; Zhang, H.; Ou, H.; Lam, J.W.Y.; Liu, Y.; Shi, L.; Ding, D.; Tang, B.Z. Molecular Motion in Aggregates: Manipulating TICT for Boosting Photothermal Theranostics. *J. Am. Chem. Soc.* **2019**, *141*, 5359–5368. [CrossRef] [PubMed]
53. Qi, J.; Duan, X.; Cai, Y.; Jia, S.; Chen, C.; Zhao, Z.; Li, Y.; Peng, H.; Kwok, R.T.K.; Lam, J.W.Y.; et al. Simultaneously boosting the conjugation, brightness and solubility of organic fluorophores by using AIEgens. *Chem. Sci.* **2020**, *11*, 8438–8447. [CrossRef]
54. Chen, J.; Qi, J.; Chen, C.; Chen, J.; Liu, L.; Gao, R.; Zhang, T.; Song, L.; Ding, D.; Zhang, P.; et al. Tocilizumab-Conjugated Polymer Nanoparticles for NIR-II Photoacoustic-Imaging-Guided Therapy of Rheumatoid Arthritis. *Adv. Mater.* **2020**, *32*, 2003399. [CrossRef] [PubMed]
55. Wang, L.H.V.; Hu, S. Photoacoustic Tomography: In Vivo Imaging from Organelles to Organs. *Science* **2012**, *335*, 1458–1462. [CrossRef] [PubMed]
56. De La Zerda, A.; Zavaleta, C.; Keren, S.; Vaithilingam, S.; Bodapati, S.; Liu, Z.; Levi, J.; Smith, B.R.; Ma, T.J.; Oralkan, O.; et al. Carbon nanotubes as photoacoustic molecular imaging agents in living mice. *Nat. Nanotechnol.* **2008**, *3*, 557–562. [CrossRef] [PubMed]
57. Xu, M.H.; Wang, L.H.V. Photoacoustic imaging in biomedicine. *Rev. Sci. Instrum.* **2006**, *77*, 041101. [CrossRef]
58. Fu, Q.; Zhu, R.; Song, J.; Yang, H.; Chen, X. Photoacoustic Imaging: Contrast Agents and Their Biomedical Applications. *Adv. Mater.* **2019**, *31*, 1805875. [CrossRef]
59. Zhou, Y.; Wang, D.; Zhang, Y.; Chitgupi, U.; Geng, J.; Wang, Y.; Zhang, Y.; Cook, T.R.; Xia, J.; Lovell, J.F. A Phosphorus Phthalocyanine Formulation with Intense Absorbance at 1000 nm for Deep Optical Imaging. *Theranostics* **2016**, *6*, 688–697. [CrossRef]
60. Wu, J.Y.Z.; You, L.Y.; Lan, L.; Lee, H.J.; Chaudhry, S.T.; Li, R.; Cheng, J.X.; Mei, J.G. Semiconducting Polymer Nanoparticles for Centimeters-Deep Photoacoustic Imaging in the Second Near-Infrared Window. *Adv. Mater.* **2017**, *29*, 1703403. [CrossRef]
61. Aoki, H.; Nojiri, M.; Mukai, R.; Ito, S. Near-infrared absorbing polymer nano-particle as a sensitive contrast agent for photo-acoustic imaging. *Nanoscale* **2015**, *7*, 337–343. [CrossRef]
62. Guo, X.; Cao, B.; Wang, C.; Lu, S.; Hu, X. In vivo photothermal inhibition of methicillin-resistant Staphylococcus aureus infection by in situ templated formulation of pathogen-targeting phototheranostics. *Nanoscale* **2020**, *12*, 7651–7659. [CrossRef] [PubMed]
63. Qi, J.; Fang, Y.; Kwok, R.T.K.; Zhang, X.; Hu, X.; Lam, J.W.Y.; Ding, D.; Tang, B.Z. Highly Stable Organic Small Molecular Nanoparticles as an Advanced and Biocompatible Phototheranostic Agent of Tumor in Living Mice. *ACS Nano* **2017**, *11*, 7177–7188. [CrossRef]
64. Song, S.; Zhao, Y.; Kang, M.; Zhang, Z.; Wu, Q.; Fu, S.; Li, Y.; Wen, H.; Wang, D.; Tang, B.Z. Side-Chain Engineering of Aggregation-Induced Emission Molecules for Boosting Cancer Photothernaostics. *Adv. Funct. Mater.* **2021**. [CrossRef]
65. Huang, X.; El-Sayed, I.H.; Qian, W.; El-Sayed, M.A. Cancer cell imaging and photothermal therapy in the near-infrared region by using gold nanorods. *J. Am. Chem. Soc.* **2006**, *128*, 2115–2120. [CrossRef] [PubMed]
66. Gao, H.; Cheng, T.; Liu, J.; Liu, J.; Yang, C.; Chu, L.; Zhang, Y.; Ma, R.; Shi, L. Self-Regulated Multifunctional Collaboration of Targeted Nanocarriers for Enhanced Tumor Therapy. *Biomacromolecules* **2014**, *15*, 3634–3642. [CrossRef]
67. Smith, B.R.; Gambhir, S.S. Nanomaterials for In Vivo Imaging. *Chem. Rev.* **2017**, *117*, 901–986. [CrossRef]
68. Weber, J.; Beard, P.C.; Bohndiek, S.E. Contrast agents for molecular photoacoustic imaging. *Nat. Methods* **2016**, *13*, 639–650. [CrossRef] [PubMed]
69. Kobayashi, H.; Ogawa, M.; Alford, R.; Choyke, P.L.; Urano, Y. New Strategies for Fluorescent Probe Design in Medical Diagnostic Imaging. *Chem. Rev.* **2010**, *110*, 2620–2640. [CrossRef]
70. Zhao, X.; Long, S.; Li, M.; Cao, J.; Li, Y.; Guo, L.; Sun, W.; Du, J.; Fan, J.; Peng, X. Oxygen-Dependent Regulation of Excited-State Deactivation Process of Rational Photosensitizer for Smart Phototherapy. *J. Am. Chem. Soc.* **2020**, *142*, 1510–1517. [CrossRef] [PubMed]
71. Yan, D.; Xie, W.; Zhang, J.; Wang, L.; Wang, D.; Tang, B.Z. Donor/pi-Bridge Manipulation for Constructing a Stable NIR-II Aggregation-Induced Emission Luminogen with Balanced Phototheranostic Performance. *Angew. Chem. Int. Ed.* **2021**, *133*, 26973–26980. [CrossRef]
72. Hawkins, M.J.; Soon-Shiong, P.; Desai, N. Protein nanoparticles as drug carriers in clinical medicine. *Adv. Drug Deliv. Rev.* **2008**, *60*, 876–885. [CrossRef]
73. Xu, W.; Wang, D.; Tang, B.Z. NIR-II AIEgens: A Win-Win Integration towards Bioapplications. *Angew. Chem. Int. Ed.* **2021**, *60*, 7476–7487. [CrossRef] [PubMed]
74. Kang, M.; Zhou, C.; Wu, S.; Yu, B.; Zhang, Z.; Song, N.; Lee, M.M.S.; Xu, W.; Xu, F.; Wang, D.; et al. Evaluation of Structure-Function Relationships of Aggregation-Induced Emission Luminogens for Simultaneous Dual Applications of Specific Discrimination and Efficient Photodynamic Killing of Gram-Positive Bacteria. *J. Am. Chem. Soc.* **2019**, *141*, 16781–16789. [CrossRef] [PubMed]

75. Zhu, W.; Kang, M.; Wu, Q.; Zhang, Z.; Wu, Y.; Li, C.; Li, K.; Wang, L.; Wang, D.; Tang, B.Z. Zwitterionic AIEgens: Rational Molecular Design for NIR-II Fluorescence Imaging-Guided Synergistic Phototherapy. *Adv. Funct. Mater.* **2021**, *31*, 2007026. [CrossRef]
76. Xu, W.; Lee, M.M.S.; Nie, J.; Zhang, Z.; Kwok, R.T.K.; Lam, J.W.Y.; Xu, F.; Wang, D.; Tang, B.Z. Three-Pronged Attack by Homologous Far-red/NIR AIEgens to Achieve 1 + 1 + 1 > 3 Synergistic Enhanced Photodynamic Therapy. *Angew. Chem. Int. Ed.* **2020**, *59*, 9610–9616. [CrossRef] [PubMed]
77. Senthilnathan, N.; Gaurav, K.; Ramana, C.V.; Radhakrishnan, T.P. Zwitterionic small molecule based fluorophores for efficient and selective imaging of bacterial endospores. *J. Mater. Chem. B* **2020**, *8*, 4601–4608. [CrossRef]
78. Lovell, J.F.; Jin, C.S.; Huynh, E.; Jin, H.L.; Kim, C.; Rubinstein, J.L.; Chan, W.C.W.; Cao, W.G.; Wang, L.V.; Zheng, G. Porphysome nanovesicles generated by porphyrin bilayers for use as multimodal biophotonic contrast agents. *Nat. Mater.* **2011**, *10*, 324–332. [CrossRef]
79. Antaris, A.L.; Chen, H.; Cheng, K.; Sun, Y.; Hong, G.S.; Qu, C.R.; Diao, S.; Deng, Z.X.; Hu, X.M.; Zhang, B.; et al. A small-molecule dye for NIR-II imaging. *Nat. Mater.* **2016**, *15*, 235–242. [CrossRef] [PubMed]
80. Jung, H.S.; Lee, J.H.; Kim, K.; Koo, S.; Verwilst, P.; Sessler, J.L.; Kang, C.; Kim, J.S. A Mitochondria-Targeted Cryptocyanine-Based Photothermogenic Photosensitizer. *J. Am. Chem. Soc.* **2017**, *139*, 9972–9978. [CrossRef] [PubMed]
81. Li, X.; Kim, C.Y.; Lee, S.; Lee, D.; Chung, H.M.; Kim, G.; Heo, S.H.; Kim, C.; Hong, K.S.; Yoon, J. Nanostructured Phthalocyanine Assemblies with Protein-Driven Switchable Photoactivities for Biophotonic Imaging and Therapy. *J. Am. Chem. Soc.* **2017**, *139*, 10880–10886. [CrossRef] [PubMed]
82. Li, X.S.; Yu, S.; Lee, Y.; Guo, T.; Kwon, N.; Lee, D.; Yeom, S.C.; Cho, Y.; Kim, G.; Huang, J.D.; et al. In Vivo Albumin Traps Photosensitizer Monomers from Self-Assembled Phthalocyanine Nanovesicles: A Facile and Switchable Theranostic Approach. *J. Am. Chem. Soc.* **2019**, *141*, 1366–1372. [CrossRef] [PubMed]
83. Zhang, Z.; Xu, W.; Kang, M.; Wen, H.; Guo, H.; Zhang, P.; Xi, L.; Li, K.; Wang, L.; Wang, D.; et al. An All-Round Athlete on the Track of Phototheranostics: Subtly Regulating the Balance between Radiative and Nonradiative Decays for Multimodal Imaging-Guided Synergistic Therapy. *Adv. Mater.* **2020**, *32*, 2003210. [CrossRef] [PubMed]
84. Xu, W.; Zhang, Z.; Kang, M.; Li, Y.; Wen, H.; Lee, M.M.S.; Wang, Z.; Kwok, R.T.K.; Lam, J.W.Y.; et al. Making the Best Use of Excited-State Energy: Multimodality Theranostic Systems Based on Second Near-Infrared (NIR-II) Aggregation-Induced Emission Luminogens (AIEgens). *ACS Mater. Lett.* **2020**, *2*, 1033–1040. [CrossRef]
85. Wen, H.; Zhang, Z.; Kang, M.; Li, H.; Xu, W.; Guo, H.; Li, Y.; Tan, Y.; Wen, Z.; Wu, Q.; et al. One-for-all phototheranostics: Single component AIE dots as multi-modality theranostic agent for fluorescence-photoacoustic imaging-guided synergistic cancer therapy. *Biomaterials* **2021**, *274*, 120892. [CrossRef] [PubMed]

Review

Design of Magnetic Nanoplatforms for Cancer Theranostics

Wangbo Jiao [1], Tingbin Zhang [1], Mingli Peng [1], Jiabao Yi [2], Yuan He [1,*] and Haiming Fan [1,*]

[1] Key Laboratory of Synthetic and Natural Functional Molecule Chemistry of the Ministry of Education, College of Chemistry and Materials Science, Northwest University, Xi'an 710069, China; jwbzzh@stumail.nwu.edu.cn (W.J.); zhangtb@nwu.edu.cn (T.Z.); mlpeng@nwu.edu.cn (M.P.)

[2] Global Innovative Centre for Advanced Nanomaterials, School of Engineering, The University of Newcastle, Newcastle, NSW 2308, Australia; jiabao.yi@newcastle.edu.au

* Correspondence: yuanhe@nwu.edu.cn (Y.H.); fanhm@nwu.edu.cn (H.F.)

Abstract: Cancer is the top cause of death globally. Developing smart nanomedicines that are capable of diagnosis and therapy (theranostics) in one–nanoparticle systems are highly desirable for improving cancer treatment outcomes. The magnetic nanoplatforms are the ideal system for cancer theranostics, because of their diverse physiochemical properties and biological effects. In particular, a biocompatible iron oxide nanoparticle based magnetic nanoplatform can exhibit multiple magnetic–responsive behaviors under an external magnetic field and realize the integration of diagnosis (magnetic resonance imaging, ultrasonic imaging, photoacoustic imaging, etc.) and therapy (magnetic hyperthermia, photothermal therapy, controlled drug delivery and release, etc.) in vivo. Furthermore, due to considerable variation among tumors and individual patients, it is a requirement to design iron oxide nanoplatforms by the coordination of diverse functionalities for efficient and individualized theranostics. In this article, we will present an up–to–date overview on iron oxide nanoplatforms, including both iron oxide nanomaterials and those that can respond to an externally applied magnetic field, with an emphasis on their applications in cancer theranostics.

Keywords: iron oxide nanoparticles; magnetotheranostics; cancer

1. Introduction

Due to the huge differences between individual patients, a tough question exists in the field of tumor diagnosis and therapy: when and where to apply what kind of treatment for a particular patient? Image–guided therapy, known as theranostics, provides a new solution for this problem. The integration of diagnosis and therapy means that treatment can be carried out under the guidance of images and monitored in real time to achieve precise and personalized medical treatment. Due to their diversified functions, nanomaterials provide a great opportunity for the integration of efficient diagnosis and therapy into a single nanoplatform [1]. There are several requirements for an ideal theranostics nanoplatform. Firstly, the nanoplatform should possess good diagnostic and/or therapeutic capabilities. Secondly, this nanoplatform must be able accumulate at the target area. Thirdly, the biocompatibility of this nanoplatform must be acceptable. Finally, it should have the ability to be integrated with other diagnostic and/or therapeutic technologies for multi–modality theranostics. Magnetotheranostics is a kind of advanced medical technology that utilizes the interaction between magnetic nanoplatforms and magnetic fields to realize the integration of therapy and diagnosis on a single nanoparticle. The magnetic nanoplatforms are known to have excellent biocompatibility [2,3], diversified diagnostic and therapeutic capabilities [4], as well as active/passive targeting capabilities [5,6], while the magnetic field is well recognized for no attenuation [7] and little damage to the tissue [8]. As a result, magnetotheranostics has received a great deal of attention in cancer research recently.

As the core of magnetotheranostics nanoplatforms, magnetic iron oxide nanoparticles (MIONs) themselves have the ability to achieve diagnosis and therapy. The classic example of MIONs for diagnosis is magnetic resonance imaging (MRI) contrast agent [9]. Some iron oxide–based MRI contrast agents have been clinically approved, such as Ferrixan and Ferumoxide [10]. MIONs can achieve T1–weighted or T2–weighted MRI enhancement by affecting the T1 or T2 relaxation rate of surrounding protons. Although MIONs are generally considered to have only single–modal imaging capabilities, the development of advanced imaging technologies based on MIONs such as magnetic particle imaging (MPI) and magnetomotive optical coherence tomography (MMOCT) has also greatly enriched the imaging prospect of MIONs [11,12]. On the other hand, therapy technologies based on MIONs are gradually being developed. Magnetic hyperthermia (MHT) is a therapy method through the ability of MIONs to convert the energy of alternating magnetic field (AMF) into heat. In Europe, Nanotherm®, which is used as a nano agent of magnetic hyperthermia for brain gliomas, has been approved for clinical use [13]. MIONs also have the ability to kill cancer cells by producing reactive oxygen species (ROS) through the Fenton reaction catalyzed by Fe^{2+}, which is known as chemodynamic therapy (CDT) [14]. Additionally, MIONs can integrate with other materials for multi–modality therapy or diagnostics (Figure 1). For example, gold–magnetic composite nanomaterials are used as optical–magnetic hybrid nano–platforms for the integration of PDT, PTT, and MRI, and even CT and PET [15–17]. Appropriate surface modification can bring better biocompatibility [18], blood circulation time [19], active targeting [20], and even additional therapeutic and diagnostic functions [21] for MIONs. When combined with functional molecules, MION could satisfy more diverse demands. Fluorescent molecules can give MIONs the ability of fluorescence imaging [22]. Another example is that MIONs can carry and deliver drugs to the tumor area through some tumor–targeting methods [23].

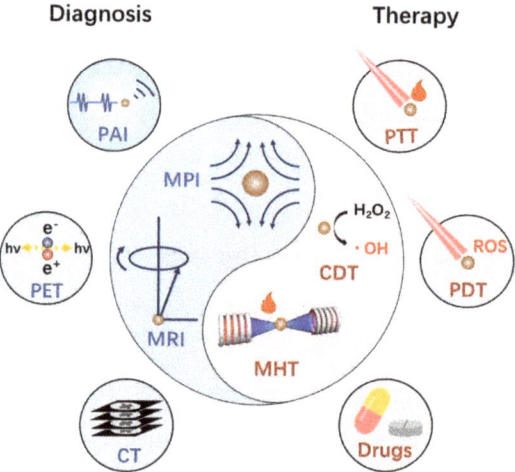

Figure 1. Diverse diagnosis and treatment technologies based on functionalized MIONs.

In this review, we summarized the progress in magnetotheranostics nanoplatforms for cancers in recent years. First, we reviewed the synthetic techniques of MIONs. Then, we introduced the design of therapy and diagnosis technologies (e.g., MRI, MHT, CDT, and others) based on MIONs. The next part of this review will focus on the design strategy of magnetotheranostics nanoplatforms combined with other therapy or diagnostic platforms, including phototheranostics, computed tomography (CT), positron emission computed tomography/single photon emission computed tomography (PET/SPECT), fluorescence imaging (FI), and drug delivery. The summary of the design strategy from magnetic function unit to external modification will help to deepen the understanding on their therapy and diagnostic capabilities. We hope this can provide some new ideas for the future design of new magnetotheranostics nanoplatforms.

2. Controlled Synthesis of Magnetic Nanoplatforms

The magnetic function unit of magnetotheranostics mainly refers to MIONs. MIONs can be synthesized by physical, biological, and chemical means. Physical methods include ball milling, vapor deposition, photolithography and other technologies, but the properties of MIONs synthesized by physical methods are difficult to control [24]. The biosynthesis of MIONs has some advantages, such as better environmental friendliness and product biocompatibility, but it also faces the problems of low crystallinity and difficulty in controlling the size and morphology [25]. Chemical synthesis of MIONs is the most commonly used method. Starting from the initial co–precipitation method [26,27], researchers have successively developed thermal decomposition methods [28], hydrothermal methods [29], solvothermal methods [30], sol–gel methods [31], Micelle methods [32], and other methods to construct MIONs.

MIONs are a type of iron–based metal oxide nanoparticles with a spinel structure, whose composition can be expressed as MFe_2O_4, and M represents divalent metal ions, including Mn^{2+}, Fe^{2+}, Co^{2+}, Ni^{2+}, Zn^{2+}, etc. (Figure 2a). In the most common Fe_3O_4 materials, M = Fe^{2+} and the Fe^{2+} occupies the octahedral (O_h) sites of the spinel structure, forming an inverse spinel structure. The antiferromagnetic coupling between Fe^{3+} makes the overall magnetic spin behave as 4 µB of Fe^{2+}. The conditions of $CoFe_2O_4$ and $NiFe_2O_4$ are similar to those of Fe_3O_4 materials, and the total magnetic spins are 3 µB of Co^{2+} and 2 µB of Ni^{2+} respectively. When M = Mn^{2+}, Mn^{2+} mainly occupies octahedral sites and partly occupies the tetrahedral (T_d) sites, forming a mixed spinel structure. However, since the magnetic spins of Mn^{2+} and Fe^{3+} are both 5 µB, they always show a total magnetic spin of 5 µB in the end [33]. In $ZnFe_2O_4$, Zn^{2+} occupies a tetrahedral position to form a normal spinel structure. The magnetic spin of Zn^{2+} is 0 µB, and the magnetic spins of two Fe^{3+} cancel each other out, showing 0 µB overall, theoretically. Interestingly, in $Zn_{0.4}Fe_{2.6}O_4$, the antiferromagnetic coupling between Fe^{3+} is broken by the configuration reversal caused by the partial doping of Zn^{2+}, and shows a higher remanent magnetic spin [34]. The composition control of MIONs can be easily achieved by adjusting the ratio of metal precursors. In the thermal decomposition method, this usually depends on the feeding amount of different metal–organic complexes. For example, Sun et al. synthesized $MnFe_2O_4$ and $CoFe_2O_4$ nanoparticles by thermal decomposition of manganese acetylacetonate, cobalt acetylacetonate with iron(III) acetylacetonate [35]. Long–chain fatty acid complexes are another type of common organic precursors. Zhang et al. used iron(III) erucate and manganese oleate, cobalt oleate to synthesize $MnFe_2O_4$ and $CoFe_2O_4$ nanoparticles [36]. In other synthesis methods, MFe_2O_4 is usually synthesized by directly adding different metal ions. Co–precipitation and solvothermal methods can be used to generate MFe_2O_4 by adding Mn^{2+}, Co^{2+}, Ni^{2+}, or Zn^{2+} [26,29].

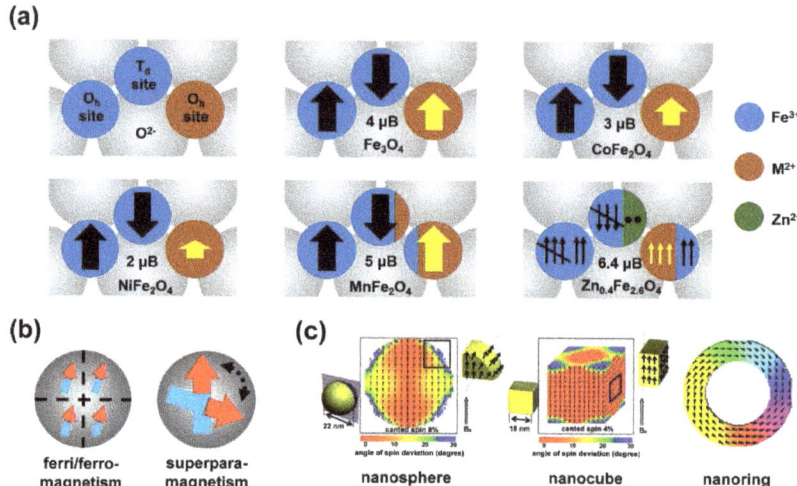

Figure 2. (a) Structure schematic of spinel structure and magnetic spin of MFe$_2$O$_4$. (b) Magnetic domain of ferrimagnetism/ferromagnetism (**left**) and superparamagnetism (**right**). (c) Magnetic spin states simulated using OOMMF program for nanosphere and nanocube [37] and vortex domain structure of nanoring simulated by the LLG Micromagnetics Simulator™ package [38]. Reprint permission from [37,38]. Copyright 2012 American Chemical Society and 2012 American Institute of Physics.

In addition to the influence of the crystal structure itself, the magnetic properties of MIONs are greatly affected by their size. Excluding the influence of other factors, the smaller the MIONs, the lower the saturation magnetization (Ms) [39]. When the size is smaller than the critical size, the magnetic anisotropy of MIONs is not enough to resist the effects of thermal disturbance, resulting in the loss of its own remanence and hysteresis, but still maintaining a high initial magnetic susceptibility. This phenomenon is called superparamagnetic (Figure 2b). Due to its zero remanence, superparamagnetic iron oxide nanoparticles (SPIOs or SPIONs) have excellent colloidal dispersion and better stability than ferromagnetic or ferrimagnetic nanoparticles, and they have been approved for clinical use [10]. The size of MIONs can be controlled by the reaction temperature and the amount of surfactant. Park et al. used several solvents with different boiling points to control the reaction temperature in thermal decomposition of iron oleate, including 1–hexadecene (b.p. 274 °C), octyl ether (b.p. 287 °C), 1–octadecene (b.p. 317 °C), 1–eicosene (b.p. 330 °C), and trioctylamine (b.p. 365 °C), and successfully synthesized Fe$_3$O$_4$ nanoparticles of 5–22 nm sizes [28]. Xu et al. [40] used iron acetylacetonate to synthesize 7–10 nm Fe$_3$O$_4$ nanoparticles by adjusting the ratios of oleylamine and benzyl ether in the high temperature thermal decomposition process.

The morphology will also affect the magnetic properties of MIONs (Figure 2c). Compared with spherical MIONs, cubic MIONs have a higher Ms due to the less distributed spin disorder layer on the surface [37,41]. The ring–shaped MIONs possess a unique vortex magnetic domain [38], enabling it to have zero remanence and zero hysteresis while maintaining ferrimagnetism [42]. To synthesize MIONs with a specified morphology, which means to achieve anisotropic growth of MIONs, it is necessary to provide a near thermodynamically stable environment during crystal growth so that the interface energy of different crystal faces dominates the process. Therefore, the usage of thermal decomposition method or hydrothermal method are more appropriate choices. In the work of Zhou et al., sodium oleate was used to control the growth of different crystal faces during the thermal decomposition of iron oleate to obtain cubes, concaves, multibranch shaped MIONs [43]. An example of hydrothermal control of the morphology of MIONs is the synthesis of α–Fe$_2$O$_3$

nanorings and nanotubes by phosphate and sulfate dianion–assisted hydrothermal method, and controlled synthesis of Fe_3O_4 and γ–Fe_2O_3 nanorings/tubes by thermal reduction and thermal oxidation [44].

The naked MIONs may not be suitable for direct biological application. During or after the preparation of MIONs, they need to be further modified to endow them with better stability for biological applications. Polymers are a kind of widely used coatings, including synthetic polyethylene glycol (PEG) [18,45], polyethyleneimine (PEI) [46], polyacrylic acid (PAA) [47], polyvinylpyrrolidone (PVP) [48] and natural dextran [19], chitosan [49], alginate [50], and so on. For example, modification of PEG not only enhances the colloidal dispersion and stability of MIONs and improves their biocompatibility, but also prolongs the average time for MIONs to be recognized and swallowed by macrophages in the liver and spleen, which will extend the blood circulation time of MIONs [18]. Dextran is shown to have a similar effect [19]. Silane is also a common coating of MIONs. Amino silane is used for surface modification of the magnetic hyperthermia agent NanoTherm® [51]. The modification with antibodies [20], targeting peptides [52] or other targeting molecules [53] can provide MIONs with active targeting capability. Meanwhile, surface modifications may also affect the therapeutic and diagnostic functions of MIONs. The thickness of the surface coating has been proven to affect the T2 relaxation performance of MIONs [54] and the performance of magnetic hyperthermia [55]. In the study of Zeng et al. [56], the difference in anchor groups can also affect its Ms and further affect its T2 relaxation properties. Connecting graphene oxide on the surface of MIONs can improve its magnetothermal effect in the form of dielectric loss [57]. Surface ligands are often used to control the spatial position between various groups [58]. PEG coatings of different molecular weights have been used to precisely adjust the distance between natural enzymes and MIONs nanozymes [59]. Self–assembled monolayers (SAMs) can also control the spatial distribution of surface functional groups at the molecular level [60]. Modification of detection molecules on the surface allows MIONs to be widely used in the diagnosis of various biomolecules in vitro, including circulating tumor cells (CTC), alpha–fetoprotein, ctDNA, and other markers. The target cell/molecules can be captured by MIONs, magnetically separated and then detected by polymerase chain reaction (PCR), enzyme–linked immunosorbent assay (ELISA), or more sensitive atomic force microscope (AFM)–based technology [61,62].

3. Basis of Magnetic Nanomaterials Mediated Diagnosis and Therapy of Cancer

The magnetic properties of MIONs magnetic core can affect the relaxation process of protons, making it useful for MRI contrast agents. MIONs can absorb the energy of the magnetic field to generate in situ heating under the alternating magnetic field, and then can realize the magnetic hyperthermia of the tumor. The Fenton reaction by Fe^{2+} enables the generation of ROS to mediate tumor chemodynamic therapy. These physicochemical properties can be applied for cancer diagnosis and therapy. Together with the low toxicity [2] and clear degradation metabolism [3], MIONs have received increasing attention for theranostics.

3.1. Biosafety of Magnetic Nanoplatforms

MION formulations are generally considered to have excellent biological safety. Naked MIONs have strong antigenicity and are prone to cause allergic reactions. Surface modification such as dextran can significantly avoid side effects. There have been a large number of studies to evaluate the possible side effects of MIONs for clinical use so far. In the cell viability studies, most MIONs reported only showed cytotoxicity at particularly high concentrations [63–65]. For example, the ability of nerve cells to lengthen neurons was reduced in a dose–dependent manner by Fe. In this example, anionic magnetic nanoparticles are applied [66]. Animal experiments showed that the LD50 of MIONs is indeed affected by its surface modification. The LD50 of naked MIONs is 300~600 mg/kg. When MIONs were coated with dextran, the LD50 are increased to 2000~6000 mg/kg [67]. When used to treat lymph node metastases from thoracic squamous cell carcinoma of the esophagus,

MIONs (ferucarbotran in this case) exhibited negligible side effects [68]. The application of dextran–coated MIONs in the diagnosis of carotid inflammatory plaques also showed no obvious side effects [69]. However, some adverse reactions in the clinical study of ferumoxtrans–10 were observed, and even one case died after the injection of undiluted MIONs [70]. The report pointed out that the safety of MIONs is highly dependent on the dose used. The death may be caused by the rapid formation of aggregated particles in undiluted MIONs, which then accumulate in the kidney and liver through phagocytes to cause acute toxicity [71].

3.2. Magnetic Resonance Imaging

Due to its high soft tissue contrast, high temporal and spatial resolution, and no ionizing radiation, MRI is widely used for imaging of soft tissues such as brain, heart, muscle, and tumor [72]. MRI signals are derived from nuclear magnetic resonance (NMR) signals from water protons in human tissues. Depending on the received proton longitudinal relaxation (T1) or transverse relaxation (T2) signals, MRI imaging methods are divided into two types: T1 weighting and T2 weighting. These relaxation signals can be affected by the magnetic properties of MIONs, thereby enhancing their signal strength and improving the contrast between diseased tissues and normal tissues. The contrast agents for these two imaging methods are thus called T1 contrast agents and T2 contrast agents, respectively, and their ability to enhance the corresponding relaxation rate is characterized by r1 and r2 values. The superparamagnetism of SPIONs makes them capable of disturbing the magnetic uniformity near itself under the high main magnetic field conditions of MRI, which can accelerate the lateral relaxation of surrounding protons, reduce the signal intensity in T2–weighted magnetic resonance images, and achieve negative image enhancement. Although some SPIONs were approved for clinical use as pure T2 contrast agents, they have been gradually withdrawn from clinical application in recent years, mainly due to shortcomings of poor imaging specificity (such as confusion with bleeding and calcification) [73]. Nevertheless, the T2 contrast enhancement of MIONs has been widely used in recent years for image tracking and therapy guidance of the spatiotemporal position of MIONs in vivo, such as the use of T2–weighted MRI to track cells marked by MIONs [74,75].

T1–weighted MRI can avoid the shortcomings of T2–weighted MRI, so it has better clinical usage. During the relaxation process, the protons can transfer energy with the T1 contrast agent to shorten its T1 relaxation time, especially for Gd^{3+}, Fe^{3+}, Mn^{2+}, and other ions containing a large number of unpaired valence electrons. In comparison to the clinically used Gd–based contrast agents with biological safety problems [76,77], MIONs are well recognized for better biocompatibility, and Fe^{3+} grants them potential T1 imaging capabilities. However, large–sized MIONs have high Ms and T2 enhanced imaging performance. The high r2/r1 ratio limits their application in T1 contrast imaging. The emergence of ultrasmall SPIONs provides an opportunity to solve this problem. When the size of MIONs is reduced, their Ms decreases sharply. This reduces the r2 value, meanwhile the increased surface area increases its r1 value, leading to declined r2/r1 ratio to the range that allows T1 imaging. With the progress in new technologies for the large–scale synthesis of ultrasmall SPIONs [36,78], the clinical application of ultrasmall SPIONs as T1 contrast agents has also rapidly developed. Wei et al. [79] reported the synthesis of a zwitterion (ZES) coated ultrasmall SPIONs with a magnetic core diameter of about 3 nm, a hydrophilic shell thickness of about 1 nm, a low r2/r1 value and a long internal circulation time, for high resolution T1 MRI imaging of vessels with a spatial resolution of about 0.2 mm. Miao et al. [80] studied the effect of different doping of the core–shell structure on the T1 imaging performance of ultrasmall SPIONs. The optimized 3.8 nm $Zn_xFe_{3-x}O_4@Zn_xMn_yFe_{3-x-y}O_4$ core–shell ultrasmall SPIONs has an r1 relaxation rate of 20.22 $mM^{-1}s^{-1}$, which is 5.2–fold and 6.5–fold larger than that of the undoped ultrasmall SPIONs and the clinically used Gd–DTPA. Such nanoagent was then shown to be able to detect micro metastases in the lungs of mice. In order to extend the circulation time in vivo, some MIONs are designed to form larger aggregates in the form of clusters before entering the target area, showing

T2 enhanced performance. After reaching the target area, it disintegrates into ultrasmall SPIONs responsively, realizing the conversion of T2 to T1 contrast enhancement [81]. This mode conversion provides a new strategy to monitor the in vivo behavior of these MIONs.

3.3. Other Diagnosis Applications

Magnetic particle imaging (MPI) has the advantages of no tissue signal attenuation, a linear correlation between the signal and tracer concentration, and no ionizing radiation during detection. It has become an emerging tomographic imaging technology that is expected to enter clinical applications, especially in lung and other organs that are difficult to be imaged by MRI [82]. The earliest tracer used in MPI technology is SPIONs, and so far it is still the dominant tracer materials [82]. The Ms of the MPI tracer is decisive for its imaging performance. The Ms of Fe@Fe_3O_4 NPs are reported to be as high as 176 emu/g, which enables good MPI performance [83]. Changes in the crystallinity of MIONs can also alter the Ms, thereby affecting their MPI performance [84]. Magnetomotive optical coherence tomography (MMOCT) is another type of imaging technology based on MIONs, and the image contrast is derived from dynamic magnetomotive force. Unlike MRI and MPI, MMOCT requires a magnetic field as low as 0.08 T [85] and can detect ultra–low concentrations of tracers. MMOCT has been shown to be able to image tumor models in animals [12]. Due to its ability to detect the movement state of MIONs particles, it has also recently been used for real–time monitoring of magnetic hyperthermia [86].

3.4. Magnetic Hyperthermia

Hyperthermia is a treatment with a long history. MIONs have the ability to convert the energy of an alternating magnetic field (AMF) into heat. Hyperthermia using this magnetothermal effect is called magnetic hyperthermia (MHT). As a means of in situ hyperthermia, magnetic hyperthermia can kill tumor tissues more accurately, and it is not limited by the depth of tissue penetration. It has developed many application scenarios in the field of tumor therapy [87–89]. The improvement of the efficiency of tumor magnetic hyperthermia depends on the improvement of specific absorption rate (SAR), and SAR is directly proportional to the Ms of MIONs [90]. It has been proven that the Ms of MIONs are strongly related to their size [39]. SPIONs with a diameter of less than 20 nm have a SAR in the range of hundreds of W/g [42]. In theory, the SAR of MIONs could be improved by increasing the size of MIONs. Paradoxically, larger–sized MIONs also exhibit ferromagnetic/ferrimagnetic properties. The existence of remanence disfavors colloidal stability of the MIONs, which in turn can decrease its SAR. In recent years, some MIONs with special magnetic domain structures have gradually shown their advantages. In one example, iron oxide nanorings of a specific size will exhibit vortex magnetic domains. The magnetic domains of this structure are closed loops connected end to end. While maintaining high ferrous hysteresis loss, the residual magnetization is kept at zero, achieving a SAR exceeding 2000 W/g while having excellent colloidal dispersion [42]. The FePt@IONP synthesized by Yang et al. [91] also showed magnetic domains in the vortex state. This structure reduced the magnetic dipole–dipole interaction between FePt@IONP nanoparticles, prevented them from gathering in the remaining state, and improved the colloidal stability. Exchange coupling between FePt core and iron oxide shells can enhance the magnetic anisotropy of FePt@IONP, thereby improving its SAR and enabling more efficient magnetic hyperthermia. In addition, during the MRI scan of this FePt@IONP, the presence of a high main magnetic field induced the formation of NP chains and introduces an increase in the local uneven dipole field that ultimately enhances the T2 relaxation performance. Zhang et al. [92] grew a multi–domain eMION through the biomineralization process of capsules. The encapsulin–produced magnetic IONs (eMIONs) consist of FeO subdomains containing Fe_3O_4 with ~100% crystallinity. The SAR of eMIONs can reach 2390 W/g, and their nanozyme activity was also enhanced under the action of AMF, showing excellent tumor therapy capabilities, and the entire therapy process could be monitored in real time using MRI.

3.5. Chemodynamic Therapy

In an acidic environment, Fe^{2+} in MIONs can catalyze the Fenton reaction:

$$Fe^{3+} + H_2O_2 = Fe^{2+} + HO_2\bullet + H^+$$

$$Fe^{2+} + H_2O_2 = Fe^{3+} + \bullet OH + OH^-$$

Due to its similar behavior to peroxidases such as horseradish peroxidase (HRP), the ability of MIONs to catalyze the Fenton reaction is also called peroxidase–like activity of a nanozyme. This reaction can generate a large amount of ROS in the tumor cells, break the redox balance of tumor cells, and then damage tumor cells [93]. This phenomenon is known as chemodynamic therapy (CDT). However, since the pH in the tumor microenvironment does not meet the optimal conditions for the Fenton reaction [94,95], the direct application of MIONs in tumor CDT is limited. Hence, strategies to improve the Fenton reaction activity of MIONs are vital for their CDT efficacy. Liang et al. [81] synthesized a porous yolk–shell Fe/Fe_3O_4 nanoparticles (PYSNP, Figure 3a,b), which used Fe_3O_4 shell to protect Fe^0 from oxidation and to deliver it to the tumor microenvironment. Disintegration of the nanoparticles into fragments results in the transformation of its MRI imaging performance from T2 to T1, realizing image tracking of the delivery process. Finally, the relatively higher Fenton reactivity of the exposed Fe^0 was used to achieve high–efficiency tumor CDT (Figure 3c). Aiming at the problem that the concentration of H_2O_2 in cancer cells is not sufficient for effective CDT [96], Du et al. [97] used mesoporous silica nanoshell to connect ultrasmall SPIONs with Au nanorods and constructed a core–shell–satellite nanomaces (Au @ MSN@IONP), AuNR can convert near infrared light into heat to cause heat stress in cancer cells and to generate a large amount of H_2O_2. The H_2O_2 then acted as a substrate of MIONs for Fenton reaction, achieving a highly specific anticancer effect, and inhibited PI3K/Akt/FoxO axis which is closely related to the redox regulation and survival of breast cancer cells. MIONs–enhanced MRI provided imaging guidance for the photodynamic therapy. Low–intensity focused ultrasound (LIFU) has also recently been used to enhance CDT by Deng and coworkers [98]. They loaded vitamin C (Vc) and SPIO together inside the PLGA nanospheres to fabricate PLGA–SPIO&Vc. The PLGA–SPIO&Vc were delivered to tumor cells through magnetic targeting and EPR effect. Subsequently, the cavitation and oscillation effects of LIFU were used to promote the release of Vc and to lower the environmental pH. Vc also worked as a H_2O_2 precursor to provide the materials for Fenton reaction. Photoacoustic imaging was used by them to detect the progress of the Fenton reaction.

Interestingly, the magnetothermal effect of MIONs can also be used to enhance its intrinsic catalytic activity. Exposing MIONs to AMF was shown to effectively enhance their peroxidase nanozyme activity without changing the bulk temperature of the solution, and the degree of rate enhancement has a linear dependence on the SAR of MIONs [99]. Zhang et al. studied the influence of the magnetothermal effect of MIONs on the activity of the natural enzyme glucose oxidase (GOx) [59]. They found that the degree to which GOx is enhanced by AMF stimulation is related to the distance between the MIONs and GOx. With an optimal distance of 1 nm, the hybrid MIONs–GOx catalyst shows the highest cascade activity to produce a large amount of ROS and achieves the best tumor inhibition effect in a mouse breast cancer model.

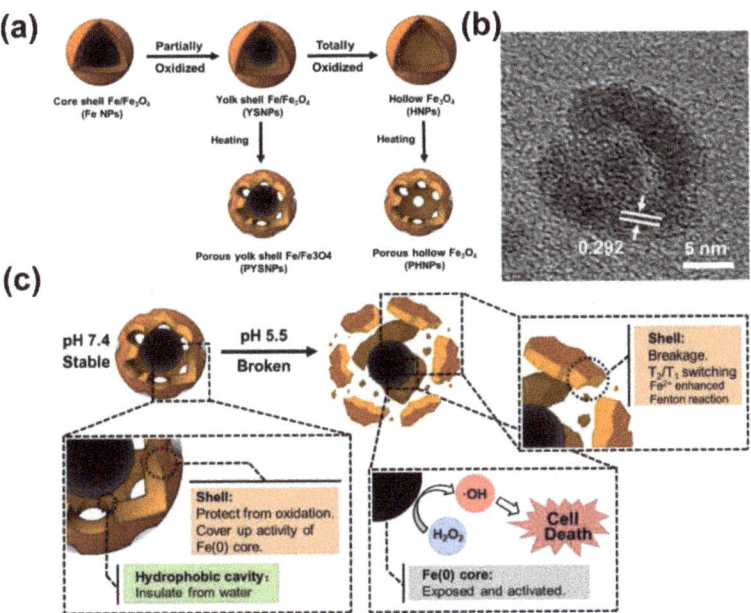

Figure 3. (**a**) Schematic illustration of PYSNP preparation in [64]. (**b**) HRTEM images of PYSNP. (**c**) Schematic illustration of the pH activated Fe release in PYSNPs. Reprint with permission from [81]. Copyright 2020 Elsevier Ltd., Amsterdam, The Netherlands.

4. Implementation of Magnetotheranostic Based on Magnetic Nanoplatforms

MIONs can be coupled with a variety of nanocomposites, so that multiple diagnosis and therapy technologies can be integrated on a single nanoplatform. This includes magnetotheranostics based on magnetic field, magnetoptical theranostics, PET or CT, fluorescence imaging, drug delivery, etc. MIONs can be used as a nanoplatform to integrate almost all existing diagnosis and therapy methods. The following will describe functional modifications of MIONs for integration of magnetic diagnosis and treatment.

4.1. Magnetotheranostics Based on Magnetic Nanoplatforms Only

As mentioned in Section 3, MIONs have their own therapeutic and diagnostic functions such as MRI, MPI, MHT, and CDT. Therefore, improvements on MIONs are beneficial for their magnetotheranostics performances. The Ms of MIONs received great attention from researchers at an earlier time [34], because the improvement of Ms will simultaneously enhance the T2–weighted imaging performance of MIONs and the thermal conversion efficiency of MHT. Recent work in this area has become more diversified, and one direction is T1–T2 dual–modality imaging combined with treatment. Liu et al. [90] synthesized wüstite $Fe_{0.6}Mn_{0.4}O$ nanoflowers. Unlike the antiferromagnetic bulk wüstite, $Fe_{0.6}Mn_{0.4}O$ nanoflowers exhibit ferromagnetism, which may be due to exchange coupling effect. The as–prepared nanoflowers exhibit excellent magnetic induction heating effects (SAR can reach 535 W/g), which could induce tumor regression in breast cancer through MHT. The longitudinal relaxation rate r1 and lateral relaxation rate r2 of $Fe_{0.6}Mn_{0.4}O$ nanoflowers are as high as 4.9 and 61.2 $mM^{-1}_{[Fe]+[Mn]} \cdot s^{-1}$, respectively. These nanoflowers showed both T1 and T2 enhancing properties in the mouse glioma model. Different from this static T1–T2 dual–modal contrast agent, another type is dynamic T1–T2 dual–modal contrast agents. They can present two states of T1 or T2 contrast, and certain events will prompt the transition between the two states. This is generally accomplished by disintegrating large particles into small particles (T2 to T1) or aggregation of small particles into large particles (T1 to T2). An example of the conversion of T2 enhancement to T1 enhancement is listed in Section 3.2 [81]. In the work by Zhou et al. [100], the ultrasmall SPIONs aggregated into clusters in the tumor in situ, resulting in the conversion of T1 enhancement to T2 enhancement. They used hyaluronic acid (HA) to encapsulate ultrasmall SPIONs, which showed T1 enhanced performance before penetrating into the tumor. After entering the tumor area, the surface–modified HA was degraded by the abundant hyaluronidase, which decreased the colloidal stability of ultrasmall SPIONs and caused aggregation of the nanoparticles into clusters, resulting in enhanced T2 imaging performance and weakened T1 imaging performance. Although the therapy performance of the designed ultrasmall SPIONs was not investigated in this work, its penetration–aggregation design still provided a strategy for future magnetotheranostics.

Another direction that has received widespread attention is the tumor immune effect caused by ROS produced in CDT. The FDA–approved iron supplement ferumoxytol (FMX) was confirmed to be able to polarize tumor–associated macrophages from the anti–inflammatory M2 phenotype to the pro–inflammatory M1 phenotype through the induced ROS [101]. Further studies have shown that the ROS generated by iron oxide–loaded nanovaccines (IONVs) can promote the presentation of tumor antigens, mediate tumor immune cell infiltration, and stimulate non–toxic long–term protective antitumor immunity [102]. MRI can provide real–time location of IONVs in this work. Liu et al. [57] reported that the Fenton reaction activity of MIONs (FVIOs–GO, Figure 4a) could be enhanced under the action of AMF, causing the massive production of ROS (Figure 4b,c), which in turn lead to calreticulin (CRT) of cancer cells migration from the endoplasmic reticulum to the outside of the plasma membrane (Figure 4d), causing immunogenic cell death (ICD). MRI could be also used to monitor the location of MIONs.

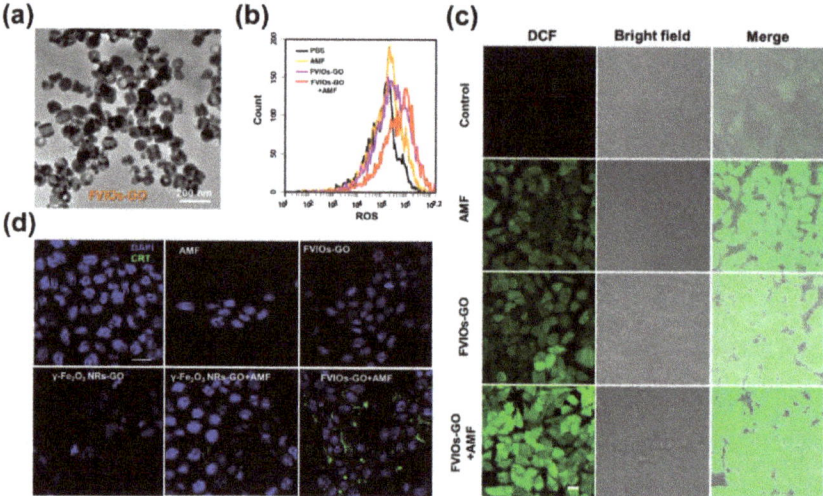

Figure 4. (a) TEM image of FVIOs–GO. (b) ROS generation of 4T1 breast cancer cells after treatment in a flow cytometry assay. (c) Confocal laser scanning microscope (CLSM) images of 4T1 cells treated with complete media (control), AMF, FVIOs–GO only, and FVIOs–GO–containing media under AMF. Generation of ROS was monitored using DCFH–DA (green). Scale bars for all images are 20 μm. (d) Confocal images showing the CRT exposure on 4T1 tumor cells in vitro after treatment with γ–Fe$_2$O$_3$ nanorings+AMF or FVIOs–GO–mediated magnetothermodynamic (MTD) therapy. Scale bars for all images are 20 μm. Reprint with permission from [57]. Copyright 2020 American Chemical Society.

4.2. Integration of Magnetic Nanoplatforms with Phototheransotics

Similar to magnetotheranostics, comprehensive application of the optical properties of nanomaterials in therapeutic diagnostics are called phototheranostics, covering technologies such as photothermal therapy (PTT), photodynamic therapy (PDT), and photoacoustic imaging (PAI). Corresponding to the magnetic core in magnetotheranostics, the phototheranostics nanoplatform needs a photosensitizer as its core. Among the various photosensitizers, noble metal nanoparticles, especially Au nanoparticles [103] can efficiently complete energy conversion through the localized surface plasmon resonance (LSPR) effect, and are often used to form a hybrid magnetoptical theranostics nanoplatform with MIONs for therapeutic diagnostics. Liu et al. [104] assembled SiO_2–coated Au nanowreaths (AuNWs) with ultrasmall SPIONs through molecules containing polycystamine blocks (Figure 5a). After sensing the GSH in the tumor cells, the disulfide bond of polycystamine was cleaved, causing the disassembly of ultrasmall SPIONs from the surface of AuNWs, and the MRI contrast performance of the ultrasmall SPIONs changed from T2 enhancement to T1 enhancement (Figure 5b). The released AuNWs were used for photothermal therapy and photoacoustic imaging (Figure 5c,d). Amphiphilic Janus nanoparticles with hydrophilic PEG–modified AuNPs and hydrophobic poly(lipid hydro–peroxide)–co–poly(4–vinylpyrene) (PLHPVP) modified MIONs were reported to form a double–layered vesicle [17]. Its MRI and PAI performance could be enhanced through magnetic dipole interaction and strong plasma coupling. The inner cavity of this vesicle could be loaded with DOX to deliver the drug into tumor cells, and the outer side was modified with the radioactive isotope ^{64}Cu for PET imaging. After entering the tumor cells, the acidic environment disassembled the vesicles, and PLHPVP became hydrophilic under the influence of H^+ and allowed Fe^{2+} to contact the environment. Fe^{2+} further reacted with LHP to generate ROS and cooperates with DOX to kill tumor cells. Metal sulfide is also one of the common photosensitizers. IONPs anchored on titanium disulfide (TiS_2) nanosheets have strong absorption and excellent magnetic properties in the second near–infrared (NIR–II) window and had been developed as NIR–II PAI and MRI–guided photothermal therapy, combined with immunotherapy to prevent tumor recurrence [105]. Besides MIONs, iron(II) sulfide nanoparticles could also exhibit superparamagnetism, and their strong absorption in the near–infrared region enables them to be used as PTT agents. In addition, its ultra–high r2 relaxivity makes it an excellent T2 contrast agent [106]. In addition to these inorganic materials, some organic materials having light–to–heat conversion capabilities are also used in magnetotheranostics. Polydopamine (PDA) has traditionally been used to modify MIONs [21] for better colloidal stability and biocompatibility. PDA can also be used as a photothermal agent to mediate photothermal therapy. T2–weighted MRI was used for image–guided therapy. Porphyrin is a bioinspired organic photosensitizer. The porphyrin derivative meso–tetrakis(4–carboxyphenyl)porphyrin (TCPP) could be excited by the Cerenkov luminescence of ^{89}Zr connected to the MIONs platform to generate ROS, which was used for PDT without external light source [107]. In this example, MIONs were designed as $Zn_{0.4}Mn_{0.6}Fe_2O_4$ to obtain optimized Ms for magnetic targeting. In another study, protoporphyrin IX (PpIX) was used to coat SPIONs to form clusters, which could realize the integration of diagnosis and therapy of MRI and PDT [108].

4.3. Integration of Magnetic Nanoplatforms with Fluorescence Imaging

Fluorescence imaging (FI) relies heavily on the penetration of light in tissue, which limits its application in the field of in vivo diagnosis and therapy. However, because fluorescent molecules can be designed to achieve highly specific responsiveness, fluorescence imaging is often used as an auxiliary imaging method for the magnetotheranostics nanoplatform to monitor its response behavior in the body. Zhou et al. [109] designed a nanoplatform capable of detecting tumor hypoxic environment, composed of ultrasmall SPIONs and assembly–responsive fluorescent dyes (NBD), and used nitroimidazole derivatives as hypoxia–sensitive detectors. Zhou et al. [97] designed a nano–platform capable of detecting the hypoxic environment of tumors, consisting of USION and assembly–responsive fluorescent dyes (NBD), and used nitroimidazole derivatives as hypoxia–sensitive detectors. In an oxygen–rich environment, ultrasmall SPIONs showed enhanced T1 performance. Under hypoxic conditions, NBD cross–linked irreversibly, leading to self–assembly of ultrasmall SPIONs, and thus its contrast performance changed from T1 MRI to T2. At the same time, the cross–linking of NBD increased its fluorescence intensity, indicating that the oxygen environment level in this area had decreased. In another example, combination of MRI and fluorescence imaging was used to monitor the progress of CDT [110]. NQ–Cy, MIONs, and GOx were mixed and loaded in the micelles of DSPE–PEG–FA. The MRI capability of MIONs provided information to monitor the delivery of the micelles. Next, the increase of fluorescence at 830 nm indicated successful release of NQ–Cy into the cytoplasm. Subsequently, the released MIONs and GOx produced a large amount of ROS in the cytoplasm, causing cell oxidative stress and an increase in NQO1 enzyme expression. Then NQ–Cy was decomposed by NQO1, and its emission wavelength shifted from 830 to 670 nm, indicating that CDT had effectively activated the oxidative stress of cells.

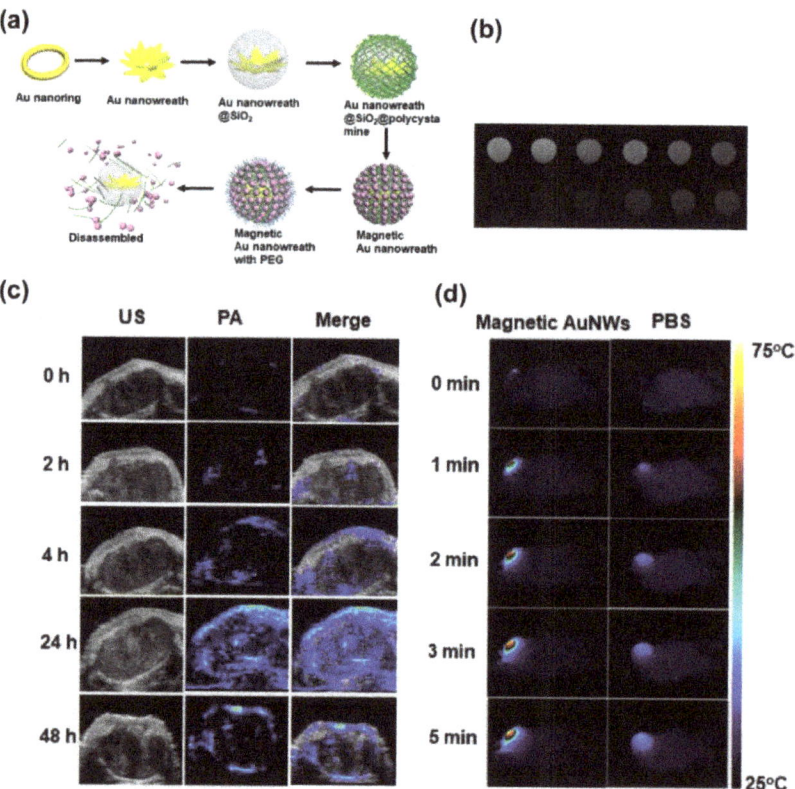

Figure 5. (a) Schematic illustration of the synthesis of magnetic gold nanowreath. (b) Corresponding T1–weighted images (top, disassembled; bottom, assembled) of magnetic AuNWs. Concentrations of Fe are 0.5, 0.25, 0.125, 0.0625, 0.03125, and 0.01563 mM (from left to right). (c) Ultrasonic (US), photoacoustic imaging (PA), and merged images of tumor before injection (0 h) and at 2, 4, 24, and 48 h after intravenous injection of magnetic AuNWs upon irradiation by an 808 nm pulsed laser. (d) Representative thermal images of U87MG tumor bearing mice after injection of magnetic AuNWs and PBS. The tumors were irradiated by an 808 nm CW laser at 0.75 W/cm^2. Reprint with permission from [104]. Copyright 2018 American Chemical Society.

4.4. Integration of Magnetic Nanoplatforms with CT&PET/SPECT

Although iodine–based contrast agents for CT have been well developed, certain components of the magnetotheranostics nanoplatforms also have the ability to act as CT contrast agents. The use of CT to guide the diagnosis and therapy of the magnetic nanoplatforms has remarkable application prospects. A relatively common example is the gold–magnetic composites, wherein the high CT value of gold enables it to be traced by CT [111]. Liu et al. developed an ultrasonication–triggered interfacial assembly approach (Figure 6a,b) to synthesize magnetic Janus amphiphilic nanoparticles (MJANPs) for image–guided cancer MHT (Figure 6c) [16]. Au NPs–MIONs MJANPs made of Au NPs and MIONs could achieve MRI/CT dual–modality imaging and could be used to guide MHT (Figure 6d). Similarly, if CuInS/ZnS NPs and MIONs were used to make CuInS/ZnS NPs–MIONs MJANPs, MRI/FI dual–modality imaging can be used to guide MHT (Figure 6e). In the bismuth ferrite (BFO) nanoplatform designed by Feng et al., BFO nanoparticles could achieve CT contrast effects similar to iohexol [112]. The 2D–Ta$_3$C$_4$–MIONs designed for PTT and T2–weighted MRI also have a CT value exceeds that of the clinically used CT contrast agent iopromide [113]. PET/SPECT imaging relies on the labeling of radioisotopes. Combining radioisotopes with the magnetotheranostic nanoplatforms is a common method for multimodal imaging. These radioisotopes include ^{18}F, ^{59}Fe, ^{64}Cu, ^{68}Ga, ^{89}Zr, ^{99m}Tc [114–117]. In a comparative study by Zhang et al. [114], two modification strategies using radioisotopes (^{59}Fe) in the core of MIONs and radioisotopes (^{64}Cu) labeling in the shell were systematically compared. They believed that the shell labeling was relatively more attractive due to its flexible design, easy operation, and low radiation risk, but the core labeling had better stability for in vitro tests.

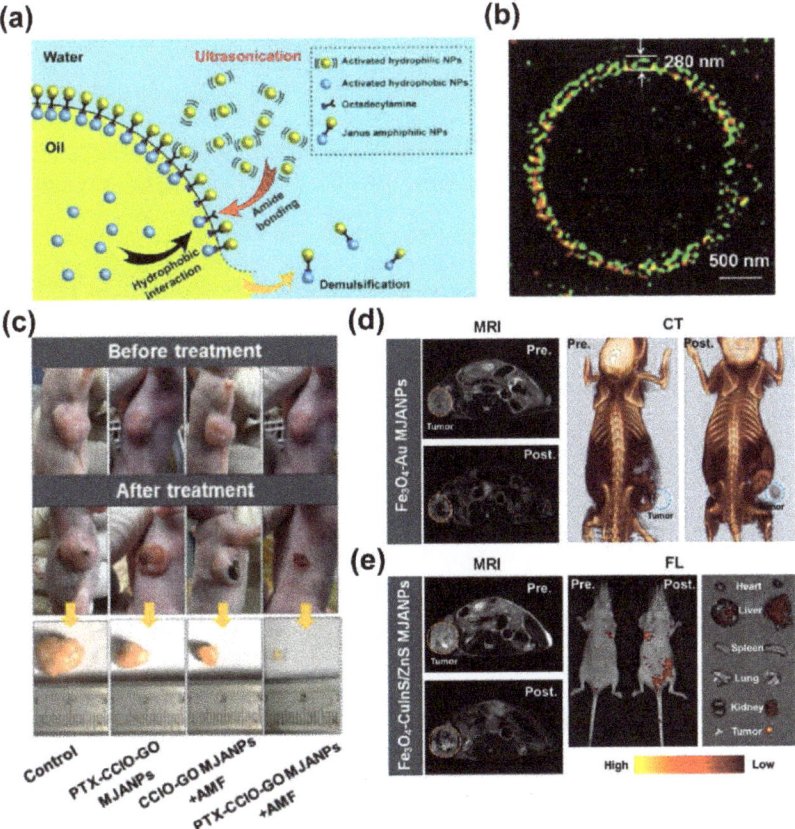

Figure 6. (a) Schematics of the ultrasonically triggered interfacial assembly process. (b) SIM image of the Pickering emulsion. Fluoresceinamine was used to label graphene oxide nanosheets (green), and rhodamine B isothiocyanate (RITC) was used to label Co cluster-embedded IONPs (red). (c) Photographs

of pre- and post-MHT treatment mice. (**d**) In vivo T2-weighted MRI and CT images of tumor-bearing mice pre- and post-injection of Au NPs−MIONs MJANPs. Arrowheads indicate the tumor. (**e**) In vivo T2-weighted MRI and fluorescence images of tumor-bearing nude mice pre- and postinjection of CuInS/ZnS−MIONs MJANPs. Arrowheads indicate the tumor. Reprint with permission from [16]. Copyright 2019 American Chemical Society.

4.5. Magnetic Nanoplatforms Carrier Based Drug Delivery

Chemotherapy is one of the main methods in current tumor therapy. There are currently about 80 chemotherapeutic drugs in clinical service, but these drugs can inevitably cause damage to normal tissues at therapeutic doses [118]. In order to improve the chemotherapy efficiency of drugs on tumors and to reduce their toxicity to normal tissues, it is necessary to develop an efficient drug delivery system. The magnetotheranostic nanoplatform has become a representative delivery system due to its adjustable size, easy image tracking, and clear metabolic pathway. The delivery mechanism of MIONs can be divided into passive targeting and active targeting. Passive targeting mainly depends on the enhanced permeability and retention (EPR) effect of MIONs. Although the mechanism still needs to be further investigated [119,120], the EPR effect can indeed enhance the enrichment of nanoparticles (not just MIONs) in the tumor area [5]. In addition, the ligands of tumor characteristic markers can be employed to functionalize MIONs in order to give them the ability to actively target tumors [6], and for more efficient drug delivery. In a recent work, paclitaxel (PTX) and cisplatin (CDDP) have been loaded into the carboxymethyl dextran coating of the clinical iron supplement FMX, and actively target gliomas through HMC, which was an organic anion transport polypeptide targeting agent with near–infrared fluorescence. This system was used for MRI/FI visualized drug delivery of glioblastoma multiforme (GBM) [121]. Liu et al. [122] designed a delivery system with a Yolk–shell structure. The vesicles were composed of PEG–PPS–SS–PEG and loaded with ultrasmall SPIONs and DOX, which were encapsulated together by a polyacrylic acid coating. In tumor cells, vesicles were disintegrated under the influence of GSH. The complex of ultrasmall SPIONs and DOX was then separated, so that the drug release process could be monitored by an enhancement in T1 MRI. The complex microenvironment of the tumor tissue could limit the penetration of the nanoplatforms [123,124]. In the work of Zhang et al., hyaluronidase was used to reduce the viscosity of the tumor ECM and to improve the tumor penetration of the magnetotheranostic nanoplatform [125]. In this platform, ultrasmall SPIONs were stabilized with layered double hydroxide (LDH), hyaluronic acid (HA) was modified on the outside of LDH and DOX was loaded inside the LDH. LDH–Fe_3O_4–HA/DOX could efficiently penetrate into the tumor pretreated with hyaluronidase and enter the lysosome in the tumor cell through the HA–CD44 pathway. The LDH sensed the pH decrease in the lysosome and released DOX to kill tumor cells. The entire delivery process was monitored by T1–weighted MRI. On the other hand, the magnetothermal effect of MIONs was designed to remotely control drug release [126]. Temperature–sensitive poly(lactic–co–glycolic acid) (PLGA) was used to coat SPIONs and DOX (Figure 7a). With the help of T2–weighted MRI (Figure 7b) to monitor the enrichment of the nanoplatforms in the tumor area, AMF was then applied to exert magnetothermal effect to the system. When the temperature rose above 42 °C, PLGA underwent a phase change, releasing DOX for chemotherapy (Figure 7c). Liu et al. combined tumor penetration and drug controlled release into one magnetotheranostics platform [127]. They modified temperature–sensitive hyperbranched PEI on ferrimagnetic vortex–domain iron oxide nanorings (FVIOs) with DOX loaded inside the PEI. The 0.1 kHz low–frequency magnetic field induced the magnetic force effect of FVIOs to effectively penetrate the magnetotheranostics nanoplatform deep into the tumor tissue. The cells then took up the positively charged magnetotheranostics platform. After entering the cells, the 360 kHz intermediate frequency magnetic field was turned on to raise the surface temperature of FVIOs, causing phase transition of the temperature–sensitive PEI. PEI shrank violently and then DOX was released. The sudden increase in intracellular DOX concentration during this release process was sufficient to effectively kill DOX–resistant MCF–7 breast cancer cells. Some smart micro–nano platforms with more complex structures are often called "micro/nano–robots" because their behaviors can be precisely manipulated under the external magnetic fields. Park et al. developed a degradable hyperthermia micro–robot (DHM) with a three–dimensional spiral structure [7], which contains MIONs and 5–fluorouracil (5–FU). The movement of DHM was controlled by a rotating magnetic field (RMF), and the AMF was then applied for magnetic hyperthermia. Upon AMF stimulation, 5–FU can be released in different modes for precise chemotherapy. MIONs can carry multiple types of drugs for collaborative treatment. In the work by Li et al. [128], PEI–PEG–coated MIONs were used to load gemcitabine and miRNA for pancreatic cancer treatment. They further installed the CD44v6 targeting molecule on the magnetic nanoplatform to improve the delivery efficiency, while the delivery process can be monitored using MRI.

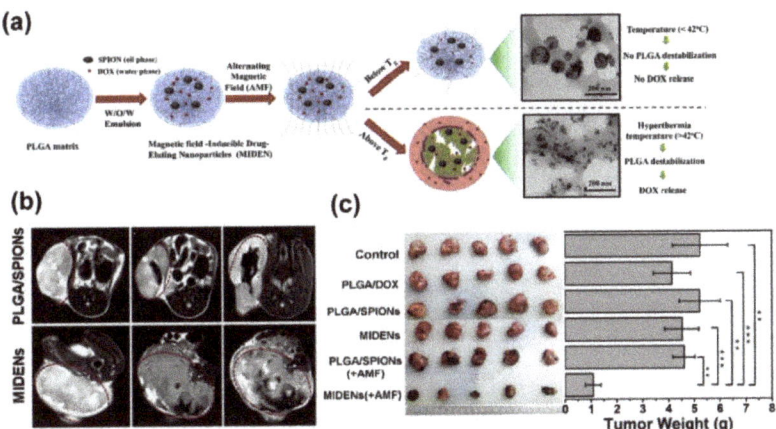

Figure 7. (a) Schematic illustration showing the formation of magnetic field inducible drug–eluting nanoparticles (MIDENs). (b) T2 MRI of CT26 xenografted mice administered intratumorally with PLGA/SPIONs and MIDENs (10 mg$_{[Fe]}$/kg body weight). Tumor regions are marked with a red circle. (c) Photographs of tumor tissues and weights of tumors collected 15 days post treatment (n = 5). The statistical significance was evaluated using Student's t–test. ** $p < 0.01$, *** $p < 0.001$. Reprint with permission from [126]. Copyright 2018 Elsevier Ltd., Amsterdam, The Netherlands.

5. Summary and Perspectives

In summary, we focused on reviewing the design of the magnetic nanoplatforms in the integration of tumor diagnosis and therapy in recent years. In the past, magnetotheranostics has mostly referred to the combination of MRI and other therapy technologies [95], which is a mixture of the MRI–enhancing properties of its magnetic core, MIONs, and materials with other properties, such as gold magnetic materials. With the in–depth study of the MIONs by researchers, the therapeutic function of MIONs (such as MHT and CDT) has gradually been integrated to realize magnetotheranostics. Some new imaging methods based on MIONs, such as MPI, MMOCT, and even magnetoacoustics ultrasound imaging [129], are also emerging and are expected to enrich future magnetotheranostics nanoplatforms.

MIONs as MRI contrast agents, magnetothermal agents, and iron supplements have been approved for clinical use. Although the magnetotheranostics nanoplatforms based on MIONs have potential for clinical implementation, there are still several issues to be solved. First of all, the current preparation cost of the magnetotheranostics nanoplatforms is relatively high, the stability of large–scale synthesis is questionable, and the quality control lacks evaluation standards. Secondly, the EPR effect of nanoparticles has not been effectively proved. Finally, in some complex magnetotheranostics nanoplatform systems, the potential toxicity of each component has not yet been resolved. As more researchers focus on elucidating these matters, we foresee that magnetotheranostics nanoplatforms will serve as an important new theranostics technology in future clinical practice.

Author Contributions: Conceptualization, W.J. and H.F.; investigation, W.J. and T.Z.; data curation, W.J.; writing—original draft preparation, W.J. and T.Z.; writing—review and editing, Y.H. and H.F.; supervision, M.P. and J.Y.; project administration, Y.H. and H.F.; funding acquisition, T.Z., M.P., J.Y., Y.H. and H.F. All authors have read and agreed to the published version of the manuscript.

Funding: This project was funded by National Natural Science Foundation of China, grant numbers 32001005, 31400663, 61874088, 81771981 and 22072115, and the Shaanxi Province Funds for Distinguished Young Scholars, grant number 202031900097.

Institutional Review Board Statement: Not applicable.

Informed Consent Statement: Not applicable.

Data Availability Statement: Data are contained within the article.

Acknowledgments: We sincerely thank the supports from Ning Gu. Youwei Du gave us a lot of help in the field of magnetic theory. We appreciate the guidance from Sishen Xie.

Conflicts of Interest: The authors declare no conflict of interest.

References

1. Chen, H.; Zhang, W.; Zhu, G.; Xie, J.; Chen, X. Rethinking cancer nanotheranostics. *Nat. Rev. Mater.* **2017**, *2*, 17024. [CrossRef] [PubMed]
2. McCormack, P.L. Ferumoxytol In Iron Deficiency Anaemia in Adults with Chronic Kidney Disease. *Drugs* **2012**, *72*, 2013–2022. [CrossRef]
3. Lartigue, L.; Alloyeau, D.; Kolosnjaj–Tabi, J.; Javed, Y.; Guardia, P.; Riedinger, A.; Péchoux, C.; Pellegrino, T.; Wilhelm, C.; Gazeau, F. Biodegradation of Iron Oxide Nanocubes: High–Resolution In Situ Monitoring. *ACS Nano* **2013**, *7*, 3939–3952. [CrossRef]
4. Tong, S.; Zhu, H.; Bao, G. Magnetic iron oxide nanoparticles for disease detection and therapy. *Mater. Today* **2019**, *31*, 86–99. [CrossRef]
5. Fang, J.; Islam, W.; Maeda, H. Exploiting the dynamics of the EPR effect and strategies to improve the therapeutic effects of nanomedicines by using EPR effect enhancers. *Adv. Drug Deliv. Rev.* **2020**, *157*, 142–160. [CrossRef]
6. Mitchell, M.J.; Billingsley, M.M.; Haley, R.M.; Wechsler, M.E.; Peppas, N.A.; Langer, R. Engineering precision nanoparticles for drug delivery. *Nat. Rev. Drug Discov.* **2020**, *20*, 101–124. [CrossRef]
7. Park, J.; Jin, C.; Lee, S.; Kim, J.; Choi, H. Magnetically Actuated Degradable Microrobots for Actively Controlled Drug Release and Hyperthermia Therapy. *Adv. Health Mater.* **2019**, *8*, e1900213. [CrossRef] [PubMed]
8. Cazares–Cortes, E.; Cabana, S.; Boitard, C.; Nehlig, E.; Griffete, N.; Fresnais, J.; Wilhelm, C.; Abou–Hassan, A.; Ménager, C. Recent insights in magnetic hyperthermia: From the "hot–spot" effect for local delivery to combined magneto–photo–thermia using magneto–plasmonic hybrids. *Adv. Drug Deliv. Rev.* **2018**, *138*, 233–246. [CrossRef] [PubMed]
9. Ni, D.; Bu, W.; Ehlerding, E.B.; Cai, W.; Shi, J. Engineering of inorganic nanoparticles as magnetic resonance imaging contrast agents. *Chem. Soc. Rev.* **2017**, *46*, 7438–7468. [CrossRef]
10. Dadfar, S.M.; Roemhild, K.; Drude, N.; von Stillfried, S.; Knüchel, R.; Kiessling, F.; Lammers, T. Iron oxide nanoparticles: Diagnostic, therapeutic and theranostic applications. *Adv. Drug Deliv. Rev.* **2019**, *138*, 302–325. [CrossRef]
11. Lu, C.; Han, L.; Wang, J.; Wan, J.; Song, G.; Rao, J. Engineering of magnetic nanoparticles as magnetic particle imaging tracers. *Chem. Soc. Rev.* **2021**, *50*, 8102–8146. [CrossRef] [PubMed]
12. John, R.; Rezaeipoor, R.; Adie, S.G.; Chaney, E.J.; Oldenburg, A.L.; Marjanovic, M.; Haldar, J.P.; Sutton, B.P.; Boppart, S.A. In vivo magnetomotive optical molecular imaging using targeted magnetic nanoprobes. *Proc. Natl. Acad. Sci. USA* **2010**, *107*, 8085–8090. [CrossRef] [PubMed]
13. Teijeiro–Valiño, C.; Gómez, M.A.G.; Yañez–Villar, S.; García–Acevedo, P.; Arnosa–Prieto, A.; Belderbos, S.; Gsell, W.; Himmelreich, U.; Piñeiro, Y.; Rivas, J. Biocompatible magnetic gelatin nanoparticles with enhanced MRI contrast performance prepared by single–step desolvation method. *Nano Express* **2021**, *2*, 020011. [CrossRef]
14. Zhou, Y.; Fan, S.; Feng, L.; Huang, X.; Chen, X. Manipulating Intratumoral Fenton Chemistry for Enhanced Chemodynamic and Chemodynamic-Synergized Multimodal Therapy. *Adv. Mater.* **2021**, *33*, 2104223. [CrossRef] [PubMed]
15. Li, C.; Cuichen, W.; Ocsoy, I.; Zhu, G.; Yasun, E.; You, M.; Wu, C.; Zheng, J.; Song, E.; Huang, C.Z.; et al. Gold–Coated Fe3O4Nanoroses with Five Unique Functions for Cancer Cell Targeting, Imaging, and Therapy. *Adv. Funct. Mater.* **2013**, *24*, 1772–1780. [CrossRef]
16. Liu, X.; Peng, M.L.; Li, G.; Miao, Y.Q.; Luo, H.; Jing, G.; He, Y.; Zhang, C.; Zhang, F.; Fan, H. Ultrasonication–Triggered Ubiquitous Assembly of Magnetic Janus Amphiphilic Nanoparticles in Cancer Theranostic Applications. *Nano Lett.* **2019**, *19*, 4118–4125. [CrossRef]

17. Song, J.; Lin, L.; Yang, Z.; Zhu, R.; Zhou, Z.; Li, Z.-W.; Wang, F.; Chen, J.; Yang, H.-H.; Chen, X. Self–Assembled Responsive Bilayered Vesicles with Adjustable Oxidative Stress for Enhanced Cancer Imaging and Therapy. *J. Am. Chem. Soc.* **2019**, *141*, 8158–8170. [CrossRef]
18. Tromsdorf, U.I.; Bruns, O.; Salmen, S.C.; Beisiegel, U.; Weller, H. A Highly Effective, Nontoxic T1 MR Contrast Agent Based on Ultrasmall PEGylated Iron Oxide Nanoparticles. *Nano Lett.* **2009**, *9*, 4434–4440. [CrossRef]
19. Harisinghani, M.G.; Barentsz, J.; Hahn, P.F.; Deserno, W.M.; Tabatabaei, S.; Van De Kaa, C.H.; De La Rosette, J.; Weissleder, R. Noninvasive Detection of Clinically Occult Lymph–Node Metastases in Prostate Cancer. *N. Engl. J. Med.* **2003**, *348*, 2491–2499. [CrossRef]
20. Tang, L.; Casas, J.; Venkataramasubramani, M. Magnetic Nanoparticle Mediated Enhancement of Localized Surface Plasmon Resonance for Ultrasensitive Bioanalytical Assay in Human Blood Plasma. *Anal. Chem.* **2013**, *85*, 1431–1439. [CrossRef]
21. Li, B.; Gong, T.; Xu, N.; Cui, F.; Yuan, B.; Yuan, Q.; Sun, H.; Wang, L.; Liu, J. Improved Stability and Photothermal Performance of Polydopamine-Modified Fe3O4 Nanocomposites for Highly Efficient Magnetic Resonance Imaging-Guided Photothermal Therapy. *Small* **2020**, *16*, e2003969. [CrossRef] [PubMed]
22. Wang, P.; Shi, Y.; Zhang, S.; Huang, X.; Zhang, J.; Zhang, Y.; Si, W.; Dong, X. Hydrogen Peroxide Responsive Iron–Based Nanoplatform for Multimodal Imaging–Guided Cancer Therapy. *Small* **2018**, *15*, e1803791. [CrossRef] [PubMed]
23. Shen, Z.; Chen, T.; Ma, X.; Ren, W.; Zhou, Z.; Zhu, G.; Zhang, A.; Liu, Y.; Song, J.; Li, Z.; et al. Multifunctional Theranostic Nanoparticles Based on Exceedingly Small Magnetic Iron Oxide Nanoparticles for T1–Weighted Magnetic Resonance Imaging and Chemotherapy. *ACS Nano* **2017**, *11*, 10992–11004. [CrossRef]
24. Sodipo, B.K.; Aziz, A.A. Recent advances in synthesis and surface modification of superparamagnetic iron oxide nanoparticles with silica. *J. Magn. Magn. Mater.* **2016**, *416*, 275–291. [CrossRef]
25. Wu, W.; Wu, Z.; Yu, T.; Jiang, C.; Kim, W.-S. Recent progress on magnetic iron oxide nanoparticles: Synthesis, surface functional strategies and biomedical applications. *Sci. Technol. Adv. Mater.* **2015**, *16*, 023501. [CrossRef]
26. Pereira, C.; Pereira, A.M.; Fernandes, C.; Rocha, M.; Mendes, R.; Fernández–García, M.P.; Guedes, A.; Tavares, P.B.; Grenèche, J.-M.; Araújo, J.P.; et al. Superparamagnetic MFe2O4 (M = Fe, Co, Mn) Nanoparticles: Tuning the Particle Size and Magnetic Properties through a Novel One–Step Coprecipitation Route. *Chem. Mater.* **2012**, *24*, 1496–1504. [CrossRef]
27. Massart, R. Preparation of aqueous magnetic liquids in alkaline and acidic media. *IEEE Trans. Magn.* **1981**, *17*, 1247–1248. [CrossRef]
28. Park, J.; An, K.; Hwang, Y.; Park, J.G.; Noh, H.J.; Kim, J.Y.; Park, J.H.; Hwang, N.M.; Hyeon, T. Ultra–large–scale syntheses of mono disperse nanocrystals. *Nat Mater* **2004**, *3*, 891–895. [CrossRef] [PubMed]
29. Deng, H.; Li, X.; Peng, Q.; Wang, X.; Chen, J.; Li, Y. Monodisperse magnetic single–crystal ferrite microspheres. *Angew. Chem.* **2005**, *117*, 2842–2845. [CrossRef]
30. Hu, P.; Yu, L.; Zuo, A.; Guo, C.; Yuan, F. Fabrication of Monodisperse Magnetite Hollow Spheres. *J. Phys. Chem. C* **2008**, *113*, 900–906. [CrossRef]
31. Niederberger, M. Nonaqueous Sol–Gel Routes to Metal Oxide Nanoparticles. *Accounts Chem. Res.* **2007**, *40*, 793–800. [CrossRef] [PubMed]
32. Lee, Y.; Lee, J.; Bae, C.J.; Park, J.-G.; Noh, H.-J.; Hyeon, T. Large–Scale Synthesis of Uniform and Crystalline Magnetite Nanoparticles Using Reverse Micelles as Nanoreactors under Reflux Conditions. *Adv. Funct. Mater.* **2005**, *15*, 503–509. [CrossRef]
33. Lee, J.-H.; Huh, Y.-M.; Jun, Y.-W.; Seo, J.-W.; Jang, J.-T.; Song, H.-T.; Kim, S.; Cho, E.-J.; Yoon, H.-G.Y.; Suh, J.-S.; et al. Artificially engineered magnetic nanoparticles for ultra–sensitive molecular imaging. *Nat. Med.* **2006**, *13*, 95–99. [CrossRef] [PubMed]
34. Jang, J.-T.; Nah, H.; Lee, J.-H.; Moon, S.H.; Kim, M.G.; Cheon, J. Critical Enhancements of MRI Contrast and Hyperthermic Effects by Dopant–Controlled Magnetic Nanoparticles. *Angew. Chem. Int. Ed.* **2009**, *48*, 1234–1238. [CrossRef]
35. Sun, S.; Zeng, H.; Robinson, D.B.; Raoux, S.; Rice, P.M.; Wang, S.X.; Li, G. Monodisperse MFe2O4 (M = Fe, Co, Mn) Nanoparticles. *J. Am. Chem. Soc.* **2004**, *126*, 273–279. [CrossRef]
36. Zhang, H.; Li, L.; Liu, X.L.; Jiao, J.; Ng, C.-T.; Yi, J.; E Luo, Y.; Bay, B.-H.; Zhao, L.Y.; Peng, M.L.; et al. Ultrasmall Ferrite Nanoparticles Synthesized via Dynamic Simultaneous Thermal Decomposition for High–Performance and Multifunctional T1 Magnetic Resonance Imaging Contrast Agent. *ACS Nano* **2017**, *11*, 3614–3631. [CrossRef]
37. Noh, S.-H.; Na, W.; Jang, J.-T.; Lee, J.-H.; Lee, E.J.; Moon, S.H.; Lim, Y.; Shin, J.-S.; Cheon, J. Nanoscale Magnetism Control via Surface and Exchange Anisotropy for Optimized Ferrimagnetic Hysteresis. *Nano Lett.* **2012**, *12*, 3716–3721. [CrossRef]
38. Yang, Y.; Liu, X.-L.; Yi, J.-B.; Yang, Y.; Fan, H.-M.; Ding, J. Stable vortex magnetite nanorings colloid: Micromagnetic simulation and experimental demonstration. *J. Appl. Phys.* **2012**, *111*, 044303. [CrossRef]
39. Ling, D.; Lee, N.; Hyeon, T. Chemical Synthesis and Assembly of Uniformly Sized Iron Oxide Nanoparticles for Medical Applications. *Accounts Chem. Res.* **2015**, *48*, 1276–1285. [CrossRef]
40. Xu, Z.; Shen, C.; Hou, Y.; Gao, H.; Sun, S. Oleylamine as Both Reducing Agent and Stabilizer in a Facile Synthesis of Magnetite Nanoparticles. *Chem. Mater.* **2009**, *21*, 1778–1780. [CrossRef]
41. Liu, X.L.; Fan, H.M. Innovative magnetic nanoparticle platform for magnetic resonance imaging and magnetic fluid hyperthermia applications. *Curr. Opin. Chem. Eng.* **2014**, *4*, 38–46. [CrossRef]
42. Liu, X.L.; Yang, Y.; Ng, C.T.; Zhao, L.Y.; Zhang, Y.; Bay, B.H.; Fan, H.M.; Ding, J. Magnetic Vortex Nanorings: A New Class of Hyperthermia Agent for Highly Efficient In Vivo Regression of Tumors. *Adv. Mater.* **2015**, *27*, 1939–1944. [CrossRef] [PubMed]

43. Zhou, Z.; Zhu, X.; Wu, D.; Chen, Q.; Huang, D.; Sun, C.; Xin, J.; Ni, K.; Gao, J. Anisotropic Shaped Iron Oxide Nanostructures: Controlled Synthesis and Proton Relaxation Shortening Effects. *Chem. Mater.* **2015**, *27*, 3505–3515. [CrossRef]
44. Jia, C.-J.; Sun, L.-D.; Luo, F.; Han, X.-D.; Heyderman, L.J.; Yan, Z.-G.; Yan, C.-H.; Zheng, K.; Zhang, Z.; Takano, M.; et al. Large–Scale Synthesis of Single–Crystalline Iron Oxide Magnetic Nanorings. *J. Am. Chem. Soc.* **2008**, *130*, 16968–16977. [CrossRef] [PubMed]
45. Cole, A.J.; David, A.E.; Wang, J.; Galbán, C.J.; Yang, V.C. Magnetic brain tumor targeting and biodistribution of long–circulating PEG–modified, cross–linked starch–coated iron oxide nanoparticles. *Biomaterials* **2011**, *32*, 6291–6301. [CrossRef]
46. Liu, G.; Xie, J.; Zhang, F.; Wang, Z.-Y.; Luo, K.; Zhu, L.; Quan, Q.-M.; Niu, G.; Lee, S.; Ai, H.; et al. N–Alkyl–PEI–functionalized iron oxide nanoclusters for efficient siRNA delivery. *Small* **2011**, *7*, 2742–2749. [CrossRef]
47. Kang, X.-J.; Dai, Y.-L.; Ma, P.-A.; Yang, D.-M.; Li, C.-X.; Hou, Z.-Y.; Cheng, Z.-Y.; Lin, J. Poly(acrylic acid)–Modified Fe_3O_4 Microspheres for Magnetic–Targeted and pH–Triggered Anticancer Drug Delivery. *Chem. Eur. J.* **2012**, *18*, 15676–15682. [CrossRef]
48. Riedinger, A.; Leal, M.P.; Deka, S.R.; George, C.; Franchini, I.R.; Falqui, A.; Cingolani, R.; Pellegrino, T. "Nanohybrids" Based on pH–Responsive Hydrogels and Inorganic Nanoparticles for Drug Delivery and Sensor Applications. *Nano Lett.* **2011**, *11*, 3136–3141. [CrossRef]
49. KC, R.B.; Lee, S.M.; Yoo, E.S.; Choi, J.H.; Ghim, H.D. Glycoconjugated chitosan stabilized iron oxide nanoparticles as a multifunctional nanoprobe. *Mater. Sci. Eng. C* **2009**, *29*, 1668–1673. [CrossRef]
50. Kim, J.; Arifin, D.R.; Muja, N.; Kim, T.; Gilad, A.A.; Kim, H.; Arepally, A.; Hyeon, T.; Bulte, J.W.M. Multifunctional Capsule–in–Capsules for Immunoprotection and Trimodal Imaging. *Angew. Chem. Int. Ed.* **2011**, *50*, 2317–2321. [CrossRef]
51. Nanotherm®. Available online: https://www.magforce.com/en/home/our_therapy/ (accessed on 13 December 2021).
52. Wadajkar, A.S.; Menon, J.U.; Tsai, Y.-S.; Gore, C.; Dobin, T.; Gandee, L.; Kangasniemi, K.; Takahashi, M.; Manandhar, B.; Ahn, J.-M.; et al. Prostate cancer–specific thermo–responsive polymer–coated iron oxide nanoparticles. *Biomaterials* **2013**, *34*, 3618–3625. [CrossRef]
53. Yang, M.; Cheng, K.; Qi, S.; Liu, H.; Jiang, Y.; Jiang, H.; Li, J.; Zhang, H.; Cheng, Z. Affibody modified and radiolabeled gold–Iron oxide hetero–nanostructures for tumor PET, optical and MR imaging. *Biomaterials* **2013**, *34*, 2796–2806. [CrossRef]
54. Tong, S.; Hou, S.; Zheng, Z.; Zhou, J.; Bao, G. Coating Optimization of Superparamagnetic Iron Oxide Nanoparticles for High T2 Relaxivity. *Nano Lett.* **2010**, *10*, 4607–4613. [CrossRef]
55. Liu, X.L.; Fan, H.M.; Yi, J.B.; Yang, Y.; Choo, E.S.G.; Xue, J.M.; Di Fan, D.; Ding, J. Optimization of surface coating on Fe_3O_4 nanoparticles for high performance magnetic hyperthermia agents. *J. Mater. Chem.* **2012**, *22*, 8235–8244. [CrossRef]
56. Zeng, J.; Jing, L.; Hou, Y.; Jiao, M.; Qiao, R.; Jia, Q.; Liu, C.; Fang, F.; Lei, H.; Gao, M. Anchoring Group Effects of Surface Ligands on Magnetic Properties of Fe_3O_4 Nanoparticles: Towards High Performance MRI Contrast Agents. *Adv. Mater.* **2014**, *26*, 2694–2698. [CrossRef]
57. Liu, X.; Yan, B.; Li, Y.; Ma, X.; Jiao, W.; Shi, K.; Zhang, T.; Chen, S.; He, Y.; Liang, X.-J.; et al. Graphene Oxide–Grafted Magnetic Nanorings Mediated Magnetothermodynamic Therapy Favoring Reactive Oxygen Species–Related Immune Response for Enhanced Antitumor Efficacy. *ACS Nano* **2020**, *14*, 1936–1950. [CrossRef] [PubMed]
58. Roy, D.; Park, J.W. Spatially nanoscale–controlled functional surfaces toward efficient bioactive platforms. *J. Mater. Chem. B* **2015**, *3*, 5135–5149. [CrossRef] [PubMed]
59. Zhang, Y.; Wang, Y.; Zhou, Q.; Chen, X.; Jiao, W.; Li, G.; Peng, M.; Liu, X.; He, Y.; Fan, H. Precise Regulation of Enzyme–Nanozyme Cascade Reaction Kinetics by Magnetic Actuation toward Efficient Tumor Therapy. *ACS Appl. Mater. Interfaces* **2021**, *13*, 52395–52405. [CrossRef] [PubMed]
60. Iqbal, P.; Rawson, F.J.; Ho, W.K.-W.; Lee, S.-F.; Leung, K.C.-F.; Wang, X.; Beri, A.; Preece, J.A.; Ma, J.; Mendes, P.M. Surface Molecular Tailoring Using pH–Switchable Supramolecular Dendron–Ligand Assemblies. *ACS Appl. Mater. Interfaces* **2014**, *6*, 6264–6274. [CrossRef] [PubMed]
61. Roy, D.; Kwon, S.H.; Kwak, J.W.; Park, J.W. "Seeing and counting" individual antigens captured on a microarrayed spot with Force–Based Atomic Force Microscopy. *Anal. Chem.* **2010**, *82*, 5189–5194. [CrossRef] [PubMed]
62. Lee, Y.; Kim, Y.; Lee, D.; Roy, D.; Park, J.W. Quantification of Fewer than Ten Copies of a DNA Biomarker without Amplification or Labeling. *J. Am. Chem. Soc.* **2016**, *138*, 7075–7081. [CrossRef] [PubMed]
63. Li, Y.; Liu, J.; Zhong, Y.; Zhang, D.; Wang, Z.; An, Y.-L.; Lin, M.; Gao, Z.; Zhang, J. Biocompatibility of Fe_3O_4@Au composite magnetic nanoparticles in vitro and in vivo. *Int. J. Nanomed.* **2011**, *6*, 2805–2819. [CrossRef] [PubMed]
64. Khan, M.I.; Mohammad, A.; Patil, G.; Naqvi, S.; Chauhan, L.; Ahmad, I. Induction of ROS, mitochondrial damage and autophagy in lung epithelial cancer cells by iron oxide nanoparticles. *Biomaterials* **2012**, *33*, 1477–1488. [CrossRef] [PubMed]
65. Huang, D.-M.; Hsiao, J.-K.; Chen, Y.-C.; Chien, L.-Y.; Yao, M.; Chen, Y.-K.; Ko, B.-S.; Hsu, S.-C.; Tai, L.-A.; Cheng, H.-Y.; et al. The promotion of human mesenchymal stem cell proliferation by superparamagnetic iron oxide nanoparticles. *Biomaterials* **2009**, *30*, 3645–3651. [CrossRef]
66. Pisanic, T.R.; Blackwell, J.D.; Shubayev, V.I.; Fiñones, R.R.; Jin, S. Nanotoxicity of iron oxide nanoparticle internalization in growing neurons. *Biomaterials* **2007**, *28*, 2572–2581. [CrossRef]
67. Wada, S.; Yue, L.; Tazawa, K.; Furuta, I.; Nagae, H.; Takemori, S.; Minamimura, T. New local hyperthermia using dextran magnetite complex (DM) for oral cavity: Experimental study in normal hamster tongue. *Oral Dis.* **2001**, *7*, 192–195. [CrossRef]

68. Motoyama, S.; Ishiyama, K.; Maruyama, K.; Narita, K.; Minamiya, Y.; Ogawa, J.-I. Estimating the Need for Neck Lymphadenectomy in Submucosal Esophageal Cancer Using Superparamagnetic Iron Oxide–Enhanced Magnetic Resonance Imaging: Clinical Validation Study. *World J. Surg.* **2011**, *36*, 83–89. [CrossRef]
69. Howarth, S.; Tang, T.; Trivedi, R.; Weerakkody, R.; U-King-Im, J.; Gaunt, M.; Boyle, J.; Li, Z.-Y.; Miller, S.; Graves, M.; et al. Utility of USPIO-enhanced MR imaging to identify inflammation and the fibrous cap: A comparison of symptomatic and asymptomatic individuals. *Eur. J. Radiol.* **2009**, *70*, 555–560. [CrossRef]
70. Bernd, H.; De Kerviler, E.; Gaillard, S.; Bonnemain, B. Safety and Tolerability of Ultrasmall Superparamagnetic Iron Oxide Contrast Agent. *Investig. Radiol.* **2009**, *44*, 336–342. [CrossRef] [PubMed]
71. Yildirimer, L.; Thanh, N.T.; Loizidou, M.; Seifalian, A.M. Toxicology and clinical potential of nanoparticles. *Nano Today* **2011**, *6*, 585–607. [CrossRef] [PubMed]
72. Brito, B.; Price, T.W.; Gallo, J.; Bañobre-López, M.; Stasiuk, G.J. Smart magnetic resonance imaging–based theranostics for cancer. *Theranostics* **2021**, *11*, 8706–8737. [CrossRef]
73. Wahsner, J.; Gale, E.M.; Rodríguez-Rodríguez, A.; Caravan, P. Chemistry of MRI Contrast Agents: Current Challenges and New Frontiers. *Chem. Rev.* **2018**, *119*, 957–1057. [CrossRef]
74. Wu, C.; Xu, Y.; Yang, L.; Wu, J.; Zhu, W.; Li, D.; Cheng, Z.; Xia, C.; Guo, Y.; Gong, Q.; et al. Negatively Charged Magnetite Nanoparticle Clusters as Efficient MRI Probes for Dendritic Cell Labeling and In Vivo Tracking. *Adv. Funct. Mater.* **2015**, *25*, 3581–3591. [CrossRef]
75. Karimian-Jazi, K.; Münch, P.; Alexander, A.; Fischer, M.; Pfleiderer, K.; Piechutta, M.; Karreman, M.A.; Solecki, G.M.; Berghoff, A.S.; Friedrich, M.; et al. Monitoring innate immune cell dynamics in the glioma microenvironment by magnetic resonance imaging and multiphoton microscopy (MR–MPM). *Theranostics* **2020**, *10*, 1873–1883. [CrossRef] [PubMed]
76. Marckmann, P.; Skov, L.; Rossen, K.; Dupont, A.; Damholt, M.B.; Heaf, J.G.; Thomsen, H.S. Nephrogenic systemic fibrosis: Suspected causative role of gadodiamide used for contrast-enhanced magnetic resonance imaging. *J. Am. Soc. Nephrol.* **2006**, *17*, 2359–2362. [CrossRef] [PubMed]
77. Sieber, M.A.; Lengsfeld, P.; Walter, J.; Schirmer, H.; Frenzel, T.; Siegmund, F.; Weinmann, H.-J.; Pietsch, H. Gadolinium-based contrast agents and their potential role in the pathogenesis of nephrogenic systemic fibrosis: The role of excess ligand. *J. Magn. Reson. Imaging* **2008**, *27*, 955–962. [CrossRef] [PubMed]
78. Kim, B.H.; Lee, N.; Kim, H.; An, K.; Park, Y.I.; Choi, Y.; Shin, K.; Lee, Y.; Kwon, S.G.; Bin Na, H.; et al. Large-Scale Synthesis of Uniform and Extremely Small-Sized Iron Oxide Nanoparticles for High-Resolution T1 Magnetic Resonance Imaging Contrast Agents. *J. Am. Chem. Soc.* **2011**, *133*, 12624–12631. [CrossRef]
79. Wei, H.; Bruns, O.T.; Kaul, M.G.; Hansen, E.C.; Barch, M.; Wisniowska, A.E.; Chen, O.; Chen, Y.; Li, N.; Okada, S.; et al. Exceedingly small iron oxide nanoparticles as positive MRI contrast agents. *Proc. Natl. Acad. Sci. USA* **2017**, *114*, 2325–2330. [CrossRef]
80. Miao, Y.; Zhang, H.; Cai, J.; Chen, Y.; Ma, H.; Zhang, S.; Yi, J.B.; Liu, X.; Bay, B.-H.; Guo, Y.; et al. Structure–Relaxivity Mechanism of an Ultrasmall Ferrite Nanoparticle T1 MR Contrast Agent: The Impact of Dopants Controlled Crystalline Core and Surface Disordered Shell. *Nano Lett.* **2021**, *21*, 1115–1123. [CrossRef] [PubMed]
81. Liang, H.; Guo, J.; Shi, Y.; Zhao, G.; Sun, S.; Sun, X. Porous yolk–shell Fe/Fe$_3$O$_4$ nanoparticles with controlled exposure of highly active Fe(0) for cancer therapy. *Biomaterials* **2020**, *268*, 120530. [CrossRef]
82. Chandrasekharan, P.; Tay, Z.W.; Hensley, D.; Zhou, X.Y.; Fung, B.K.; Colson, C.; Lu, Y.; Fellows, B.D.; Huynh, Q.; Saayujya, C.; et al. Using magnetic particle imaging systems to localize and guide magnetic hyperthermia treatment: Tracers, hardware, and future medical applications. *Theranostics* **2020**, *10*, 2965–2981. [CrossRef] [PubMed]
83. Gloag, L.; Mehdipour, M.; Ulanova, M.; Mariandry, K.; Nichol, M.A.; Hernández-Castillo, D.J.; Gaudet, J.; Qiao, R.; Zhang, J.; Nelson, M.; et al. Zero valent iron core–iron oxide shell nanoparticles as small magnetic particle imaging tracers. *Chem. Commun.* **2020**, *56*, 3504–3507. [CrossRef] [PubMed]
84. Song, G.; Chen, M.; Zhang, Y.; Cui, L.; Qu, H.; Zheng, X.; Wintermark, M.; Liu, Z.; Rao, J. Janus Iron Oxides @ Semiconducting Polymer Nanoparticle Tracer for Cell Tracking by Magnetic Particle Imaging. *Nano Lett.* **2017**, *18*, 182–189. [CrossRef]
85. Oldenburg, A.L.; Crecea, V.; Rinne, S.A.; Boppart, S.A. Phase-resolved magnetomotive OCT for imaging nanomolar concentrations of magnetic nanoparticles in tissues. *Opt. Express* **2008**, *16*, 11525–11539. [CrossRef] [PubMed]
86. Huang, P.-C.; Chaney, E.J.; Aksamitiene, E.; Barkalifa, R.; Spillman, D.R.; Bogan, B.J.; Boppart, S.A. Biomechanical sensing of in vivo magnetic nanoparticle hyperthermia–treated melanoma using magnetomotive optical coherence elastography. *Theranostics* **2021**, *11*, 5620–5633. [CrossRef]
87. Xu, C.; Zheng, Y.; Gao, W.; Xu, J.; Zuo, G.; Chen, Y.; Zhao, M.; Li, J.; Song, J.; Zhang, N.; et al. Magnetic Hyperthermia Ablation of Tumors Using Injectable Fe3O4/Calcium Phosphate Cement. *ACS Appl. Mater. Interfaces* **2015**, *7*, 13866–13875. [CrossRef]
88. Yin, P.; Shah, S.; Pasquale, N.J.; Garbuzenko, O.B.; Minko, T.; Lee, K. –B. Stem cell–based gene therapy activated using magnetic hyperthermia to enhance the treatment of cancer. *Biomaterials* **2015**, *81*, 46–57. [CrossRef]
89. Moise, S.; Byrne, J.M.; El Haj, A.J.; Telling, N.D. The potential of magnetic hyperthermia for triggering the differentiation of cancer cells. *Nanoscale* **2018**, *10*, 20519–20525. [CrossRef]
90. Liu, X.L.; Ng, C.T.; Chandrasekharan, P.; Yang, H.T.; Zhao, L.Y.; Peng, E.; Lv, Y.B.; Xiao, W.; Fang, J.; Yi, J.; et al. Synthesis of Ferromagnetic Fe0.6Mn0.4O Nanoflowers as a New Class of Magnetic Theranostic Platform for In Vivo T1–T2Dual–Mode Magnetic Resonance Imaging and Magnetic Hyperthermia Therapy. *Adv. Health Mater.* **2016**, *5*, 2092–2104. [CrossRef]

91. Yang, M.; Ho, C.; Ruta, S.; Chantrell, R.; Krycka, K.; Hovorka, O.; Chen, F.-R.; Lai, P.; Lai, C. Magnetic Interaction of Multifunctional Core–Shell Nanoparticles for Highly Effective Theranostics. *Adv. Mater.* **2018**, *30*, e1802444. [CrossRef]
92. Zhang, Y.; Wang, X.; Chu, C.; Zhou, Z.; Chen, B.; Pang, X.; Lin, G.; Lin, H.; Guo, Y.; Ren, E.; et al. Genetically engineered magnetic nanocages for cancer magneto–catalytic theranostics. *Nat. Commun.* **2020**, *11*, 5421. [CrossRef]
93. Zhou, Z.; Song, J.; Tian, R.; Yang, Z.; Yu, G.; Lin, L.; Zhang, G.; Fan, W.; Zhang, F.; Niu, G.; et al. Activatable Singlet Oxygen Generation from Lipid Hydroperoxide Nanoparticles for Cancer Therapy. *Angew. Chem. Int. Ed.* **2017**, *56*, 6492–6496. [CrossRef]
94. Du, J.; Bao, J.; Fu, X.; Lu, C.; Kim, S.H. Mesoporous sulfur–modified iron oxide as an effective Fenton–like catalyst for degradation of bisphenol A. *Appl. Catal. B Environ.* **2016**, *184*, 132–141. [CrossRef]
95. Zhao, S.; Yu, X.; Qian, Y.; Chen, W.; Shen, J. Multifunctional magnetic iron oxide nanoparticles: An advanced platform for cancer theranostics. *Theranostics* **2020**, *10*, 6278–6309. [CrossRef] [PubMed]
96. Huo, M.; Wang, L.; Chen, Y.; Shi, J. Tumor–selective catalytic nanomedicine by nanocatalyst delivery. *Nat. Commun.* **2017**, *8*, 357. [CrossRef] [PubMed]
97. Du, Y.; Yang, C.; Li, F.; Liao, H.; Chen, Z.; Lin, P.; Wang, N.; Zhou, Y.; Lee, J.Y.; Ding, Q.; et al. Core–Shell–Satellite Nanomaces as Remotely Controlled Self-Fueling Fenton Reagents for Imaging-Guided Triple-Negative Breast Cancer-Specific Therapy. *Small* **2020**, *16*, e2002537. [CrossRef]
98. Deng, L.; Liu, M.; Sheng, D.; Luo, Y.; Wang, D.; Yu, X.; Wang, Z.; Ran, H.; Li, P. Low–intensity focused ultrasound–augmented Cascade chemodynamic therapy via boosting ROS generation. *Biomaterials* **2021**, *271*, 120710. [CrossRef]
99. He, Y.; Chen, X.; Zhang, Y.; Wang, Y.; Cui, M.; Li, G.; Liu, X.; Fan, H. Magnetoresponsive nanozyme: Magnetic stimulation on the nanozyme activity of iron oxide nanoparticles. *Sci. China Life Sci.* **2022**, *65*, 184–192. [CrossRef]
100. Zhou, H.; Tang, J.; Li, J.; Li, W.; Liu, Y.; Chen, C. In vivo aggregation–induced transition between T1 and T2 relaxations of magnetic ultra–small iron oxide nanoparticles in tumor microenvironment. *Nanoscale* **2017**, *9*, 3040–3050. [CrossRef]
101. Zanganeh, S.; Hutter, G.; Spitler, R.; Lenkov, O.; Mahmoudi, M.; Shaw, A.; Pajarinen, J.S.; Nejadnik, H.; Goodman, S.; Moseley, M.; et al. Iron oxide nanoparticles inhibit tumour growth by inducing pro–inflammatory macrophage polarization in tumour tissues. *Nat. Nanotechnol.* **2016**, *11*, 986–994. [CrossRef] [PubMed]
102. Ruiz-De-Angulo, A.; Bilbao-Asensio, M.; Cronin, J.; Evans, S.J.; Clift, M.J.; Llop, J.; Feiner, I.V.; Beadman, R.; Bascarán, K.Z.; Mareque-Rivas, J.C. Chemically Programmed Vaccines: Iron Catalysis in Nanoparticles Enhances Combination Immunotherapy and Immunotherapy–Promoted Tumor Ferroptosis. *iScience* **2020**, *23*, 101499. [CrossRef]
103. Xu, C.; Pu, K. Second near–infrared photothermal materials for combinational nanotheranostics. *Chem. Soc. Rev.* **2020**, *50*, 1111–1137. [CrossRef]
104. Liu, Y.; Yang, Z.; Huang, X.; Yu, G.; Wang, S.; Zhou, Z.; Shen, Z.; Fan, W.; Liu, Y.; Davisson, M.; et al. Glutathione–Responsive Self–Assembled Magnetic Gold Nanowreath for Enhanced Tumor Imaging and Imaging–Guided Photothermal Therapy. *ACS Nano* **2018**, *12*, 8129–8137. [CrossRef]
105. Fu, Q.; Li, Z.; Ye, J.; Li, Z.; Fu, F.; Lin, S.-L.; Chang, C.A.; Yang, H.; Song, J. Magnetic targeted near–infrared II PA/MR imaging guided photothermal therapy to trigger cancer immunotherapy. *Theranostics* **2020**, *10*, 4997–5010. [CrossRef]
106. Yang, K.; Yang, G.; Chen, L.; Cheng, L.; Wang, L.; Ge, C.; Liu, Z. FeS nanoplates as a multifunctional nano–theranostic for magnetic resonance imaging guided photothermal therapy. *Biomaterials* **2014**, *38*, 1–9. [CrossRef]
107. Ni, D.; Ferreira, C.A.; Barnhart, T.E.; Quach, V.; Yu, B.; Jiang, D.; Wei, W.; Liu, H.; Engle, J.W.; Hu, P.; et al. Magnetic Targeting of Nanotheranostics Enhances Cerenkov Radiation–Induced Photodynamic Therapy. *J. Am. Chem. Soc.* **2018**, *140*, 14971–14979. [CrossRef]
108. Yan, L.; Amirshaghaghi, A.; Huang, D.; Miller, J.; Stein, J.M.; Busch, T.M.; Cheng, Z.; Tsourkas, A. Protoporphyrin IX (PpIX)–Coated Superparamagnetic Iron Oxide Nanoparticle (SPION) Nanoclusters for Magnetic Resonance Imaging and Photodynamic Therapy. *Adv. Funct. Mater.* **2018**, *28*, 1707030. [CrossRef]
109. Zhou, H.; Guo, M.; Li, J.; Qin, F.; Wang, Y.; Liu, T.; Liu, J.; Sabet, Z.F.; Wang, Y.; Liu, Y.; et al. Hypoxia–Triggered Self–Assembly of Ultrasmall Iron Oxide Nanoparticles to Amplify the Imaging Signal of a Tumor. *J. Am. Chem. Soc.* **2021**, *143*, 1846–1853. [CrossRef] [PubMed]
110. Ma, Y.; Yan, C.; Guo, Z.; Tan, G.; Niu, D.; Li, Y.; Zhu, W. Spatio-Temporally Reporting Dose-Dependent Chemotherapy via Uniting Dual-Modal MRI/NIR Imaging. *Angew. Chem. Int. Ed.* **2020**, *59*, 21143–21150. [CrossRef] [PubMed]
111. Ma, J.; Li, P.; Wang, W.; Wang, S.; Pan, X.; Zhang, F.; Li, S.; Liu, S.; Wang, H.; Gao, G.; et al. Biodegradable Poly(amino acid)–Gold–Magnetic Complex with Efficient Endocytosis for Multimodal Imaging-Guided Chemo-photothermal Therapy. *ACS Nano* **2018**, *12*, 9022–9032. [CrossRef] [PubMed]
112. Feng, L.; Gai, S.; He, F.; Yang, P.; Zhao, Y. Multifunctional Bismuth Ferrite Nanocatalysts with Optical and Magnetic Functions for Ultrasound–Enhanced Tumor Theranostics. *ACS Nano* **2020**, *14*, 7245–7258. [CrossRef]
113. Liu, Z.; Lin, H.; Zhao, M.; Dai, C.; Zhang, S.; Peng, W.; Chen, Y. 2D Superparamagnetic Tantalum Carbide Composite MXenes for Efficient Breast–Cancer Theranostics. *Theranostics* **2018**, *8*, 1648–1664. [CrossRef] [PubMed]
114. Zhang, J.; Ma, Y.; Yang, W.; Xue, J.; Ding, Y.; Xie, C.; Luo, W.; Gao, F.-P.; Zhang, Z.; Zhao, Y.; et al. Comparative study of core– and surface–radiolabeling strategies for the assembly of iron oxide nanoparticle–based theranostic nanocomposites. *Nanoscale* **2019**, *11*, 5909–5913. [CrossRef] [PubMed]

115. Thorek, D.L.J.; Ulmert, D.; Diop, N.-F.M.; Lupu, M.E.; Doran, M.G.; Huang, R.; Abou, D.S.; Larson, S.; Grimm, J. Non–invasive mapping of deep–tissue lymph nodes in live animals using a multimodal PET/MRI nanoparticle. *Nat. Commun.* **2014**, *5*, 3097. [CrossRef] [PubMed]
116. de Rosales, R.T.M.; Tavaré, R.; Glaria, A.; Varma, G.; Protti, A.; Blower, P.J. 99mTc–Bisphosphonate–Iron Oxide Nanoparticle Conjugates for Dual–Modality Biomedical Imaging. *Bioconj. Chem.* **2011**, *22*, 455–465. [CrossRef]
117. Stelter, L.; Pinkernelle, J.G.; Michel, R.; Schwartländer, R.; Raschzok, N.; Morgul, M.H.; Koch, M.; Denecke, T.; Ruf, J.; Bäumler, H.; et al. Modification of Aminosilanized Superparamagnetic Nanoparticles: Feasibility of Multimodal Detection Using 3T MRI, Small Animal PET, and Fluorescence Imaging. *Mol. Imaging Biol.* **2009**, *12*, 25–34. [CrossRef] [PubMed]
118. McNerney, M.E.; Godley, L.A.; Le Beau, M.M. Therapy–related myeloid neoplasms: When genetics and environment collide. *Nat. Cancer* **2017**, *17*, 513–527. [CrossRef]
119. Shi, Y.; Van Der Meel, R.; Chen, X.; Lammers, T. The EPR effect and beyond: Strategies to improve tumor targeting and cancer nanomedicine treatment efficacy. *Theranostics* **2020**, *10*, 7921–7924. [CrossRef]
120. Wilhelm, S.; Tavares, A.J.; Dai, Q.; Ohta, S.; Audet, J.; Dvorak, H.F.; Chan, W.C.W. Analysis of nanoparticle delivery to tumours. *Nat. Rev. Mater.* **2016**, *1*, 16014. [CrossRef]
121. Reichel, D.; Sagong, B.; Teh, J.; Zhang, Y.; Wagner, S.; Wang, H.; Chung, L.W.K.; Butte, P.; Black, K.L.; Yu, J.S.; et al. Near Infrared Fluorescent Nanoplatform for Targeted Intraoperative Resection and Chemotherapeutic Treatment of Glioblastoma. *ACS Nano* **2020**, *14*, 8392–8408. [CrossRef]
122. Liu, D.; Zhou, Z.; Wang, X.; Deng, H.; Sun, L.; Lin, H.; Kang, F.; Zhang, Y.; Wang, Z.; Yang, W.; et al. Yolk–shell nanovesicles endow glutathione–responsive concurrent drug release and T1 MRI activation for cancer theranostics. *Biomaterials* **2020**, *244*, 119979. [CrossRef]
123. Zhou, Q.; Shao, S.; Wang, J.; Xu, C.; Xiang, J.; Piao, Y.; Zhou, Z.; Yu, Q.; Tang, J.; Liu, X.; et al. Enzyme–activatable polymer–drug conjugate augments tumour penetration and treatment efficacy. *Nat. Nanotechnol.* **2019**, *14*, 799–809. [CrossRef]
124. Dai, Q.; Wilhelm, S.; Ding, D.; Syed, A.; Sindhwani, S.; Zhang, Y.; Chen, Y.Y.; MacMillan, P.; Chan, W.C.W. Quantifying the Ligand–Coated Nanoparticle Delivery to Cancer Cells in Solid Tumors. *ACS Nano* **2018**, *12*, 8423–8435. [CrossRef] [PubMed]
125. Zhang, N.; Wang, Y.; Zhang, C.; Fan, Y.; Li, D.; Cao, X.; Xia, J.; Shi, X.; Guo, R. LDH–stabilized ultrasmall iron oxide nanoparticles as a platform for hyaluronidase–promoted MR imaging and chemotherapy of tumors. *Theranostics* **2020**, *10*, 2791–2802. [CrossRef]
126. Thirunavukkarasu, G.K.; Cherukula, K.; Lee, H.; Jeong, Y.Y.; Park, I.-K.; Lee, J.Y. Magnetic field–inducible drug–eluting nanoparticles for image–guided thermo–chemotherapy. *Biomaterials* **2018**, *180*, 240–252. [CrossRef] [PubMed]
127. Liu, X.; Zhang, Y.; Guo, Y.; Jiao, W.; Gao, X.; Lee, W.S.V.; Wang, Y.; Deng, X.; He, Y.; Jiao, J.; et al. Electromagnetic Field-Programmed Magnetic Vortex Nanodelivery System for Efficacious Cancer Therapy. *Adv. Sci.* **2021**, *8*, 2100950. [CrossRef] [PubMed]
128. Li, Y.; Li, Y.; Chen, Y.; Chen, Y.; Li, J.; Li, J.; Zhang, Z.; Zhang, Z.; Huang, C.; Huang, C.; et al. Co–delivery of microRNA–21 antisense oligonucleotides and gemcitabine using nanomedicine for pancreatic cancer therapy. *Cancer Sci.* **2017**, *108*, 1493–1503. [CrossRef]
129. Li, X.; Yu, K.; He, B. Magnetoacoustic tomography with magnetic induction (MAT–MI) for imaging electrical conductivity of biological tissue: A tutorial review. *Phys. Med. Biol.* **2016**, *61*, R249–R270. [CrossRef]

Review

Activatable Second Near-Infrared Fluorescent Probes: A New Accurate Diagnosis Strategy for Diseases

Dong Li *,†, Jie Pan †, Shuyu Xu, Shiying Fu, Chengchao Chu and Gang Liu *

State Key Laboratory of Molecular Vaccinology and Molecular Diagnostics, Center for Molecular Imaging, Translational Medicine School of Public Health, Xiamen University, Xiamen 361102, China; 32620201150737@stu.xmu.edu.cn (J.P.); xushuyu@stu.xmu.edu.cn (S.X.); 21620191152631@stu.xmu.edu.cn (S.F.); chuchengchao@xmu.edu.cn (C.C.)
* Correspondence: lidong@xmu.edu.cn (D.L.); gangliu.cmitm@xmu.edu.cn (G.L.)
† These authors contributed equally to this work.

Abstract: Recently, second near-infrared (NIR-II) fluorescent imaging has been widely applied in biomedical diagnosis, due to its high spatiotemporal resolution and deep tissue penetration. In contrast to the "always on" NIR-II fluorescent probes, the activatable NIR-II fluorescent probes have specific targeting to biological tissues, showing a higher imaging signal-to-background ratio and a lower detection limit. Therefore, it is of great significance to utilize disease-associated endogenous stimuli (such as pH values, enzyme existence, hypoxia condition and so on) to activate the NIR-II probes and achieve switchable fluorescent signals for specific deep bioimaging. This review introduces recent strategies and mechanisms for activatable NIR-II fluorescent probes and their applications in biosensing and bioimaging. Moreover, the potential challenges and perspectives of activatable NIR-II fluorescent probes are also discussed.

Keywords: NIR-II fluorescent probes; activatable strategy; NIR-II fluorescence imaging; biomarker; biosensing

1. Introduction

Fluorescence imaging techniques are widely used in the fields of disease detection and diagnosis, surgical navigation, and drug delivery, due to their high sensitivity and noninvasiveness, and the absence of ionizing radiation [1–3]. According to different emission wavelengths, fluorescence imaging can be divided into three regions: the visible region (400–700 nm), the first near-infrared region (NIR-I, 700–900 nm), and the second near-infrared region (NIR-II, 1000–1700 nm) [4–6]. In comparison with NIR-I fluorescence imaging, NIR-II fluorescence imaging has the characteristics of less scattering, minimal tissue absorption, and low autofluorescence, which can produce deeper tissue penetration and a higher signal-to-background ratio (SBR) [7–9]. To date, varieties of NIR-II fluorescent probes, including benzobisthiadiazole (BBTD) dyes, semiconducting polymer nanoparticles (SPNPs), cyanine dyes, quantum dots (QDs), and single-wall carbon nanotubes (SWCNTs) have been developed for applications in NIR-II fluorescence imaging [10–12].

Recently, accurate in vivo diagnosis based on specific biomarkers produced by diseases has attracted extensive attention [13,14]. Specifically, the pathological microenvironment of diseased tissue is obviously different from that of normal tissue, and the occurrence and development of many diseases produce specific biomarkers [15–17]. For example, the pH value and the reactive oxygen species (ROS) concentration are abnormal in the tumor microenvironment [18–20], hepatotoxicity induced by drugs has an abnormal ONOO⁻ concentration [21], and liver injury induced by diabetes leads to viscosity changes [22]. Encouragingly, the specific biomarkers generated by the disease can be exploited to activate NIR-II fluorescence probes, realizing the accurate diagnosis of the underlying disease [23]. Compared with the "always on" types of NIR-II fluorescent probes, activatable NIR-II fluorescent probes can be specifically recognized by the target tissue and exhibit switchable fluorescence emission [13,14]. In fact, activatable NIR-II fluorescent probes cannot produce fluorescence, or produce only weak fluorescence; however, the special structure reacts with certain molecules existing in the microenvironment of the diseased tissue, thereby producing effective NIR-II fluorescence emission [24–26]. Consequently, the "off" state in normal tissues and the "on" state in the diseased tissues can significantly improve the sensitivity and resolution of the fluorescence imaging [27–29]. Therefore, it is of great significance to utilize disease-associated endogenous stimuli to specifically activate NIR-II fluorescent probes for highly sensitive and high-resolution bioimaging.

Activatable NIR-II fluorescent probes (e.g., small organic molecular fluorophores, inorganic nanoparticles, and semiconducting polymer nanoparticles) for specific fluorescence imaging have made rapid progress. Here, recent strategies for activatable NIR-II fluorescent probes and their applications in biosensing and bioimaging are summarized (Scheme 1). In addition, according to the characteristics of highly expressed biomarkers in diseases, the activation mechanism is comprehensively analyzed and discussed. Finally, the prospects are also analyzed.

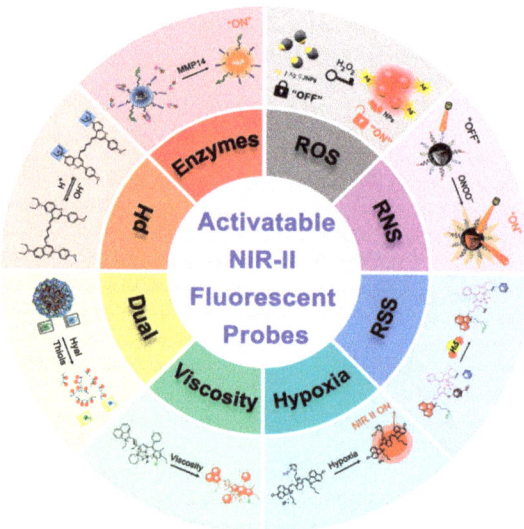

ROS: Reactive Oxygen Species RNS: Reactive Nitrogen Species
RSS: Reactive Sulfur Species

Scheme 1. Schematic illustration of strategies and mechanisms for activatable NIR-II fluorescent probes. Image for pH: reproduced from publication of Wang, S. et al. with permission, Copyright 2019, Nature Publishing Group. Image for enzymes: reproduced from publication of Zhan, Y. et al. with permission, Copyright 2021, WILEY-VCH. Image for ROS: reproduced from publication of Zhang, X. et al. with permission, Copyright 2021, American Chemical Society. Image for RNS: reproduced from publication of Yang, X. et al. with permission, Copyright 2021, WILEY-VCH. Image for RSS: reproduced from publication of Dou, K. et al. with permission, Copyright 2021, American Chemical Society. Image for hypoxia: reproduced from publication of Meng, X. et al. with permission, Copyright 2018, Creative Commons. Image for viscosity: reproduced from publication of Dou, K. et al. with permission, Copyright 2021, American Chemical Society. Image for dual: reproduced from publication of Tang, Y. et al. with permission, Copyright 2018, WILEY-VCH.

2. NIR-II Fluorescent Probes

2.1. Organic NIR-II Fluorescent Probes

Organic NIR-II fluorescent probes are applied in biological imaging for disease diagnosis, and mainly include BBTD dyes, aggregation-induced emission luminogens (AIEgens), SPNPs, cyanine dyes, rhodamine analogs, and boron dipyrromethenes (BOD-IPYs), with representative structures CH1055, TQ-BPN, poly (benzo [1,2-b:3,4-b'] difuran-alt-fluorothieno-[3,4-b] thiophene) (pDA), FD-1080, RhIndz, and NJ1060, respectively (Figure 1) [30,31]. Specifically, BBTD derivatives exhibit donor–acceptor–donor (D–A–D) characteristics with large Stokes shifts and high imaging quality [32]. In 2016, Antaris et al. reported the BBTD core structure CH1055 with fluorescence emission at 1055 nm, which outperformed the clinically used indocyanine green (ICG) for sentinel lymphatic imaging in the vicinity of mouse tumors [2]. In addition, a high enrichment of PEGylated CH1055 dye was observed in the deep tissues of mouse brain tumors at approximately 4 mm. Moreover, the tumor-to-normal tissue ratio of NIR-II imaging mediated by the CH1055 dye was 5 times higher than that of NIR-I imaging, enabling it to be used for in vivo imaging to guide tumor resection. In 2019, Zhou et al. developed two NIR-II fluorescent probes, CH1055-PEG-PT and CH1055-PEG-Affibody, which showed great potential in the fluorescence imaging of osteosarcoma and lung metastasis, respectively. In addition, CH1055-PEG-PT surpassed the imaging capability of computed tomography for a 143B tumor in vivo, and could therefore be used to guide surgical resection of 143B tumors. CH1055-PEG-Affibody could be used to visualize osteosarcoma and lung metastasis [33]. In contrast to the traditional supramolecular dyes, AIE molecules overcame the aggregation-caused quenching (ACQ) induced by intermolecular π–π stacking. Qi et al. reported a crab-shaped AIEgen, TQ-BPN, with fluorescence emission at 700–1200 nm for high-resolution microangiography and imaging in the NIR-II window [34]. The excellent imaging performance of TQ-BPN ensured visualization of the anatomy of high-depth brain capillaries (800 µm) with high spatial resolution (~3 µm). Therefore, TQ-BPN could be used to dynamically evaluate vascular diseases for the diagnosis of blood–brain barrier damage in the brain. Samanta et al. developed a highly bright, highly water-soluble aggregation-induced emission (AIE)-active two-photon (TP) (AIETP) NIR-II probe, which could form hydrophilic nanoparticles, AIETP NPs, with the polymer Pluronic F127. Encouragingly, AIETP NPs not only exhibited excellent cell permeability and biocompatibility but also exhibited good two-photon imaging properties in vivo. Moreover, due to the superior penetration depth (800 µm) and excellent spatial resolution (1.92 µm), AIETP NPs were applied to deep brain imaging in vivo [35]. SPNPs are organic semiconducting macromolecules whose backbone consists of alternating single and double bonds. Among these, pDA is a typical SPNP, with an emission wavelength of 1050 nm and a Stokes shift of approximately 400 nm, which can be used for vascular imaging in vivo [5]. Importantly, pDA has a frame rate of >25 frames per second, which can be applied to deep tissue and ultrafast imaging of mouse arterial blood flow. Cyanine dyes are based on a polymethylene skeleton and contain a unique extended conjugated system. Specifically, FD-1080 is a polymethine cyanine dye with an emission wavelength of 1080 nm, which can achieve noninvasive high-resolution angiography of deep-tissue brain and hindlimb vessels [36]. In addition to the abovementioned typical dyes, traditional NIR-I dyes modified with specific groups can achieve NIR-II fluorescence emission. RhIndz and NJ1060 are derivatives of rhodamine and BODIPY, respectively [37,38]. With the modification of special functional groups, both RhIndz and NJ1060 achieved NIR-II fluorescence emission. It is worth noting that organic NIR-II fluorescent probes show great potential in the application of surgical navigation. Zeng et al. designed a small-molecule fluorescent probe, H3-PEG2k, with excellent aqueous solubility, high brightness, and high photostability [39]. When H3-PEG2k was intravenously injected into rats with mammary carcinoma, a strong fluorescence signal could be observed within 8 h after injection. In addition, the tumor and the surrounding normal tissues could be clearly distinguished 8.5 h after injection. Encouragingly, H3-PEG2k was applied to rat mammary carcinoma imaging as well as image-guided tumor resection surgery. This research provided important guidance for the use of NIR-II fluorescence probes in clinical breast cancer imaging in vivo and in surgical navigation.

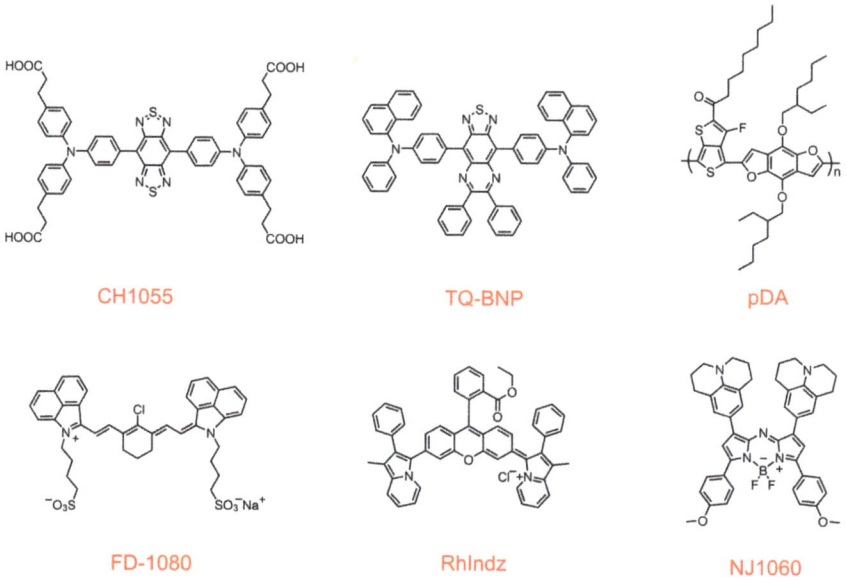

Figure 1. Chemical structures of CH1055, TQ-BNP, pDA, FD-1080, RhIndz, and NJ1060.

2.2. Inorganic NIR-II Fluorophores

Compared with organic NIR-II fluorescent probes, inorganic NIR-II fluorescent probes, including QDs, SWCNTs, and rare-earth-doped nanoparticles (RENPs) have excellent quantum yield and high stability [40,41]. Among the multiple NIR-II fluorescent probes reported thus far, QDs have attracted much attention due to their excellent fluorescence quantum yield [42]. Recently, a series of metal–sulfur QDs have been reported to possess excellent NIR-II fluorescence properties, including Ag_2S, PbS, and ZnS [32]. Specifically, Ag_2S QDs are widely used in preclinical research due to the excellent tunability of their optical properties, their excellent biocompatibility, and their low cytotoxicity. The NIR-II fluorescence imaging of Ag_2S QDs has high spatial resolution (~40 μm) and can track angiogenesis mediated by tumors (2–3 mm in diameter) in vivo [43]. On the other hand, due to their narrow band gap, SWCNTs can generate a fluorescence emission spectrum with wavelengths from 1000 nm to 1800 nm, which can be used for in vivo NIR-II fluorescence imaging of tumor vessels in deep tissues [44]. Ghosh et al. reported a SWCNT modified with a SPARC binding peptide (SBP) and M13 phage, which can be used for the detection of tumor nodules on multiple abdominal viscera and the mesentery [45]. With the NIR-II fluorescence reflectance imaging system, SBP-M13-SWNT-mediated fluorescence imaging can detect tumors at a depth of 9.7–18.2 mm. Notably, compared with the fluorescent probes fluorescein isothiocyanate (FITC) and AlexaFluor750 dye (AF750), the tumor-to-muscle ratio of SBP-M13-SWNT was 5.5 ± 1.2, which was significantly higher than that of SBP-M13-FITC (0.96 ± 0.10) or SBP-M13-AF750 (3.1 ± 0.42). RENPs were also applied in NIR-II fluorescence imaging where rare earth (RE) metals were embedded in an inorganic crystalline host matrix (for example, $NaYF_4$ or CaF_2). In addition, RENPs possess an adjustable emission spectrum, long luminescence life and good photostability; hence, they are expected to replace traditional organic fluorescent materials for fluorescence imaging [7]. Lei et al. doped cerium ions (Ce^{3+}) into the $NaYbF_4$:Er^{3+} nanostructure and achieved fluorescence emission at 1550 nm [46], and further in vivo experiments showed that $NaCeF_4$:Er/Yb could be used for deep-tissue NIR-II fluorescence imaging of mouse hind limbs. Xue et al. developed polyacrylic acid (PAA)-modified NaYF$_4$:Gd/Yb/Er nanorods (PAA-NRs), which could be used for the visual detection of microscopic tumors via NIR-II fluorescence imaging. Notably, non-invasive high-resolution and highly spatial (down to 43.65 μm) NIR-II brain vasculature imaging was achieved using PAA-NRs [47]. In summary, a wide variety of NIR-II fluorescent probes have been prepared and have shown great potential in tumor imaging, deep-seated disease detection, surgical navigation, and therapeutic effect evaluation.

3. Activatable NIR-II Fluorescent Probes

Although NIR-II fluorescent probes have excellent optical properties for bioimaging, the "always on" fluorescent probes produce nonspecific signals in normal tissues, reducing the detection sensitivity [48]. In comparison with the "always on" fluorescent probes, activatable fluorescent probes achieve high specificity by increasing the target signal intensity and reducing the background signal [49]. According to the activation modes, these strategies mainly include eight categories: pH, enzymes, ROS, reactive nitrogen species (RNS), reactive sulfur species (RSS), hypoxia, viscosity and dual-responsive.

3.1. pH

pH plays an important role in the physiological homeostasis of the living body [50,51]. Abnormal pH values affect the physiological balance, which is related to the occurrence and development of a variety of diseases [13,52]. For example, the pH value in gastric juice can affect the activity of digestive enzymes in gastric juice and the utilization of oral drugs [53]. The pH value in gastric juice can be monitored using a pH-sensing fluorescent probe. In 2019, Wang et al. reported a benzothiopyrylium pentamethine cyanine substituted by diethylamino (BTC1070) probe with high-penetration NIR-II imaging properties, which exhibited superior pH-responsive properties (Figure 2a) [54]. Interestingly, when the pH value was reduced from 5 to 2, the maximum absorption peak at 1015 nm decreased and a new absorption peak at 600–900 nm appeared (Figure 2b). The values of pKa were 0.29 and 3.81 by Boltzmann curve fitting, which clearly indicated that BTC1070 had a double protonation feature. The protonation process led to the inhibition of the intramolecular charge transfer (ICT) effect, which realized the fluorescence response ratio (Figure 2c). In addition, mice were treated with simulated gastric juice of pH 1.3 or 2.5 to simulate the acidic environment of the human stomach. The fluorescence imaging of BTC1070 in gastric juices at different pH values was monitored by noninvasive ratiometric imaging at different tissue depths, and the results indicated that BTC1070 could be used for high-contrast fluorescence imaging of deep tissues, to noninvasively detect the pH value in gastric juice (Figure 2d). Overall, this investigation showed that pentamethine fluorophore could be used in pH-activated NIR-II fluorescence imaging for accurate detection of pH in gastric juice.

Cancer is a heterogeneous disease which differs from normal tissue in morphology and growth mode [55–57]. The rapid proliferation of tumor cells leads to the characteristic changes in energy metabolism [58,59]. Tumor cells are vigorous with respect to energy metabolism, consuming large amounts of glucose and converting it to lactate, resulting in an increase in extracellular H$^+$ concentration and a decrease in pH value [18,60,61]. Inorganic NIR-II fluorescent probes have a high degree of stability, which can be controlled by modifying the pH-responsive groups on their surfaces. In 2020, Ling et al. reported a NIR-II nanodrug system (FEAD1) for precise tumor theranostics via tumor acid activation, which was self-assembled by the Fmoc-His peptide, mercaptopropionic-modified Ag$_2$S QDs (MPA-Ag$_2$S QDs), NIR absorber A1094, and doxorubicin (DOX) [62]. The NIR-II fluorescence of FEAD1 was largely quenched, due to the Forster resonance energy transfer (FRET) between Ag$_2$S QDs and A1094. However, under the acidic conditions of the tumor, the disassembly of FEAD1 by protonation of the imidazole groups led to a loss of FRET, enabling NIR-II fluorescence recovery. In vivo experiments showed that FEAD1 not only formed specific NIR-II fluorescence signals in the breast cancer sites of mice but also illuminated peritoneal metastatic tumor nodules over a long period, endowing this novel pH-activatable NIR-II fluorescent probe with excellent potential for clinical applications. In 2021, Liu et al. reported novel pH-sensitive nanovesicles assembled by thiolated polystyrene-co-poly(4-vinylpyridine)-modified Ag$_2$S QDs (Ag$_2$S Ve) for precisely activating NIR-II fluorescence imaging (Figure 2e) [63]. The pyridine group of 4-vinylpyridine was easily protonated at a lower pH value, demonstrating that Ag$_2$S Ve had the capacity for pH responsiveness. As shown in Figure 2f, under acidic conditions the fluorescence intensity of Ag$_2$S Ve gradually recovered in the range of 1000 to 1400 nm within 90 min. In vivo NIR-II fluorescence imaging indicated that Ag$_2$S Ve showed obvious fluorescence in the tumor site at 12 and 24 h, while the control group showed no fluorescence, demonstrating excellent pH-responsive properties (Figure 2g).

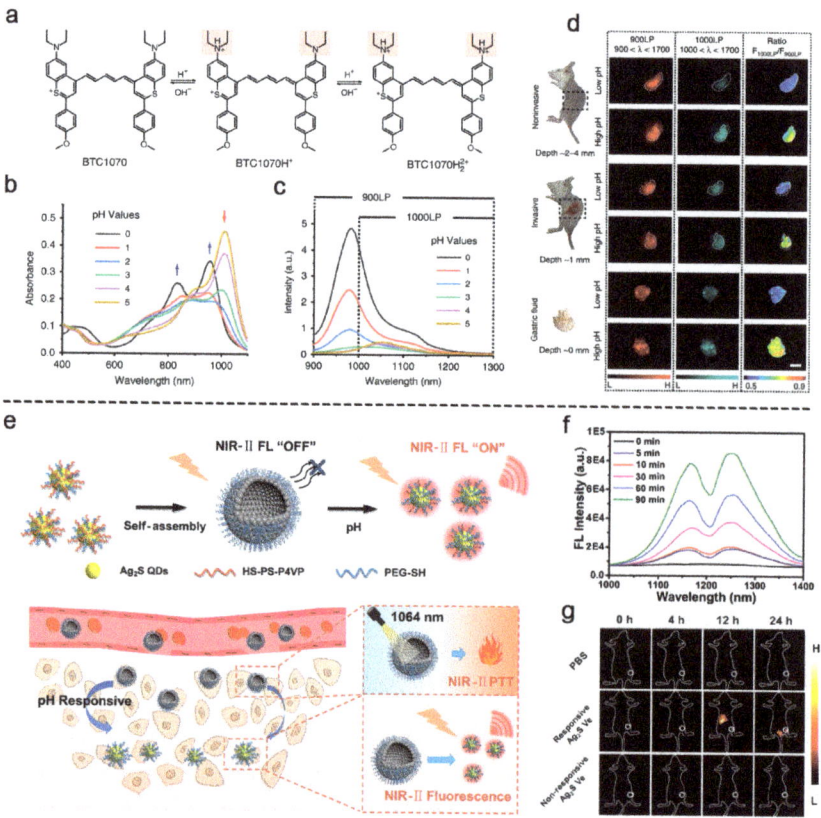

Figure 2. (a) Chemical structure of BTC1070 and its protonation mechanism. (b) Electronic absorption and (c) fluorescence spectra of BTC1070 in PBS at different pH conditions (excited at 808 nm). (d) Digital photographs of three imaging modes distinguished based on different tissue depths (left) and fluorescence imaging and ratiometric fluorescence imaging of BTC1070 in gastric juice at different pH under three imaging modes (right). Reproduced with permission [54]. Copyright 2019, Nature Publishing Group. (e) Schematic diagram of self-assembly process of Ag$_2$S QDs into pH-activatable Ag$_2$S Ve and the mechanism of pH-activatable Ag$_2$S Ve for NIR-II fluorescence-imaging-guided photothermal cancer therapy in vivo. (f) Fluorescence spectra of pH-activatable Ag$_2$S Ve triggered by low pH condition at different time points under the irradiation of 808 nm laser. (g) Time-dependent NIR-II fluorescence imaging of tumors in vivo after treatment with PBS and pH-activatable and non-activatable Ag$_2$S Ve (10 mg/mL). Reproduced with permission [63]. Copyright 2021, WILEY-VCH.

3.2. Enzymes

Enzymes are an extremely important class of biocatalysts which are involved in a great number of biological reactions in living organisms, and abnormalities in their concentrations are often associated with the development of various diseases [64–66]. For example, β-galactosidase (β-Gal), an enzyme that can hydrolyze substances containing a β-glycoside bond, exists widely in living organisms [67,68]. An increase in the content of β-Gal in the human body usually induces the occurrence of ovarian cancer, and sensitive detection of β-Gal is therefore important for the early prediction and diagnosis of ovarian cancer [69–71]. Chen et al. developed an activatable galactose-modified BODIPY BOD-M-βGal NIR-II fluorescent probe (Figure 3a) [72]. The transition of BODIPY from NIR-I to NIR-II was achieved by providing elongation of the π-conjugation to BODIPY via a vinylene unit. As shown in Figure 3b, β-Gal enhanced the fluorescence of BOD-M-βGal significantly within 20 min, showing enzyme-responsive fluorescence emission in the NIR-II region. In vivo fluorescence imaging indicated that BOD-M-βGal achieved excellent imaging of tumor regions, while the additional inhibitor D-galactose significantly inhibited the fluorescence intensity, showing the specificity of the enzyme response (Figure 3c). Moreover, at a depth of 2 mm, the bright NIR-II signal of BOD-M-βGal could be observed, while the NIR-I fluorescence signal was barely detected (Figure 3d).

Matrix metalloproteinase (MMP) is a type of matrix-degrading enzyme which plays an important role in atherosclerosis, rheumatoid arthritis, enteritis, cancer, and other inflammatory diseases [73–75]. Jeong et al. reported a protease-activatable QD (PA-NIR QD) NIR-II probe, which was obtained by combining a PbS/CdS/ZnS multishell fluorescent probe with activatable modulators (AcMs) [76]. When PA-NIR QD was loaded with the photosensitizer methylene blue (MB), the PA-NIR QD was quenched by photoinduced electron transfer (PET) and was activated specifically in the presence of MMP. After simultaneous incubation of MMP2 with PA-NIR QD, the fluorescence recovery efficiency of PA-NIR QD increased with an increase in MMP2. Furthermore, when the PA-NIR QD was injected in a tumor mouse model, the tumor area showed a stronger fluorescence signal than the normal tissue, indicating that MMP-activated PA-NIR QD achieved real-time, high-resolution fluorescence imaging. Recently, Zhan et al. reported the NIR-II fluorescent probe A&MMP@Ag$_2$S-AF7P, specifically activated by the MMP14 enzyme for rapid diagnosis of neuroblastoma (NB) (Figure 3e) [77]. A&MMP@Ag$_2$S-AF7P is modified by the MMP14-targeting peptide AF7P, polycationic peptide R9 and poly-anionic fragments E8, which are taken up by NB cells overexpressing MMP14, and FRET is disrupted, resulting in rapid activation of NIR-II fluorescence. As shown in Figure 3f, the extra MMP14 recovered the fluorescence of A&MMP@Ag$_2$S-AF7P, and the fluorescent probe was lit up within 45 min, indicating that this enzymatic reaction was fast and efficient. In vitro and ex vivo experiments showed that A&MMP@Ag$_2$S-AF7P could be activated by MMP14 to produce obvious NIR-II fluorescence signals (Figure 3g,h).

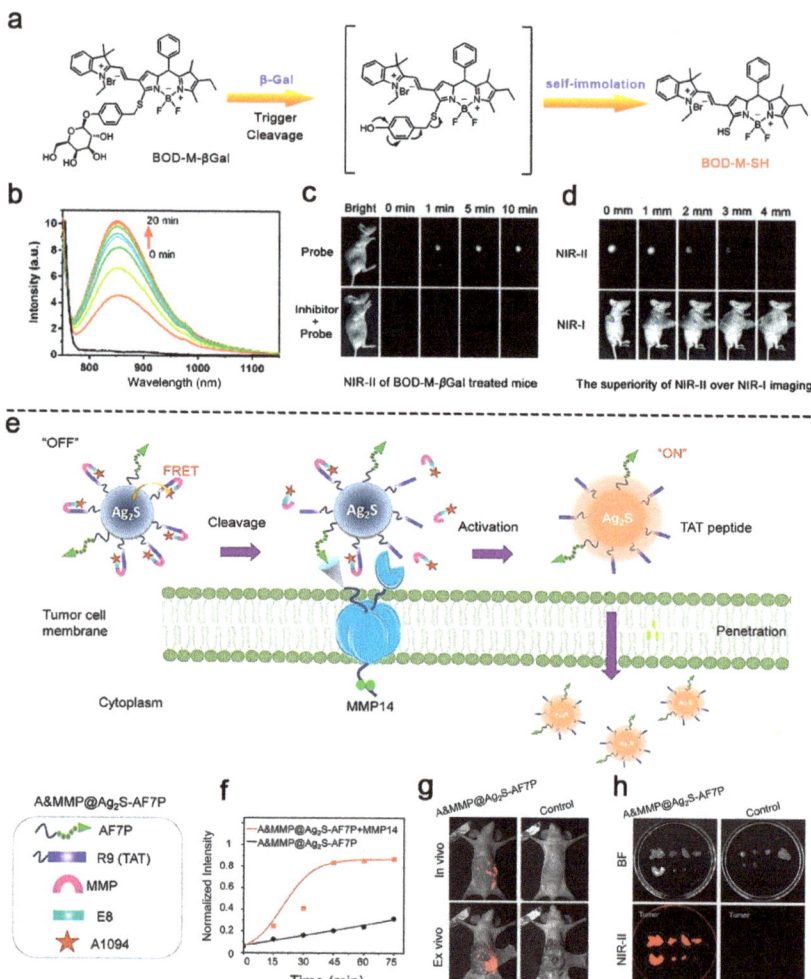

Figure 3. (a) Schematic illustration of the mechanism for β-Gal enzymatic activation process of BOD-M-βGal. (b) Time-dependent NIR-II fluorescence spectra of BOD-M-βGal upon addition of β-Gal in mixed aqueous solution. (c) NIR-II fluorescence imaging of mice at different time points after treatment with BOD-M-βGal and BOD-M-βGal + inhibitor D-galactose. (d) Comparison of NIR-II and NIR-I fluorescence imaging of mice after treatment with BOD-K-βGal and BOD-M-βGal in various deep-tissue environments. Reproduced with permission [72]. Copyright 2020, Royal Society of Chemistry. (e) Schematic diagram of A&MMP@Ag₂S-AF7P for NB detection. (f) Validation of MMP14-mediated fluorescence recovery of A&MMP@Ag₂S-AF7P. (g) NIR-II fluorescence and bright field (BF) imaging of peritoneal tumors in vivo and ex vivo after intraperitoneal injection of A&MMP@Ag₂S-AF7P. (h) BF and NIR-II fluorescence imaging of resected tumor nodules (scale bar = 5 mm). Reproduced with permission [77]. Copyright 2021, WILEY-VCH.

3.3. ROS

ROS, mainly including hydrogen peroxide (H_2O_2), superoxide anion ($O_2^{\bullet-}$), hypochlorite (ClO^-), and hydroxyl radicals ($\cdot OH$), can affect many physiological and pathological processes, such as cell apoptosis, cell signal transduction, and cancer [78–80]. Some investigations have shown that the H_2O_2 concentration in tumor tissue or at inflammatory sites was much higher than that in normal tissue [18,81]. In 2021, Zhang et al. reported a size-tunable Ag/Ag$_2$S Janus NP (JNP) NIR-II fluorescent nanoprobe activated by endogenous H$_2$O$_2$ (Figure 4a) [82]. Plasma electron transfer led to the fluorescence quenching of Ag/Ag$_2$S JNP; however, H$_2$O$_2$ could etch the surface Ag and thus activate its fluorescent property. As shown in Figure 4b,c, the Ag in Ag/Ag$_2$S JNP was gradually etched in the presence of H$_2$O$_2$ within 30 h, resulting in a slight change in absorbance at 808 nm, while the NIR-II fluorescence intensity at 1250 nm was significantly enhanced. In vivo and ex vivo NIR-II fluorescence imaging indicated that mice injected with Ag/Ag$_2$S JNP generated obvious fluorescence at the tumor site which could be inhibited by N-acetyl cysteine (NAC) (Figure 4d,e). However, ·OH is the ROS with the strongest oxidation activity in living organisms, and it can attack biological substrates such as biological proteins and DNA, causing severe oxidative damage [83,84]. Compared with conventional electron paramagnetic resonance (EPR), detection of ·OH, fluorescence imaging is noninvasive and highly sensitive, whereas NIR-II fluorescence imaging has deeper tissue penetrability and lower spontaneous background fluorescence [85–87]. In 2019, Feng et al. reported a NIR-II fluorescent probe, Hydro-1080, which could be activated by ·OH (Figure 4f) [88]. Due to the decrease in conjugation and coplanarity in Hydro-1080, there was no obvious NIR-II fluorescence intensity. After the formation of Et-1080 by ·OH oxidation, the fluorescence recovered with the recovery of the conjugation system. As shown in Figure 4g,h, with a concentration of ·OH from 0.01 to 1.60 µM, the absorption intensity at 1021 nm was enhanced, and the fluorescence intensity at 1044 nm was also enhanced, showing the responsiveness of Hydro-1080 to ·OH. The production of ·OH in mouse liver could be induced by injecting acetaminophen (APAP) into mice. After different doses of APAP were injected into mice, and they were then injected with Hydro-1080, the mice showed a dose-dependent fluorescence increase in NIR-IIa (1300–1400 nm), and inhibitor 1-aminobenzotriazole (ABT) significantly inhibited the process (Figure 4i,j).

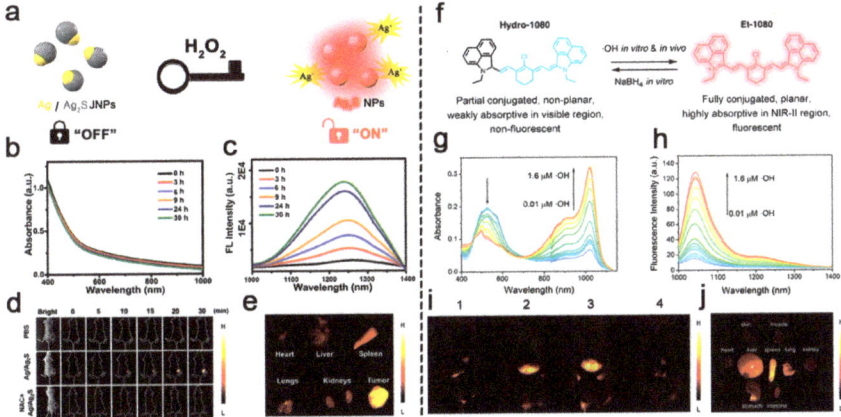

Figure 4. (**a**) Schematic representation of the mechanism of H$_2$O$_2$-activatable nanoprobe Ag/Ag$_2$S JNPs for NIR-II fluorescence imaging in vivo. (**b**) Electronic absorption and (**c**) fluorescence spectra of Ag/Ag$_2$S JNPs after treatment with H$_2$O$_2$ at different time points. (**d**) NIR-II fluorescence imaging of tumor-bearing mice at different time points after Ag/Ag$_2$S JNP injection in vivo. (**e**) NIR-II fluorescence imaging of tumor and main organs excised from the Ag/Ag$_2$S-treated mice group. Reproduced with permission [82]. Copyright 2021, American Chemical Society. (**f**) Chemical structures and mutual conversion processes of Hydro-1080 and Et-1080. (**g**) Electronic absorption and (**h**) fluorescence spectra of Hydro-1080 after reaction with different concentrations of ·OH. (**i**) NIR-IIa fluorescence imaging of mice injected first with different doses of APAP (0, 300, 500, 500 mg/kg) and then with Hydro-1080 (1 mM, 5 mL/kg). In addition, the fourth mouse in the picture was intraperitoneally pre-injected with ABT (100 mg/kg) 24 h before injecting APAP. 1: Hydro-1080; 2: APAP (300 mg/kg) + Hydro-1080; 3: APAP (500 mg/kg) + Hydro-1080; 4: APAP (500 mg/kg) + Hydro-1080 + ABT (100 mg/kg). (**j**) NIR-IIa fluorescence imaging of main organs and tissues of mice injected with APAP (500 mg/kg) and Hydro-1080. Reproduced with permission [88]. Copyright 2019, American Chemical Society.

Hypochlorous acid (HClO) is a highly oxidation-active oxo-acid of chlorine, which is endogenously produced from chloride ions and H_2O_2 catalyzed by myeloperoxidase (MPO) in neutrophils [89–91]. As a signal molecule, HClO is involved in regulating a variety of physiological processes [89–91]. However, excessive production of HClO will cause tissue damage and the formation of a variety of diseases, including cancer, arthritis, and lymphadenitis [92,93]. Ge et al. synthesized a novel NIR-II organic fluorescent probe, SETT, which could be activated by HClO-specific oxidation and showed a highly sensitive response [94]. Furthermore, SETT was loaded on the surface of down-conversion nanoparticles (DCNP) doped with Er^{3+}, obtaining the NIR-II ratiometric nanoprobe DCNP@SeTT. Interestingly, upon light irradiation at a wavelength of 980 nm, the fluorescence intensity of SETT at 1150 nm decreased with an increasing concentration of HClO, while that of DCNP at 1550 nm was unchanged. Therefore, the ratiometric fluorescence signal (I1150 nm/I1550 nm) was linearly correlated with the concentration of HClO. In addition, the DCNP@SeTT probe could be used for HClO-responsive NIR-II fluorescence imaging of rabbit osteoarthritis and mouse tumors. Tang et al. reported that a non-fullerene acceptor (ITTC) was blended with a semiconducting polymer donor (PDF) to construct a NIR-II fluorescent probe (SPNPs) that could be activated by ClO^- (Figure 5a) [95]. Upon close contact of ITTC and PDF, the donor–acceptor interaction caused PET to quench the fluorescence of PDF. As shown in Figure 5b, with an increase in ITTC content, the NIR-II fluorescence quenching degree of SPNPs also increased. Encouragingly, ClO^- could oxidize and degrade ITTC, allowing the fluorescence of SPNP25 (25% doping amount) to recover, while other reactive oxygen species and reducing species could not produce this change, indicating the specificity of this activation process (Figure 5c,d). After intravenous injection of SPNP25, the fluorescence at the inflammatory sites of the mice was significantly enhanced, while there was no fluorescence at the normal sites (Figure 5e). It is worth noting that NAC significantly inhibited fluorescence at sites of inflammation in mice.

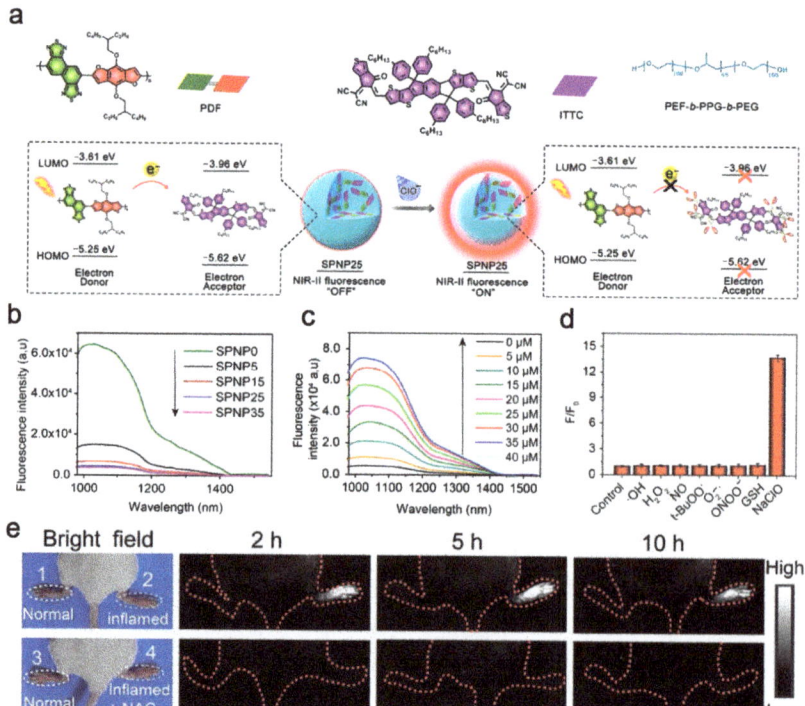

Figure 5. (a) Chemical structures of PDF, ITTC, and PEG-b-PPG-b-PEG, and the mechanism of nanoprobe

SPNP25 for endogenous ClO⁻ detection. (**b**) Fluorescence intensity of SPNPs doped with different amounts of ITTC. (**c**) Fluorescence spectra of the nanoprobe SPNP25 after reaction with various concentrations of ClO⁻. (**d**) Fluorescence intensity ratios of nanoprobe SPNP25 toward ClO⁻ (40 μM) and various biochemically related analytes (100 μM). (**e**) Time-dependent fluorescence imaging in the inflammatory and normal sites of LPS-pretreated and LPS/NAC-treated mice after injection of SPNP25. Reproduced with permission [95]. Copyright 2019, WILEY-VCH.

3.4. RNS

RNS are highly reactive biological oxidation species containing nitrogen, mainly including nitric oxide (NO), peroxynitrite (ONOO⁻), and S-nitrosothiols (RSNO) [21,96]. In common with ROS, RNS play an important role in oxidation processes under physiological conditions which are related to oxidative stress, brain diseases, cancer, and inflammation [97–99]. In fact, ONOO⁻ is not only related to the nitration of proteins in immune cells in tumors, affecting the immunosuppression of tumors, but is also related to the occurrence and development of brain injury [100,101].

Traumatic brain injury (TBI) is highly likely to cause disability and death in humans unless it can be easily diagnosed at an early stage [102]. Currently, clinical diagnosis of TBI is mainly achieved using magnetic resonance imaging (MRI) and computed tomography (CT), but these are difficult to use for real-time early diagnosis of TBI. As reported, ONOO⁻ is a biomarker of early TBI which is accompanied by the occurrence and development of TBI [103,104]. Therefore, the occurrence of early TBI can be diagnosed by monitoring the biomarker ONOO⁻. In 2020, Li et al. reported an activated NIR-II fluorescent probe (V&A@Ag$_2$S) composed of a targeting group (VCAM1 binding peptide), an A1094 chromophore NIR absorber, and Ag$_2$S QD (Figure 6a) [105]. Specifically, there is a large overlap between the NIR-II fluorescence emission spectrum of Ag$_2$S QD and the absorption spectrum of A1094, leaving the fluorescence of V&A@Ag$_2$S in the "off" state. However, when ONOO⁻ was present, A1094 was oxidized and its absorption peak disappeared, resulting in the NIR-II fluorescence signal of V&A@Ag$_2$S being turned "on". In addition, the VCAM1 binding peptide enabled V&A@Ag$_2$S to target inflamed endothelium expressing VCAM1. As shown in Figure 6b, ONOO⁻ enhanced the fluorescence intensity of V&A@Ag$_2$S at 1050 nm, and the magnitude of the enhancement had a linear relationship with the concentration of ONOO⁻. Apart from ONOO⁻/ClO⁻, other common anions and cations could not activate V&A@Ag$_2$S, showing the specificity of this activation mode (Figure 6c). Encouragingly, 3-morpholinosydnonimine hydrochloride (SIN-1) was used to induce ONOO⁻ production by the cells, and cell experiments showed that in the presence of ONOO⁻ an obvious NIR-II fluorescence signal appeared after V&A@Ag$_2$S uptake by human umbilical vein endothelial cells (HUVECs), whereas the control group did not exhibit this process (Figure 6d). In vivo experiments on TBI mice showed that V&A@Ag$_2$S could be activated by ONOO⁻, showing excellent NIR-II fluorescence imaging. In the control group, the "always on" V@Ag$_2$S probe was distributed over the whole brain region without specificity, while A@Ag$_2$S exhibited no obvious fluorescence signal. Notably, V&A@Ag$_2$S could not produce an obvious fluorescence signal in healthy mice, indicating that the "turned on" probe showed excellent specificity in biosensor and imaging applications (Figure 6e).

Ischemic stroke is also a brain disease, which is highly likely to cause death and disability in humans if it cannot be easily diagnosed at an early stage [106,107]. Similarly, ONOO⁻ is a biomarker of ischemic stroke and therefore provides the possibility of early diagnosis of the disease [103,108]. In 2021, Yang et al. reported a V&C/PbS@Ag$_2$Se nanoprobe modified by VCAM1 binding peptide and Cy7.5 fluorophores, which could be activated by ONOO⁻ to produce obvious NIR-II fluorescence intensity (Figure 6f) [109]. Meanwhile, the VCAM1 binding peptide provided V&C/PbS@Ag$_2$Se with the ability to target ischemic stroke regions, while the Cy7.5 fluorophores and PbS@Ag$_2$Se QDs engendered competitive absorption to turn off the NIR-II fluorescence. As shown in Figure 6g, the stability of PbS modified by Ag$_2$Se was significantly improved, and no significant change in fluorescence occurred within 10 days. Interestingly, both PbS@Ag$_2$Se and V/PbS@Ag$_2$Se exhibited obvious fluorescence at 1616 nm, while the fluorescence of V&C/PbS@Ag$_2$Se was quenched (Figure 6h). Absorption-competition-induced emission (ACIE) was first proposed by Zhang et al. Specifically, organic dye acted as the absorption-competition acceptor of the fluorescent probe, significantly inhibiting the fluorescence emission of the fluorescent probe. ACIE technology could be used not only for monitoring drug release but also for in situ NIR sensing of biomarkers [110]. In addition, after HUVECs were incubated with SIN-1 and V&C/PbS@Ag$_2$Se, the cells showed an obvious NIR-II fluorescence signal; conversely, in the absence of SIN-1 no NIR-II fluorescence signal was observed (Figure 6i). Furthermore, in vivo experiments showed that ONOO$^-$ activated V&C/PbS@Ag$_2$Se and lit up its NIR-II fluorescent signal (Figure 6j).

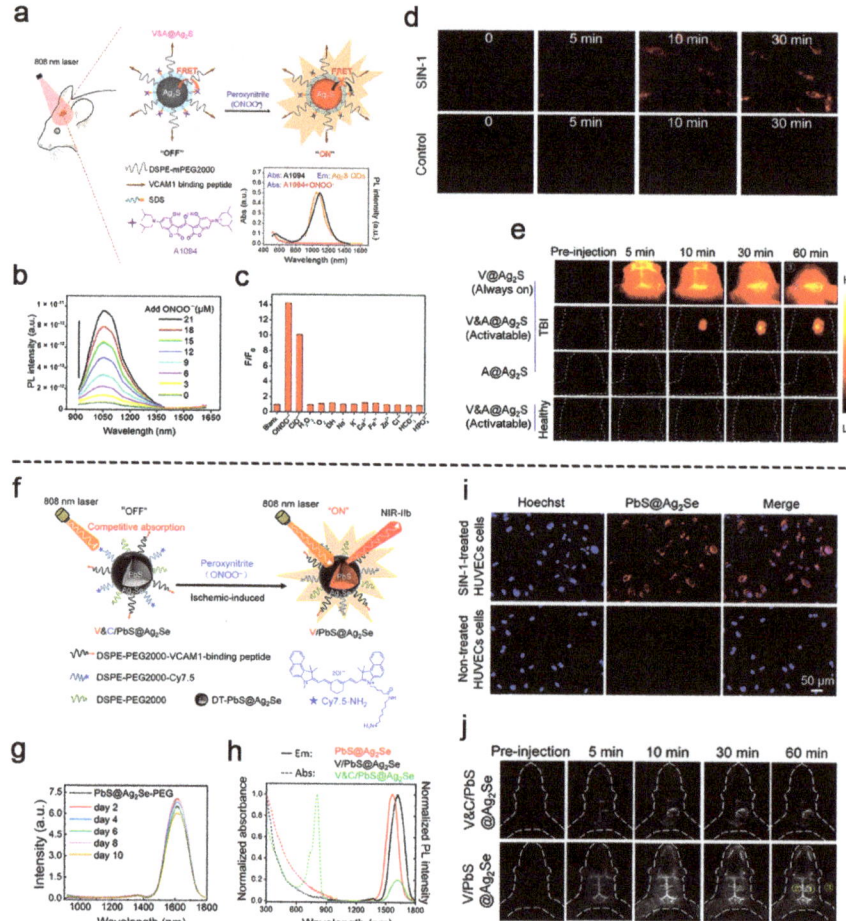

Figure 6. (a) Schematic diagram of the preparation process of the NIR-II nanoprobe V&A@Ag$_2$S and the mechanism of detecting ONOO$^-$ in vivo. (b) Photoluminescence spectra of V&A@Ag$_2$S (50 μg/mL) after reaction with the different concentrations of ONOO$^-$. (c) Fluorescence intensity ratios of V&A@Ag$_2$S (10 μg/mL) upon the addition of different ions and RNS/ROS analytes (20 μM). (d) Fluorescence microscopy images of HUVECs after incubation with V&A@Ag$_2$S in the presence and absence of SIN-1. (e) Time-dependent fluorescence imaging in TBI and healthy mice after different treatments. Reproduced with permission [105]. Copyright 2020, WILEY-VCH. (f) Schematic diagram of the design strategy of the NIR-II nanoprobe V&C/PbS@Ag$_2$Se and the mechanism of detecting ONOO$^-$ in ischemic stroke model mice. (g) Time-dependent

fluorescence spectra of PbS@Ag$_2$Se-PEG QDs in PBS of pH 7.4. (**h**) Electronic absorption and fluorescence spectra of V&C/PbS@Ag$_2$Se, V/PbS@Ag$_2$Se, and PbS@Ag$_2$Se. (**i**) Fluorescence microscopy images of HUVECs after incubation with V&C/PbS@Ag$_2$Se in the presence and absence of SIN-1. (**j**) Time-dependent NIR-II fluorescence imaging of early ischemic stroke model mice after treatment with V&C/PbS@Ag$_2$Se and V/PbS@Ag$_2$Se. Reproduced with permission [109]. Copyright 2021, WILEY-VCH.

In contrast to ROS, NO is a widely existing messenger molecule in the human body which plays an important role in cardiovascular, cerebrovascular, immune system and nervous system processes [111]. It is noteworthy that NO was considered as a biomarker of drug-induced hepatotoxicity [112,113]. Iverson et al. reported a NIR-II fluorescent probe, PEG-(AAAT)7-SWNTs, produced by covering the NO-sensitive DNA oligonucleotide ds(AAAT)7 on the surface of SWCNTs [114]. The PEG-(AAAT)7-SWNTs nanoprobe could be used to investigate the production of NO by inflammation in vivo. In 2019, Tang et al. reported an organic semiconductor nanoprobe (AOSNP) activated by NO for monitoring drug-induced hepatotoxicity, which was obtained by the amidation reaction of the NO-sensitive organic semiconducting group (FTBD) and poly (styrene-co-maleic anhydride) (PSMA) polymer [115]. Notably, in the presence of NO, FTBD could transform its receptor unit (benzo[c] [1,2,5] thiadiazole-5,6-diamine) into benzotriazole derivatives, realizing the conversion of fluorescence from NIR-I to NIR-II. NO significantly increased the NIR-II fluorescence intensity of AOSNP and changed the color of the AOSNP solution. However, other reactive oxygen species (H_2O_2, $\cdot OH$, $O_2^{\bullet-}$, and $ONOO^-$) and reducing substances (glutathione and GSH) could not enhance the fluorescence of AOSNP, indicating the specificity of the activation mode. After co-incubation with AOSNP and HepG2 hepatoma cells, the cells showed an obvious NIR-II fluorescence signal, and the additional inhibitor NAC significantly inhibited this process. Interestingly, APAP induced a large amount of NO to produce hepatotoxicity in mice. In vivo fluorescence imaging indicated that AOSNP could be activated by NO generated from APAP-treated mice to produce obvious NIR-II fluorescence signals, while this process could be inhibited by inhibitor NAC.

3.5. RSS

RSS mainly include thiols (GSH and Cys), H_2S, and persulfides (R-S-SH/H_2S_2) [116,117]. In fact, the level of intracellular GSH is higher than that of extracellular GSH, and the concentration of GSH in tumor tissue is also higher than that in normal tissue [26,118]. Li et al. reported a Ln^{3+}-doped nanoparticles (LnNPs) probe using the large spectral overlap between the absorption spectrum of Ln^{3+} and the emission spectrum of heptamethine cyanine, which could be activated by GSH for high-resolution biological imaging [119]. In contrast to ROS and RNS, RSS not only played the role of oxidation, but also played the role of reduction. H_2S is both a special biological signaling molecule and an important biomarker for early cancer [120–122]. In 2018, Shi et al. reported a H_2S-activatable nanodrug (Nano-PT) for NIR-II fluorescence imaging of colorectal cancer (CRC) [116]. Nano-PT is a monochlorinated BODIPY derivative which produces a nucleophilic reaction with H_2S. Interestingly, Nano-PT not only produced activatable NIR-II fluorescence emission, but also produced high-efficiency NIR absorption, which realized NIR-II-fluorescence-guided PTT. Therefore, PTT mediated by Nano-PT achieved highly specific CRC treatment guided by NIR-II fluorescence imaging, promoting the development of precision medicine.

In 2021, Liu et al. designed and synthesized a series of H_2S-activated NIR-II fluorescence probes, WH-X (WH-1, WH-2, WH-3, and WH-4), which were modified by a 4-nitrothiophenol fluorescence quencher (Figure 7a) [121]. The maximum emission wavelength of WH-3 was 1205 nm, which was superior to the other three compounds, and it showed excellent NIR-II fluorescence characteristics. Theoretical calculations showed that substitution of the WH-2-HS acceptor successfully reduced the gap between the highest occupied molecular orbital (HOMO) and the lowest unoccupied molecular orbital (LUMO) (Figure 7b). By extending the conjugation region in WH-3-HS and WH-4-HS, the ICT process was strengthened, leading to the gap being narrowed further. As shown in Figure 7c,

WH-3 showed a concentration-dependent fluorescence enhancement in the presence of NaHS. Using H$_2$S-overexpressing HCT-116 colon cancer cells as a model, intracellular fluorescence imaging also showed that NaHS significantly depleted WH-3, whereas the inhibitor aminooxyacetic acid (AOAA) inhibited this depletion (Figure 7d). In vivo experiments showed that WH-3 produced an NIR-II fluorescence signal at the tumor site, while the inhibitor AOAA inhibited this process, indicating that WH-3 could be activated by H$_2$S (Figure 7e).

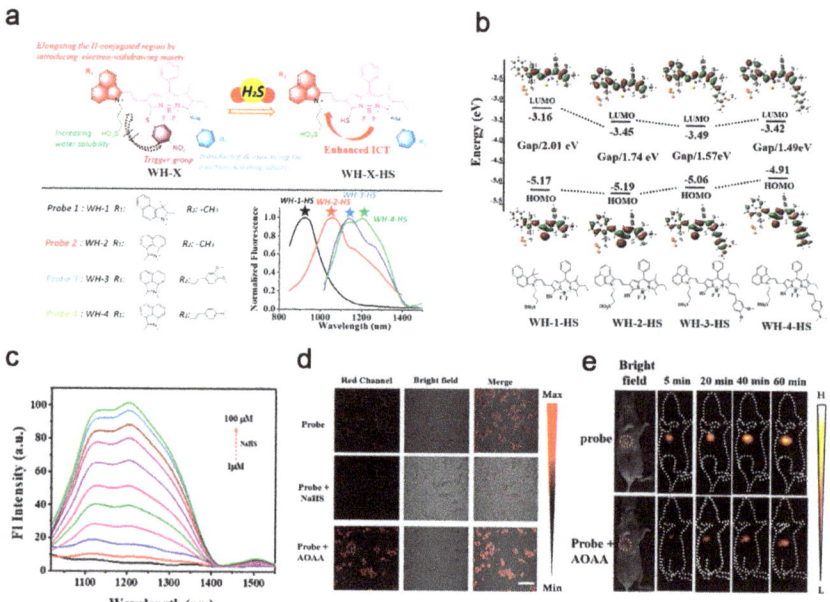

Figure 7. (a) Schematic illustration of the construction of several H$_2$S-activatable NIR-II fluorescence probes WH-X (WH-1, WH-2, WH-3, and WH-4). (b) Theoretical calculations used to verify the design strategy. The HOMO and LUMO energy levels of WH-X-HS were computed. (c) Fluorescence spectra of WH-4 in the presence of different concentrations of NaHS under the irradiation of 980 nm laser. (d) Intracellular fluorescence imaging of HCT-116 cells incubated with WH-3, WH-3 + NaHS, and WH-3 + AOAA. (e) NIR-II fluorescence imaging of tumor-bearing mice at various time points after treatment with WH-3 and WH-3 + AOAA. Reproduced with permission [121]. Copyright 2021, American Chemical Society.

3.6. Hypoxia

Hypoxia is an important characteristic of solid tumors which is related to tumor migration, invasion, and deterioration [123,124]. The occurrence of hypoxia greatly limits the therapeutic efficacy of chemotherapy, photodynamic therapy (PDT), and sonodynamic therapy (SDT) [125,126]. Therefore, it is significant to design hypoxia-activated nanomaterials for the diagnosis and treatment of tumors [127]. Meng et al. coupled IR-1048 dye and 2-(2-nitroimidazolyl) ethylamine, MZ, to develop a novel dye IR1048-MZ, achieving hypoxia-activated NIR-II/PA tumor-imaging-guided PTT (Figure 8a) [126]. Theoretical calculations showed that the HOMO and LUMO of IR1048-MZH were lower than those of IR1048-MZ, indicating enhanced fluorescence emission intensity. Nitroreductase (NTR) overexpressed by hypoxic tumor tissue reduced the nitro group of IR1048-MZ to an amine group and activated the NIR-II fluorescence signal of IR1048-MZ. As shown in Figure 8b, the fluorescence of the IR1048-MZ probe at the maximum emission wavelength (1046 nm) was suppressed, whereas it was significantly enhanced in the presence of NTR. In addition, with an increase in NTR concentration, the NIR-II fluorescence intensity also increased, and this change showed a good linear relationship (Figure 8c). Moreover, in vivo imaging showed that IR1048-MZ resulted in the A549 tumor of nude mice producing an obvious fluorescence signal, while no obvious signal was found in other parts (Figure 8d). The maximum value of the tumor background was 30, indicating that IR1048-MZ was an excellent probe for hypoxia without background.

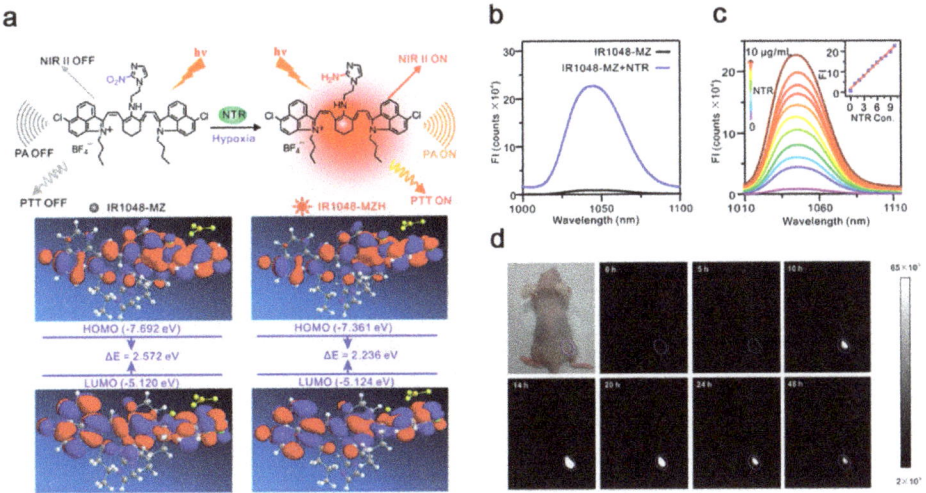

Figure 8. (a) Schematic diagram of the catalyzed mechanism of IR1048-MZ for hypoxia-activated NIR-II/PA tumor imaging. HOMO and LUMO energy levels of IR1048-MZ and IR1048-MZH were obtained by theoretical calculation. (b) Fluorescence spectra of IR1048-MZ with and without the addition of NTR. (c) NIR-II fluorescence spectra of IR1048-MZ (5 µg/mL) in response to various concentrations of NTR. (d) Time-dependent NIR-II fluorescence imaging of tumor-bearing mice after injection of IR1048-MZ (40 µg/mL, 200 µL). Reproduced with permission [126]. Copyright 2018, Creative Commons.

3.7. Viscosity

Viscosity is a major factor in the microenvironment of the living body which is related to various physiological and pathological processes [22,128]. Viscosity abnormalities are associated with many diseases, such as diabetes and Alzheimer's disease [129,130]. Consequently, a viscosity-activated NIR-II fluorescent probe can be used to analyze and detect the viscosity level of organisms to further explore the occurrence and development of related diseases. In 2020, Dou et al. developed a series of viscosity-activated NIR-II fluorescent probes using BODIPY derivatives modified by 1-ethyl-2-methyl-benz[c,d] iodolium salt as a precursor (Figure 9a) [131]. It is worth emphasizing that the substituents of WD-OCH$_3$ and WD-NME$_2$ were strong electron-donating groups and exhibited longer NIR-II fluorescence emission wavelengths (Figure 9b). Interestingly, in the ethanol–glycerol system with different viscosities, the intensities of WD-NO$_2$ at the maximum absorption wavelength and the maximum emission wavelength increased with an increase in viscosity, while the other three molecules also showed similar properties (Figure 9c,d). Briefly, at low viscosity, the conjugated structure was destroyed and caused an increase in nonradiative energy consumption, thereby leading to the weakening of the fluorescence intensity. However, at high viscosity, intramolecular rotation was restricted, causing fluorescence recovery. As shown in Figure 9e,f, monensin (Mon), nystatin (Nys), and lipopolysaccharide (LPS) could change the viscosity in mice. In vivo fluorescence imaging showed that the livers of mice treated with Mon, Nys, and LPS, which received intraperitoneal injections of WD-NO$_2$, exhibited significant fluorescence intensity, indicating that WD-NO$_2$ is an effective tool for studying the change of viscosity in vivo.

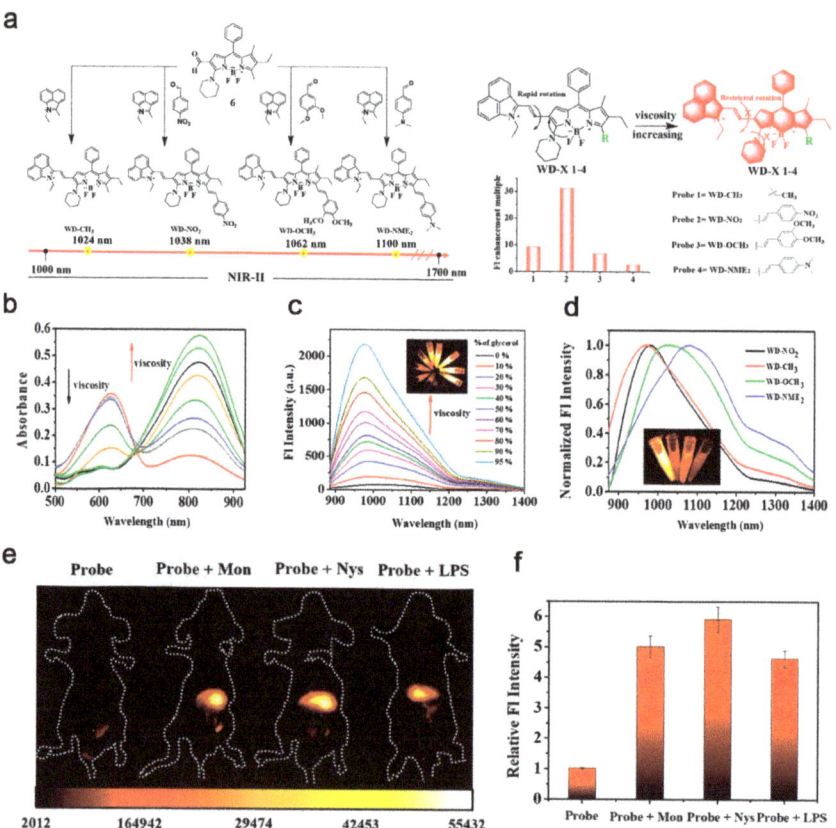

Figure 9. (**a**) Schematic illustration of the design strategy of viscosity-activatable fluorescence probes WD-X (WD-CH$_3$, WD-NO$_2$, WD-OCH$_3$, and WD-NME$_2$). Chemical structures and synthetic routes of WD-X (left) and mechanism of WD-X for viscosity-activatable NIR-II fluorescence imaging (right). (**b**) Electronic absorption and (**c**) fluorescence spectra of WD-NO$_2$ (20 μM) in the ethanol-glycerol system with different viscosities. Inset shows fluorescence images of WD-NO$_2$ in the ethanol-glycerol system with different glycerol ratios. (**d**) Normalized NIR-II fluorescence spectra of WD-X (WD-NO$_2$, WD-CH$_3$, WD-OCH$_3$, and WD-NME$_2$) in the system with 95% glycerol. Inset shows the corresponding fluorescence images under the irradiation of 808 nm (40 mW/cm^2) laser. (**e**) Fluorescence imaging and (**f**) normalized relative fluorescence intensity of mice after treatment with WD-NO$_2$, WD-NO$_2$ + Mon, WD-NO$_2$ + Nys, and WD-NO$_2$ + LPS. Reproduced with permission [131]. Copyright 2020, American Chemical Society.

3.8. Dual-Responsive

A single activation method cannot effectively deal with the complex and dynamic biological microenvironment, resulting in the phenomenon of nonspecific activation and even false-positive results [132,133]. The utilization of two activation methods in the above methods (pH, enzyme, redox, etc.) to activate the NIR-II fluorescence probe greatly improves the specificity of fluorescence imaging, which can effectively avoid the occurrence of false-positive results [134–136]. Zhang and co-workers developed a dual-activatable theranostic nanoprobe (DATN) which could output dual signals under the double stimulation of NO/acidity in inflammation-related tumors for photoacoustic and photothermal imaging in vivo [137]. DATN showed a higher photoacoustic signal under dual activation, which was 132 times that of acidity alone and 9.8 times that of NO. Under single-factor stimulation, DATN increased by about 6 °C. while under dual-factor stimulation it increased by about 27.3 °C, exhibiting the superiority of dual activation. In addition, in the tumor microenvironment, high levels of GSH (1–15 mM) and low pH (pH = 5.0–6.8) acted as a "dual key" to activate the photosensitizers. Teng et al. reported a photosensitizer, BIBCl-PAE NP, which underwent protonation under acidic conditions, further promoting the reaction of GSH with BIBCl-PAE NP to generate water-soluble BIBSG for PDT [138]. In vitro studies revealed that BIBCl-PAE NP could distinguish normal and cancer cells in cell imaging, while also exhibiting excellent PDT ability. In vivo experiments also showed that BIBCl-PAE NP was rapidly enriched in the tumor site, while specifically "lighting up" the tumor site, showing irreversible therapeutic activity with less effect on normal tissues. Tang et al. utilized NIR-II cyanine dye (IR-1061) to covalently connect hyaluronic acid (HA) and synthesized IR-1061-pendent HA polymers, which could self-assemble into single-lock-and-key-controlled HINPs in water, and then be cross-linked on the surface of HINPs through disulfide to form dual-lock-and-key-controlled HISSNPs (Figure 10a) [139]. HA and disulfide formed the "double locks" to lock HISSNPs in a quenched state, while the overexpressed hyaluronidase (Hyal) and GSH in the tumor microenvironment acted as "dual smart keys" to break HA chains and disulfide bonds enabled the fluorescence recovery of IR-1061. As shown in Figure 10b–d, when a "single smart key" (Hyal or GSH) existed, the fluorescence of HISSNPs was increased by about 3.4 times and 2.9 times, respectively, while when Hyal and GSH existed together, the fluorescence of HISSNPs was increased by about 13.3 times. In vitro cell experiments showed that MCF-7 breast cancer cells produced obvious fluorescence after uptake of HISSNPs, whereas normal cells did not show the phenomenon (Figure 10e). As we know, 6-O-palmitoyl-l-ascorbic acid (an inhibitor of Hyal) and N-ethylmalemide (a thiol scavenger) could significantly inhibit the activation process, while 6-O-palmitoyl-l-ascorbic acid could inhibit the activity of HINPs. In vivo fluorescence imaging indicated that mice injected with HISSNPs generated obvious NIR-II fluorescence at tumor sites with a higher sensitivity than that of HINPs, indicating the sensitivity and specificity of the dual lock and key (Figure 10f). Notably, although the dual-responsive NIR-II fluorescent probe showed excellent specificity, activatable NIR-II fluorescent probes with single activation methods are still in their infancy, and activatable NIR-II fluorescent probes with two activation methods have rarely been reported.

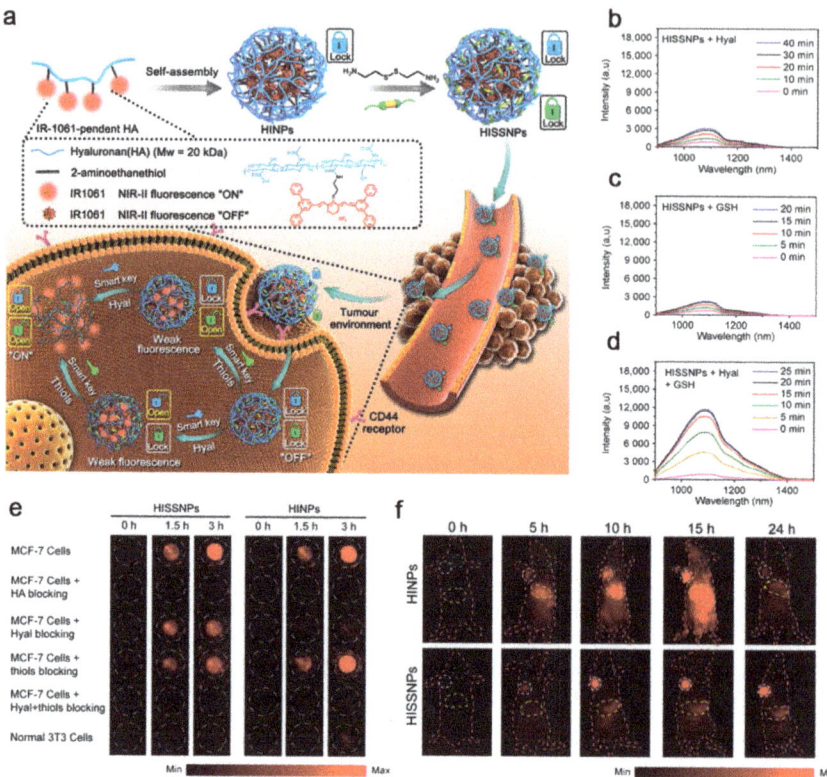

Figure 10. (a) Schematic diagram of the synthetic route of dual-lock-and-key-controlled NIR-II fluorescence probe HISSNPs and its corresponding mechanism for tumor-specific imaging. Time-dependent fluorescence (excited at 808 nm) spectra of HISSNPs incubated with (b) Hyal, (c) GSH, and (d) Hyal and GSH in PBS. (e) NIR-II fluorescence imaging of 3T3 cells and MCF-7 breast cancer cells at the time points of 0 h, 1.5 h, and 3 h after different treatments. (f) Time-dependent NIR-II fluorescence imaging of MCF-7 cancer xenografts model mice after injection with HISSNPs and HINPs (35 mg/kg). White circle: tumor site; yellow circle: abdominal liver site; green circle: muscle (refers to normal tissue). Reproduced with permission [139]. Copyright 2018, WILEY-VCH.

4. Summary and Outlook

Compared with traditional NIR-I fluorescence imaging, NIR-II fluorescence imaging has a deeper tissue penetration depth and a higher SBR. In recent years, a great number of NIR-II fluorescent materials, including SWCNTs, QDs, RENPs, BBTDs, and cyanine dyes were reported to have excellent photophysical and photochemical properties. However, these "always on" fluorescent probes lacked selectivity, showing a low SBR and poor detection sensitivity. On the other hand, the activatable NIR-II fluorescent probe is in an "off" state in normal tissues, while in diseased tissues it can be activated by a specific biomarker to present an "on" state. According to the type of activation mode, these strategies mainly include eight categories: pH, enzyme, ROS, RNS, RSS, hypoxia, viscosity, and dual-responsive. Specifically, the activation mechanisms were mainly the consumption of an inhibitor and changes in the functional group, charge, and conformation. These activatable NIR-II fluorescent probes and their activation strategies and mechanisms are summarized in Table 1.

Table 1. Representatives of activatable NIR-II fluorescent probes for biomedical applications.

Activation Mode	Probe Name	Probe Fluorophores	Activation Mechanism	Excitation/ Emission (nm)	Biological Model	Refs
pH	BTC1070	Pentamethine cyanine	Protonation, inhibition of ICT effect	808/1000–1300	Simulated gastric fluid with different pH	[54]
pH	Ag$_2$S Ve	Ag$_2$S	Protonation, increased hydrophilicity	808/1250	4T1 breast cancer	[63]
β-Gal	BOD-M-βGal	BODPY derivatives	Hydrolyzation of β-galactose residues	808/900–1300	SKOV3 ovarian cancer	[72]
MMP14	A&MMP@Ag$_2$S-AF7P	Ag$_2$S	Consumption of NIR absorber A1094	785/1050	NB	[77]
H$_2$O$_2$	Ag/Ag$_2$S JNP	Ag$_2$S	Oxidizing the plasmonic Ag parts	825/1250	MCF-7 breast cancer	[82]
·OH	Hydro-1080	cyanine dyes	Conjugated system recovery	980/1100	Liver injury	[88]
HClO	SPNPs	PDF	Degradation of the receptor ITTC	808/1010	Inflamed paws	[95]
ONOO$^-$	V&A@Ag$_2$S	Ag$_2$S	Consumption of A1094 chromophore	825/1050	TBI	[105]
ONOO$^-$	V&C/PbS@Ag$_2$Se	PbS@Ag$_2$Se	Consumption of absorber Cy7.5	808/1616	Ischemic stroke	[109]
NO	AOSNP	Ag$_2$S	Energy receptor units are converted	808/900–1150	Liver injury	[115]
H$_2$S	WH-3	BODIPY derivatives	Release fluorescence quencher	980/1140	HCT-116 colon cancer	[121]
Hypoxia (NTR)	IR-1048-MZ	IR-1048	MZ group was reduced	980/1046	A549 lung cancer	[126]
Viscosity	WD-NO$_2$	BODIPY derivatives	Intramolecular rotation of chemical bonds is limited	808/900–1400	Abnormal liver viscosity	[131]
Hyal and GSH	HISSNPs	IR-1061	Fracture of HA chains and disulfide bonds	808/1070	MCF-7 breast cancer	[139]

Although activatable NIR-II fluorescent probes have made encouraging progress, they still face some challenges, mainly reflected in the following. (1) False-positive results. This is mainly due to the limited difference in the concentration of biomarkers in diseased tissues and normal tissues, which may cause false-positive results in NIR-II fluorescence imaging. For example, NB highly expresses MMP; however, many normal cells also produce MMP. It is necessary to consider how to avoid false-positive results affecting the diagnosis of the disease. On the one hand, the probe can be activated in a dual-responsive manner, improving the specificity of imaging. On the other hand, the occurrence of false-positive results is maximally avoided by setting the limit of detection. (2) Delivery of the fluorescent probe. The effective enrichment of the activatable NIR-II fluorescent probe in target tissues is the fundamental guarantee of successful biological imaging. Therefore, drug delivery systems, including liposomes and mesoporous organosilicons, are used to deliver fluorescent probes to the target tissue and improve their concentration in the target tissue. (3) Safety. Although the activatable NIR-II fluorescent probes need special conditions to be activated, fluorescent probes in normal tissues will have toxic and side effects. Moreover, nano fluorescent probes with larger particle sizes are not easily cleared in vivo, and long-term toxicity arises. Therefore, research on safety should be enhanced to provide guarantees for clinical applications.

In conclusion, activatable NIR-II fluorescent probes have undergone rapid development, which has effectively promoted research on fluorescence imaging. This review provides a reference for the design of, and research on, activatable NIR-II fluorescent probes and their applications in biological imaging.

Author Contributions: Conceptualization, D.L. and G.L.; writing—original draft preparation, D.L. and J.P.; resources, S.X.; data curation, S.F.; project administration, D.L.; writing—review and editing, G.L. and C.C.; supervision, G.L. All authors have read and agreed to the published version of the manuscript.

Funding: This work was supported by the Major State Basic Research Development Program of China (2017YFA0205201), the National Natural Science Foundation of China (NSFC) (81925019, 81901876 and U1705281), the Fundamental Research Funds for the Central Universities (20720190088 and 20720200019), and the Program for New Century Excellent Talents in University, China (NCET-13-0502).

Institutional Review Board Statement: Not applicable.

Informed Consent Statement: Not applicable.

Data Availability Statement: Data are contained within the article.

Conflicts of Interest: The authors declare no conflict of interest.

References

1. Hong, G.; Antaris, A.L.; Dai, H. Near-infrared fluorophores for biomedical imaging. *Nat. Biomed. Eng.* **2017**, *1*, 1–22. [CrossRef]
2. Antaris, A.L.; Chen, H.; Cheng, K.; Sun, Y.; Hong, G.; Qu, C.; Diao, S.; Deng, Z.; Hu, X.; Zhang, B.; et al. A small-molecule dye for NIR-II imaging. *Nat. Mater.* **2016**, *15*, 235–242. [CrossRef] [PubMed]
3. Gao, X.; Cui, Y.; Levenson, R.M.; Chung, L.W.; Nie, S. In vivo cancer targeting and imaging with semiconductor quantum dots. *Nat. Biotechnol.* **2004**, *22*, 969–976. [CrossRef] [PubMed]
4. Smith, A.M.; Mancini, M.C.; Nie, S. Bioimaging: Second window for in vivo imaging. *Nat. Nanotechnol.* **2009**, *4*, 710–711. [CrossRef]
5. Hong, G.; Zou, Y.; Antaris, A.L.; Diao, S.; Wu, D.; Cheng, K.; Zhang, X.; Chen, C.; Liu, B.; He, Y.; et al. Ultrafast fluorescence imaging in vivo with conjugated polymer fluorophores in the second near-infrared window. *Nat. Commun.* **2014**, *5*, 4206. [CrossRef]
6. Sheng, Z.; Guo, B.; Hu, D.; Xu, S.; Wu, W.; Liew, W.H.; Yao, K.; Jiang, J.; Liu, C.; Zheng, H.; et al. Bright aggregation-induced-emission dots for targeted synergetic NIR-II fluorescence and NIR-I photoacoustic imaging of orthotopic brain tumors. *Adv. Mater.* **2018**, *30*, 1800766. [CrossRef]
7. He, S.; Song, J.; Qu, J.; Cheng, Z. Crucial breakthrough of second near-infrared biological window fluorophores: Design and synthesis toward multimodal imaging and theranostics. *Chem. Soc. Rev.* **2018**, *47*, 4258–4278. [CrossRef]
8. Yang, Y.; Wang, S.; Lu, L.; Zhang, Q.; Yu, P.; Fan, Y.; Zhang, F. NIR-II chemiluminescence molecular sensor for in vivo high-contrast inflammation imaging. *Angew. Chem. Int. Ed.* **2020**, *59*, 18380–18385. [CrossRef]
9. Antaris, A.L.; Chen, H.; Diao, S.; Ma, Z.; Zhang, Z.; Zhu, S.; Wang, J.; Lozano, A.X.; Fan, Q.; Chew, L.; et al. A high quantum yield molecule-protein complex fluorophore for near-infrared II imaging. *Nat. Commun.* **2017**, *8*, 15269. [CrossRef]
10. Hu, Z.; Fang, C.; Li, B.; Zhang, Z.; Cao, C.; Cai, M.; Su, S.; Sun, X.; Shi, X.; Li, C.; et al. First-in-human liver-tumour surgery guided by multispectral fluorescence imaging in the visible and near-infrared-I/II windows. *Nat. Biomed. Eng.* **2020**, *4*, 259–271. [CrossRef]
11. Yang, Q.; Hu, Z.; Zhu, S.; Ma, R.; Ma, H.; Ma, Z.; Wan, H.; Zhu, T.; Jiang, Z.; Liu, W.; et al. Donor engineering for NIR-II molecular fluorophores with enhanced fluorescent performance. *J. Am. Chem. Soc.* **2018**, *140*, 1715–1724. [CrossRef]
12. Lei, Z.; Zhang, F. Molecular engineering of NIR-II fluorophores for improved biomedical detection. *Angew. Chem. Int. Ed.* **2021**, *60*, 16294–16308. [CrossRef]
13. Huang, J.; Pu, K. Activatable molecular probes for second near-infrared fluorescence, chemiluminescence, and photoacoustic imaging. *Angew. Chem. Int. Ed.* **2020**, *59*, 11717–11731. [CrossRef]
14. Chen, C.; Tian, R.; Zeng, Y.; Chu, C.; Liu, G. Activatable fluorescence probes for "turn-on" and ratiometric biosensing and bioimaging: From NIR-I to NIR-II. *Bioconjug. Chem.* **2020**, *31*, 276–292. [CrossRef]
15. Zheng, H.; Ma, B.; Shi, Y.; Dai, Q.; Li, D.; Ren, E.; Zhu, J.; Liu, J.; Chen, H.; Yin, Z.; et al. Tumor microenvironment-triggered MoS_2@GA-Fe nanoreactor: A self-rolling enhanced chemodynamic therapy and hydrogen sulfide treatment for hepatocellular carcinoma. *Chem. Eng. J.* **2021**, *406*, 126888. [CrossRef]
16. Chu, C.; Yu, J.; Ren, E.; Ou, S.; Zhang, Y.; Wu, Y.; Wu, H.; Zhang, Y.; Zhu, J.; Dai, Q.; et al. Multimodal photoacoustic imaging-guided regression of corneal neovascularization: A non-invasive and safe strategy. *Adv. Sci.* **2020**, *7*, 2000346. [CrossRef]
17. Shi, X.; Zhang, Y.; Tian, Y.; Xu, S.; Ren, E.; Bai, S.; Chen, X.; Chu, C.; Xu, Z.; Liu, G. Multi-responsive bottlebrush-like unimolecules self-assembled nano-riceball for synergistic sono-chemotherapy. *Small Methods* **2020**, *5*, 2000416. [CrossRef]
18. He, T.; Jiang, C.; He, J.; Zhang, Y.; He, G.; Wu, J.; Lin, J.; Zhou, X.; Huang, P. Manganese-dioxide-coating-instructed plasmonic modulation of gold nanorods for activatable duplex-imaging-guided NIR-II photothermal-chemodynamic therapy. *Adv. Mater.* **2021**, *33*, 2008540. [CrossRef]
19. Feng, B.; Hou, B.; Xu, Z.; Saeed, M.; Yu, H.; Li, Y. Self-amplified drug delivery with light-inducible nanocargoes to enhance cancer immunotherapy. *Adv. Mater.* **2019**, *31*, 1902960. [CrossRef]

20. Li, X.; Zheng, B.Y.; Ke, M.R.; Zhang, Y.; Huang, J.D.; Yoon, J. A tumor-pH-responsive supramolecular photosensitizer for activatable photodynamic therapy with minimal in vivo skin phototoxicity. *Theranostics* **2017**, *7*, 2746–2756. [CrossRef]
21. Li, D.; Wang, S.; Lei, Z.; Sun, C.; El-Toni, A.M.; Alhoshan, M.S.; Fan, Y.; Zhang, F. Peroxynitrite activatable NIR-II fluorescent molecular probe for drug-induced hepatotoxicity monitoring. *Anal. Chem.* **2019**, *91*, 4771–4779. [CrossRef]
22. Danko, M.; Hrdlovic, P.; Martinicka, A.; Benda, A.; Cigan, M. Spectral properties of ionic benzotristhiazole based donor-acceptor NLO-phores in polymer matrices and their one- and two-photon cellular imaging ability. *Photochem. Photobiol. Sci.* **2017**, *16*, 1832–1844. [CrossRef]
23. Zhou, C.; Zhang, L.; Sun, T.; Zhang, Y.; Liu, Y.; Gong, M.; Xu, Z.; Du, M.; Liu, Y.; Liu, G.; et al. Activatable NIR-II plasmonic nanotheranostics for efficient photoacoustic imaging and photothermal cancer therapy. *Adv. Mater.* **2021**, *33*, 2006532. [CrossRef]
24. Zhao, M.; Li, B.; Zhang, H.; Zhang, F. Activatable fluorescence sensors for in vivo bio-detection in the second near-infrared window. *Chem. Sci.* **2020**, *12*, 3448–3459. [CrossRef]
25. Zheng, B.D.; Ye, J.; Zhang, X.Q.; Zhang, N.; Xiao, M.T. Recent advances in supramolecular activatable phthalocyanine-based photosensitizers for anti-cancer therapy. *Coord. Chem. Rev.* **2021**, *447*, 214155. [CrossRef]
26. Li, X.; Kolemen, S.; Yoon, J.; Akkaya, E.U. Activatable photosensitizers: Agents for selective photodynamic therapy. *Adv. Funct. Mater.* **2017**, *27*, 1604053. [CrossRef]
27. Li, D.; Wang, X.Z.; Yang, L.F.; Li, S.C.; Hu, Q.Y.; Li, X.; Zheng, B.Y.; Ke, M.R.; Huang, J.D. Size-tunable targeting-triggered nanophotosensitizers based on self-assembly of a phthalocyanine-biotin conjugate for photodynamic therapy. *ACS Appl. Mater. Interf.* **2019**, *11*, 36435–36443. [CrossRef]
28. Li, X.; Yu, S.; Lee, Y.; Guo, T.; Kwon, N.; Lee, D.; Yeom, S.C.; Cho, Y.; Kim, G.; Huang, J.D.; et al. In vivo albumin traps photosensitizer monomers from self-assembled phthalocyanine nanovesicles: A facile and switchable theranostic approach. *J. Am. Chem. Soc.* **2019**, *141*, 1366–1372. [CrossRef]
29. Li, X.; Yu, S.; Lee, D.; Kim, G.; Lee, B.; Cho, Y.; Zheng, B.Y.; Ke, M.R.; Huang, J.D.; Nam, K.T.; et al. Facile supramolecular approach to nucleic-acid-driven activatable nanotheranostics that overcome drawbacks of photodynamic therapy. *ACS Nano* **2018**, *12*, 681–688. [CrossRef]
30. Ding, F.; Zhan, Y.; Lu, X.; Sun, Y. Recent advances in near-infrared II fluorophores for multifunctional biomedical imaging. *Chem. Sci.* **2018**, *9*, 4370–4380. [CrossRef]
31. Li, L.; Dong, X.; Li, J.; Wei, J. A short review on NIR-II organic small molecule dyes. *Dyes Pigments* **2020**, *183*, 108756. [CrossRef]
32. Zhou, H.; Xiao, Y.; Hong, X. New NIR-II dyes without a benzobisthiadiazole core. *Chin. Chem. Lett.* **2018**, *29*, 1425–1428. [CrossRef]
33. Zhou, H.; Yi, W.; Li, A.; Wang, B.; Ding, Q.; Xue, L.; Zeng, X.; Feng, Y.; Li, Q.; Wang, T.; et al. Specific small-molecule NIR-II fluorescence imaging of osteosarcoma and lung metastasis. *Adv. Healthc. Mater.* **2020**, *9*, 1901224. [CrossRef] [PubMed]
34. Qi, J.; Sun, C.; Zebibula, A.; Zhang, H.; Kwok, R.T.K.; Zhao, X.; Xi, W.; Lam, J.W.Y.; Qian, J.; Tang, B.Z. Real-time and high-resolution bioimaging with bright aggregation-induced emission dots in short-wave infrared region. *Adv. Mater.* **2018**, *30*, 1706856. [CrossRef]
35. Samanta, S.; Huang, M.; Li, S.; Yang, Z.; He, Y.; Gu, Z.; Zhang, J.; Zhang, D.; Liu, L.; Qu, J. AIE-active two-photon fluorescent nanoprobe with NIR-II light excitability for highly efficient deep brain vasculature imaging. *Theranostics* **2021**, *11*, 2137–2148. [CrossRef]
36. Li, B.; Lu, L.; Zhao, M.; Lei, Z.; Zhang, F. An efficient 1064 nm NIR-II excitation fluorescent molecular dye for deep-tissue high-resolution dynamic bioimaging. *Angew. Chem. Int. Ed.* **2018**, *57*, 7483–7487. [CrossRef]
37. Bai, L.; Sun, P.; Liu, Y.; Zhang, H.; Hu, W.; Zhang, W.; Liu, Z.; Fan, Q.; Li, L.; Huang, W. Novel aza-BODIPY based small molecular NIR-II fluorophores for in vivo imaging. *Chem. Commun.* **2019**, *55*, 10920–10923. [CrossRef]
38. Rathnamalala, C.S.L.; Gayton, J.N.; Dorris, A.L.; Autry, S.A.; Meador, W.; Hammer, N.I.; Delcamp, J.H.; Scott, C.N. Donor-acceptor-donor NIR II emissive rhodindolizine dye synthesized by C-H bond functionalization. *J. Org. Chem.* **2019**, *84*, 13186–13193. [CrossRef]
39. Zeng, X.; Xue, L.; Chen, D.; Li, S.; Nong, J.; Wang, B.; Tang, L.; Li, Q.; Li, Y.; Deng, Z.; et al. A bright NIR-II fluorescent probe for breast carcinoma imaging and image-guided surgery. *Chem. Commun.* **2019**, *55*, 14287–14290. [CrossRef]
40. Yang, R.Q.; Lou, K.L.; Wang, P.Y.; Gao, Y.Y.; Zhang, Y.Q.; Chen, M.; Huang, W.H.; Zhang, G.J. Surgical navigation for malignancies guided by near-infrared-II fluorescence imaging. *Small Methods* **2021**, *5*, 2001066. [CrossRef]
41. Yang, F.; Zhang, Q.; Huang, S.; Ma, D. Recent advances of near infrared inorganic fluorescent probes for biomedical applications. *J. Mater. Chem. B* **2020**, *8*, 7856–7879. [CrossRef]
42. Zhu, S.; Tian, R.; Antaris, A.L.; Chen, X.; Dai, H. Near-infrared-II molecular dyes for cancer imaging and surgery. *Adv. Mater.* **2019**, *31*, 1900321. [CrossRef]
43. Li, C.; Zhang, Y.; Wang, M.; Zhang, Y.; Chen, G.; Li, L.; Wu, D.; Wang, Q. In vivo real-time visualization of tissue blood flow and angiogenesis using Ag_2S quantum dots in the NIR-II window. *Biomaterials* **2014**, *35*, 393–400. [CrossRef]
44. Takeuchi, T.; Iizumi, Y.; Yudasaka, M.; Kizaka-Kondoh, S.; Okazaki, T. Characterization and biodistribution analysis of oxygen-doped single-walled carbon nanotubes used as in vivo fluorescence imaging probes. *Bioconjug. Chem.* **2019**, *30*, 1323–1330. [CrossRef]

45. Ghosh, D.; Bagley, A.F.; Na, Y.J.; Birrer, M.J.; Bhatia, S.N.; Belcher, A.M. Deep, noninvasive imaging and surgical guidance of submillimeter tumors using targeted M13-stabilized single-walled carbon nanotubes. *Proc. Natl. Acad. Sci. USA* **2014**, *111*, 13948–13953. [CrossRef]
46. Lei, X.; Li, R.; Tu, D.; Shang, X.; Liu, Y.; You, W.; Sun, C.; Zhang, F.; Chen, X. Intense near-infrared-II luminescence from NaCeF4:Er/Yb nanoprobes for in vitro bioassay and in vivo bioimaging. *Chem. Sci.* **2018**, *9*, 4682–4688. [CrossRef]
47. Xue, Z.; Zeng, S.; Hao, J. Non-invasive through-skull brain vascular imaging and small tumor diagnosis based on NIR-II emissive lanthanide nanoprobes beyond 1500nm. *Biomaterials* **2018**, *171*, 153–163. [CrossRef]
48. Su, M.; Dai, Q.; Chen, C.; Zeng, Y.; Chu, C.; Liu, G. Nano-medicine for thrombosis: A precise diagnosis and treatment strategy. *Nano-Micro Lett.* **2020**, *12*, 96. [CrossRef]
49. Ke, M.R.; Chen, S.F.; Peng, X.H.; Zheng, Q.F.; Zheng, B.Y.; Yeh, C.K.; Huang, J.D. A tumor-targeted activatable phthalocyanine-tetrapeptide-doxorubicin conjugate for synergistic chemo-photodynamic therapy. *Eur. J. Med. Chem.* **2017**, *127*, 200–209. [CrossRef]
50. Zhou, K.; Liu, H.; Zhang, S.; Huang, X.; Wang, Y.; Huang, G.; Sumer, B.D.; Gao, J. Multicolored pH-tunable and activatable fluorescence nanoplatform responsive to physiologic pH stimuli. *J. Am. Chem. Soc.* **2012**, *134*, 7803–7811. [CrossRef]
51. Yue, Y.; Huo, F.; Lee, S.; Yin, C.; Yoon, J. A review: The trend of progress about pH probes in cell application in recent years. *Analyst* **2017**, *142*, 30–41. [CrossRef] [PubMed]
52. Gong, F.; Yang, N.; Wang, X.; Zhao, Q.; Chen, Q.; Liu, Z.; Cheng, L. Tumor microenvironment-responsive intelligent nanoplatforms for cancer theranostics. *Nano Today* **2020**, *32*, 100851. [CrossRef]
53. Raish, M.; Shahid, M.; Bin Jardan, Y.A.; Ansari, M.A.; Alkharfy, K.M.; Ahad, A.; Abdelrahman, I.A.; Ahmad, A.; Al-Jenoobi, F.I. Gastroprotective effect of sinapic acid on ethanol-induced gastric ulcers in rats: Involvement of Nrf2/HO-1 and NF-kappaB signaling and antiapoptotic role. *Front. Pharmacol.* **2021**, *12*, 622815. [CrossRef] [PubMed]
54. Wang, S.; Fan, Y.; Li, D.; Sun, C.; Lei, Z.; Lu, L.; Wang, T.; Zhang, F. Anti-quenching NIR-II molecular fluorophores for in vivo high-contrast imaging and pH sensing. *Nat. Commun.* **2019**, *10*, 1058. [CrossRef] [PubMed]
55. Zhao, M.; Wang, J.; Lei, Z.; Lu, L.; Wang, S.; Zhang, H.; Li, B.; Zhang, F. NIR-II pH sensor with a FRET adjustable transition point for in situ dynamic tumor microenvironment visualization. *Angew. Chem. Int. Ed.* **2021**, *60*, 5091–5095. [CrossRef]
56. Wu, J.; You, L.; Chaudhry, S.T.; He, J.; Cheng, J.X.; Mei, J. Ambient oxygen-doped conjugated polymer for pH-activatable aggregation-enhanced photoacoustic imaging in the second near-infrared window. *Anal. Chem.* **2021**, *93*, 3189–3195. [CrossRef]
57. Ren, E.; Chu, C.; Zhang, Y.; Wang, J.; Pang, X.; Lin, X.; Liu, C.; Shi, X.; Dai, Q.; Lv, P.; et al. Mimovirus vesicle-based biological orthogonal reaction for cancer diagnosis. *Small Methods* **2020**, *4*, 2000291. [CrossRef]
58. Li, J.; Zheng, L.; Li, C.; Xiao, Y.; Liu, J.; Wu, S.; Zhang, B. Mannose modified zwitterionic polyester-conjugated second near-infrared organic fluorophore for targeted photothermal therapy. *Biomater. Sci.* **2021**, *9*, 4648–4661. [CrossRef]
59. Bai, S.; Zhang, Y.; Li, D.; Shi, X.; Lin, G.; Liu, G. Gain an advantage from both sides: Smart size-shrinkable drug delivery nanosystems for high accumulation and deep penetration. *Nano Today* **2021**, *36*, 101038. [CrossRef]
60. Wang, X.; Li, C.; Qian, J.; Lv, X.; Li, H.; Zou, J.; Zhang, J.; Meng, X.; Liu, H.; Qian, Y.; et al. NIR-II responsive hollow magnetite nanoclusters for targeted magnetic resonance imaging-guided photothermal/chemo-therapy and chemodynamic therapy. *Small* **2021**, *17*, 2100794. [CrossRef]
61. Zhao, P.H.; Ma, S.T.; Hu, J.Q.; Zheng, B.Y.; Ke, M.R.; Huang, J.D. Artesunate-based multifunctional nanoplatform for photothermal/photoinduced thermodynamic synergistic anticancer therapy. *ACS Appl. Bio. Mater.* **2020**, *3*, 7876–7885. [CrossRef]
62. Ling, S.; Yang, X.; Li, C.; Zhang, Y.; Yang, H.; Chen, G.; Wang, Q. Tumor microenvironment-activated NIR-II nanotheranostic system for precise diagnosis and treatment of peritoneal metastasis. *Angew. Chem. Int. Ed.* **2020**, *59*, 7219–7223. [CrossRef]
63. Liu, T.; Zhang, X.; Liu, D.; Chen, B.; Ge, X.; Gao, S.; Song, J. Self-assembled Ag$_2$S-QD vesicles for in situ responsive NIR-II fluorescence imaging-guided photothermal cancer therapy. *Adv. Opt. Mater.* **2021**, *9*, 2100233. [CrossRef]
64. Zhang, Y.; Wang, X.; Chu, C.; Zhou, Z.; Chen, B.; Pang, X.; Lin, G.; Lin, H.; Guo, Y.; Ren, E.; et al. Genetically engineered magnetic nanocages for cancer magneto-catalytic theranostics. *Nat. Commun.* **2020**, *11*, 5421. [CrossRef]
65. Shi, Y.; Wang, J.; Liu, J.; Lin, G.; Xie, F.; Pang, X.; Pei, Y.; Cheng, Y.; Zhang, Y.; Lin, Z.; et al. Oxidative stress-driven DR5 upregulation restores TRAIL/Apo2L sensitivity induced by iron oxide nanoparticles in colorectal cancer. *Biomaterials* **2020**, *233*, 119753. [CrossRef]
66. Gong, L.; Shan, X.; Zhao, X.H.; Tang, L.; Zhang, X.B. Activatable NIR-II fluorescent probes applied in biomedicine: Progress and perspectives. *ChemMedChem* **2021**, *16*, 2426–2440. [CrossRef]
67. Suzuki, H.; Ohto, U.; Higaki, K.; Mena-Barragan, T.; Aguilar-Moncayo, M.; Ortiz Mellet, C.; Nanba, E.; Garcia Fernandez, J.M.; Suzuki, Y.; Shimizu, T. Structural basis of pharmacological chaperoning for human beta-galactosidase. *J. Biol. Chem.* **2014**, *289*, 14560–14568. [CrossRef]
68. Ohto, U.; Usui, K.; Ochi, T.; Yuki, K.; Satow, Y.; Shimizu, T. Crystal structure of human beta-galactosidase: Structural basis of Gm1 gangliosidosis and morquio B diseases. *J. Biol. Chem.* **2012**, *287*, 1801–1812. [CrossRef]
69. Asanuma, D.; Sakabe, M.; Kamiya, M.; Yamamoto, K.; Hiratake, J.; Ogawa, M.; Kosaka, N.; Choyke, P.L.; Nagano, T.; Kobayashi, H.; et al. Sensitive beta-galactosidase-targeting fluorescence probe for visualizing small peritoneal metastatic tumours in vivo. *Nat. Commun.* **2015**, *6*, 6463. [CrossRef]
70. Kim, E.J.; Kumar, R.; Sharma, A.; Yoon, B.; Kim, H.M.; Lee, H.; Hong, K.S.; Kim, J.S. In vivo imaging of beta-galactosidase stimulated activity in hepatocellular carcinoma using ligand-targeted fluorescent probe. *Biomaterials* **2017**, *122*, 83–90. [CrossRef]

71. Zhen, X.; Zhang, J.; Huang, J.; Xie, C.; Miao, Q.; Pu, K. Macrotheranostic probe with disease-activated near-infrared fluorescence, photoacoustic, and photothermal signals for imaging-guided therapy. *Angew. Chem. Int. Ed.* **2018**, *57*, 7804–7808. [CrossRef] [PubMed]
72. Chen, J.A.; Pan, H.; Wang, Z.; Gao, J.; Tan, J.; Ouyang, Z.; Guo, W.; Gu, X. Imaging of ovarian cancers using enzyme activatable probes with second near-infrared window emission. *Chem. Commun.* **2020**, *56*, 2731–2734. [CrossRef] [PubMed]
73. Gimeno, A.Z.; Santana, A.; Jimenez, A.; Parra, A.D.; Nicolas, G.; Paz, C.; Diaz, F.; Medina, C.; Diaz, L.; Quintero, E. Up-regulation of gelatinases in the colorectal adenoma-carcinoma sequence. *Eur. J. Cancer* **2006**, *42*, 3246–3252. [CrossRef] [PubMed]
74. Xiang, X.; Zhao, X.; Qu, H.; Li, D.; Yang, D.; Pu, J.; Mei, H.; Zhao, J.; Huang, K.; Zheng, L.; et al. Hepatocyte nuclear factor 4 alpha promotes the invasion, metastasis and angiogenesis of neuroblastoma cells via targeting matrix metalloproteinase 14. *Cancer Lett.* **2015**, *359*, 187–197. [CrossRef] [PubMed]
75. Esposito, M.R.; Binatti, A.; Pantile, M.; Coppe, A.; Mazzocco, K.; Longo, L.; Capasso, M.; Lasorsa, V.A.; Luksch, R.; Bortoluzzi, S.; et al. Somatic mutations in specific and connected subpathways are associated with short neuroblastoma patients' survival and indicate proteins targetable at onset of disease. *Int. J. Cancer* **2018**, *143*, 2525–2536. [CrossRef] [PubMed]
76. Jeong, S.; Song, J.; Lee, W.; Ryu, Y.M.; Jung, Y.; Kim, S.Y.; Kim, K.; Hong, S.C.; Myung, S.J.; Kim, S. Cancer-microenvironment-sensitive activatable quantum dot probe in the second near-infrared window. *Nano Lett.* **2017**, *17*, 1378–1386. [CrossRef]
77. Zhan, Y.; Ling, S.; Huang, H.; Zhang, Y.; Chen, G.; Huang, S.; Li, C.; Guo, W.; Wang, Q. Rapid unperturbed-tissue analysis for intraoperative cancer diagnosis using an enzyme-activated NIR-II nanoprobe. *Angew. Chem. Int. Ed.* **2021**, *60*, 2637–2642. [CrossRef]
78. Li, D.; Hu, Q.Y.; Wang, X.Z.; Li, X.; Hu, J.Q.; Zheng, B.Y.; Ke, M.R.; Huang, J.D. A non-aggregated silicon(IV) phthalocyanine-lactose conjugate for photodynamic therapy. *Bioorg. Med. Chem. Lett.* **2020**, *30*, 127164. [CrossRef]
79. Zhao, Y.Y.; Chen, J.Y.; Hu, J.Q.; Zhang, L.; Lin, A.L.; Wang, R.; Zheng, B.Y.; Ke, M.R.; Li, X.; Huang, J.D. The substituted zinc(II) phthalocyanines using "sulfur bridge" as the linkages. synthesis, red-shifted spectroscopic properties and structure-inherent targeted photodynamic activities. *Dyes Pigments* **2021**, *189*, 109270. [CrossRef]
80. He, X.J.; Chen, H.; Xu, C.C.; Fan, J.Y.; Xu, W.; Li, Y.H.; Deng, H.; Shen, J.L. Ratiometric and colorimetric fluorescent probe for hypochlorite monitor and application for bioimaging in living cells, bacteria and zebrafish. *J. Hazard. Mater.* **2020**, *388*, 122029. [CrossRef]
81. Ye, J.; Li, Z.; Fu, Q.; Li, Q.; Zhang, X.; Su, L.; Yang, H.; Song, J. Quantitative photoacoustic diagnosis and precise treatment of inflammation in vivo using activatable theranostic nanoprobe. *Adv. Funct. Mater.* **2020**, *30*, 2001771. [CrossRef]
82. Zhang, X.; Wang, W.; Su, L.; Ge, X.; Ye, J.; Zhao, C.; He, Y.; Yang, H.; Song, J.; Duan, H. Plasmonic-fluorescent janus Ag/Ag$_2$S nanoparticles for in situ H$_2$O$_2$-activated NIR-II fluorescence imaging. *Nano Lett.* **2021**, *21*, 2625–2633. [CrossRef]
83. Khojah, H.M.; Ahmed, S.; Abdel-Rahman, M.S.; Hamza, A.B. Reactive oxygen and nitrogen species in patients with rheumatoid arthritis as potential biomarkers for disease activity and the role of antioxidants. *Free Radic. Biol. Med.* **2016**, *97*, 285–291. [CrossRef]
84. Sayre, L.M.; Perry, G.; Smith, M.A. Oxidative stress and neurotoxicity. *Chem. Res. Toxicol.* **2008**, *21*, 172–188. [CrossRef]
85. Wang, X.; Li, P.; Ding, Q.; Wu, C.; Zhang, W.; Tang, B. Observation of acetylcholinesterase in stress-induced depression phenotypes by two-photon fluorescence imaging in the mouse brain. *J. Am. Chem. Soc.* **2019**, *141*, 2061–2068. [CrossRef]
86. Oka, T.; Yamashita, S.; Midorikawa, M.; Saiki, S.; Muroya, Y.; Kamibayashi, M.; Yamashita, M.; Anzai, K.; Katsumura, Y. Spin-trapping reactions of a novel gauchetype radical trapper G-CYPMPO. *Anal. Chem.* **2011**, *83*, 9600–9604. [CrossRef]
87. Sun, W.; Guo, S.; Hu, C.; Fan, J.; Peng, X. Recent development of chemosensors based on cyanine platforms. *Chem. Rev.* **2016**, *116*, 7768–7817. [CrossRef]
88. Feng, W.; Zhang, Y.; Li, Z.; Zhai, S.; Lv, W.; Liu, Z. Lighting up NIR-II fluorescence in vivo: An activable probe for noninvasive hydroxyl radical imaging. *Anal. Chem.* **2019**, *91*, 15757–15762. [CrossRef]
89. Zhu, H.; Fan, J.; Wang, J.; Mu, H.; Peng, X. An "enhanced PET"-based fluorescent probe with ultrasensitivity for imaging basal and elesclomol-induced HClO in cancer cells. *J. Am. Chem. Soc.* **2014**, *136*, 12820–12823. [CrossRef]
90. Wu, L.; Wu, I.C.; DuFort, C.C.; Carlson, M.A.; Wu, X.; Chen, L.; Kuo, C.T.; Qin, Y.; Yu, J.; Hingorani, S.R.; et al. Photostable ratiometric pdot probe for in vitro and in vivo imaging of hypochlorous acid. *J. Am. Chem. Soc.* **2017**, *139*, 6911–6918. [CrossRef]
91. Li, J.; Rao, J.; Pu, K. Recent progress on semiconducting polymer nanoparticles for molecular imaging and cancer phototherapy. *Biomaterials* **2018**, *155*, 217–235. [CrossRef]
92. Wang, S.; Liu, L.; Fan, Y.; El-Toni, A.M.; Alhoshan, M.S.; Li, D.; Zhang, F. In vivo high-resolution ratiometric fluorescence imaging of inflammation using NIR-II nanoprobes with 1550 nm emission. *Nano Lett.* **2019**, *19*, 2418–2427. [CrossRef]
93. Miao, Q.; Xie, C.; Zhen, X.; Lyu, Y.; Duan, H.; Liu, X.; Jokerst, J.V.; Pu, K. Molecular afterglow imaging with bright, biodegradable polymer nanoparticles. *Nat. Biotechnol.* **2017**, *35*, 1102–1110. [CrossRef]
94. Ge, X.; Lou, Y.; Su, L.; Chen, B.; Guo, Z.; Gao, S.; Zhang, W.; Chen, T.; Song, J.; Yang, H. Single wavelength laser excitation ratiometric NIR-II fluorescent probe for molecule imaging in vivo. *Anal. Chem.* **2020**, *92*, 6111–6120. [CrossRef]
95. Tang, Y.; Li, Y.; Lu, X.; Hu, X.; Zhao, H.; Hu, W.; Lu, F.; Fan, Q.; Huang, W. Bio-erasable intermolecular donor–acceptor interaction of organic semiconducting nanoprobes for activatable NIR-II fluorescence imaging. *Adv. Funct. Mater.* **2019**, *29*, 1807376. [CrossRef]
96. Kwon, N.; Kim, D.; Swamy, K.M.K.; Yoon, J. Metal-coordinated fluorescent and luminescent probes for reactive oxygen species (ROS) and reactive nitrogen species (RNS). *Coord. Chem. Rev.* **2021**, *427*, 213581. [CrossRef]

97. Tang, Y.; Pei, F.; Lu, X.; Fan, Q.; Huang, W. Recent advances on activatable NIR-II fluorescence probes for biomedical imaging. *Adv. Opt. Mater.* **2019**, *7*, 1900917. [CrossRef]
98. Szabo, C.; Ischiropoulos, H.; Radi, R. Peroxynitrite: Biochemistry, pathophysiology and development of therapeutics. *Nat. Rev. Drug Discov.* **2007**, *6*, 662–680. [CrossRef]
99. Ai, X.; Wang, Z.; Cheong, H.; Wang, Y.; Zhang, R.; Lin, J.; Zheng, Y.; Gao, M.; Xing, B. Multispectral optoacoustic imaging of dynamic redox correlation and pathophysiological progression utilizing upconversion nanoprobes. *Nat. Commun.* **2019**, *10*, 1087. [CrossRef]
100. Zhao, M.; Li, B.; Wu, Y.; He, H.; Zhu, X.; Zhang, H.; Dou, C.; Feng, L.; Fan, Y.; Zhang, F. A tumor-microenvironment-responsive lanthanide-cyanine FRET sensor for NIR-II luminescence-lifetime in situ imaging of hepatocellular carcinoma. *Adv. Mater.* **2020**, *32*, 2001172. [CrossRef]
101. Zhang, J.; Zhen, X.; Upputuri, P.K.; Pramanik, M.; Chen, P.; Pu, K. Activatable photoacoustic nanoprobes for in vivo ratiometric imaging of peroxynitrite. *Adv. Mater.* **2017**, *29*, 1604764. [CrossRef] [PubMed]
102. Uteshev, V.V. Allosteric modulation of nicotinic acetylcholine receptors: The concept and therapeutic trends. *Curr. Pharm. Des.* **2016**, *22*, 1986–1997. [CrossRef] [PubMed]
103. Li, X.; Tao, R.R.; Hong, L.J.; Cheng, J.; Jiang, Q.; Lu, Y.M.; Liao, M.H.; Ye, W.F.; Lu, N.N.; Han, F.; et al. Visualizing peroxynitrite fluxes in endothelial cells reveals the dynamic progression of brain vascular injury. *J. Am. Chem. Soc.* **2015**, *137*, 12296–12303. [CrossRef] [PubMed]
104. Adibhatla, R.M.; Hatcher, J.F. Lipid oxidation and peroxidation in CNS health and disease: From molecular mechanisms to therapeutic opportunities. *Antioxid. Redox Signal.* **2010**, *12*, 125–169. [CrossRef]
105. Li, C.; Li, W.; Liu, H.; Zhang, Y.; Chen, G.; Li, Z.; Wang, Q. An activatable NIR-II nanoprobe for in vivo early real-time diagnosis of traumatic brain injury. *Angew. Chem. Int. Ed.* **2020**, *59*, 247–252. [CrossRef]
106. Zhang, K.; Tu, M.; Gao, W.; Cai, X.; Song, F.; Chen, Z.; Zhang, Q.; Wang, J.; Jin, C.; Shi, J.; et al. Hollow prussian blue nanozymes drive neuroprotection against ischemic stroke via attenuating oxidative stress, counteracting inflammation, and suppressing cell apoptosis. *Nano Lett.* **2019**, *19*, 2812–2823. [CrossRef]
107. Li, S.; Jiang, D.; Ehlerding, E.B.; Rosenkrans, Z.T.; Engle, J.W.; Wang, Y.; Liu, H.; Ni, D.; Cai, W. Intrathecal administration of nanoclusters for protecting neurons against oxidative stress in cerebral ischemia/reperfusion injury. *ACS Nano* **2019**, *13*, 13382–13389. [CrossRef]
108. Liu, Y.; Ai, K.; Ji, X.; Askhatova, D.; Du, R.; Lu, L.; Shi, J. Comprehensive insights into the multi-antioxidative mechanisms of melanin nanoparticles and their application to protect brain from injury in ischemic stroke. *J. Am. Chem. Soc.* **2017**, *139*, 856–862. [CrossRef]
109. Yang, X.; Wang, Z.; Huang, H.; Ling, S.; Zhang, R.; Zhang, Y.; Chen, G.; Li, C.; Wang, Q. A targeted activatable NIR-IIb nanoprobe for highly sensitive detection of ischemic stroke in a photothrombotic stroke model. *Adv. Healthc. Mater.* **2021**, *10*, 2001544. [CrossRef]
110. Wang, R.; Zhou, L.; Wang, W.; Li, X.; Zhang, F. In vivo gastrointestinal drug-release monitoring through second near-infrared window fluorescent bioimaging with orally delivered microcarriers. *Nat. Commun.* **2017**, *8*, 14702. [CrossRef]
111. Xie, C.; Zhen, X.; Lyu, Y.; Pu, K. Nanoparticle regrowth enhances photoacoustic signals of semiconducting macromolecular probe for in vivo imaging. *Adv. Mater.* **2017**, *29*, 1703693. [CrossRef]
112. Peng, J.; Samanta, A.; Zeng, X.; Han, S.; Wang, L.; Su, D.; Loong, D.T.; Kang, N.Y.; Park, S.J.; All, A.H.; et al. Real-time in vivo hepatotoxicity monitoring through chromophore-conjugated photon-upconverting nanoprobes. *Angew. Chem. Int. Ed.* **2017**, *56*, 4165–4169. [CrossRef]
113. Shuhendler, A.J.; Pu, K.; Cui, L.; Uetrecht, J.P.; Rao, J. Real-time imaging of oxidative and nitrosative stress in the liver of live animals for drug-toxicity testing. *Nat. Biotechnol.* **2014**, *32*, 373380. [CrossRef]
114. Iverson, N.M.; Barone, P.W.; Shandell, M.; Trudel, L.J.; Sen, S.; Sen, F.; Ivanov, V.; Atolia, E.; Farias, E.; McNicholas, T.P.; et al. In vivo biosensing via tissue-localizable near-infrared-fluorescent single-walled carbon nanotubes. *Nat. Nanotechnol.* **2013**, *8*, 873–880. [CrossRef]
115. Tang, Y.; Li, Y.; Wang, Z.; Pei, F.; Hu, X.; Ji, Y.; Li, X.; Zhao, H.; Hu, W.; Lu, X.; et al. Organic semiconducting nanoprobe with redox-activatable NIR-II fluorescence for in vivo real-time monitoring of drug toxicity. *Chem. Commun.* **2019**, *55*, 27–30. [CrossRef]
116. Shi, B.; Yan, Q.; Tang, J.; Xin, K.; Zhang, J.; Zhu, Y.; Xu, G.; Wang, R.; Chen, J.; Gao, W.; et al. Hydrogen sulfide-activatable second near-infrared fluorescent nanoassemblies for targeted photothermal cancer therapy. *Nano Lett.* **2018**, *18*, 6411–6416. [CrossRef]
117. Xu, G.; Yan, Q.; Lv, X.; Zhu, Y.; Xin, K.; Shi, B.; Wang, R.; Chen, J.; Gao, W.; Shi, P.; et al. Imaging of colorectal cancers using activatable nanoprobes with second near-infrared window emission. *Angew. Chem. Int. Ed.* **2018**, *57*, 3626–3630. [CrossRef]
118. Bai, S.; Jia, D.; Ma, X.; Liang, M.; Xue, P.; Kang, Y.; Xu, Z. Cylindrical polymer brushes-anisotropic unimolecular micelle drug delivery system for enhancing the effectiveness of chemotherapy. *Bioact. Mater.* **2021**, *6*, 2894–2904. [CrossRef]
119. Li, Z.; Wu, J.; Wang, Q.; Liang, T.; Ge, J.; Wang, P.; Liu, Z. A universal strategy to construct lanthanide-doped nanoparticles-based activable NIR-II luminescence probe for bioimaging. *iScience* **2020**, *23*, 100962. [CrossRef]
120. Yang, Z.; Luo, Y.; Hu, Y.; Liang, K.; He, G.; Chen, Q.; Wang, Q.; Chen, H. Photothermo-promoted nanocatalysis combined with H_2S-mediated respiration inhibition for efficient cancer therapy. *Adv. Funct. Mater.* **2021**, *31*, 2007991. [CrossRef]
121. Dou, K.; Feng, W.; Fan, C.; Cao, Y.; Xiang, Y.; Liu, Z. Flexible designing strategy to construct activatable NIR-II fluorescent probes with emission maxima beyond 1200 nm. *Anal. Chem.* **2021**, *93*, 4006–4014. [CrossRef]

122. Liu, Q.; Zhong, Y.; Su, Y.; Zhao, L.; Peng, J. Real-time imaging of hepatic inflammation using hydrogen sulfide-activatable second near-infrared luminescent nanoprobes. *Nano Lett.* **2021**, *21*, 4606–4614. [CrossRef]
123. Xu, M.; Wang, P.; Sun, S.; Gao, L.; Sun, L.; Zhang, L.; Zhang, J.; Wang, S.; Liang, X. Smart strategies to overcome tumor hypoxia toward the enhancement of cancer therapy. *Nanoscale* **2020**, *12*, 21519–21533. [CrossRef]
124. Brown, J.M.; Wilson, W.R. Exploiting tumour hypoxia in cancer treatment. *Nat. Rev. Cancer* **2004**, *4*, 437–447. [CrossRef] [PubMed]
125. Sundaram, A.; Peng, L.; Chai, L.; Xie, Z.; Ponraj, J.S.; Wang, X.; Wang, G.; Zhang, B.; Nie, G.; Xie, N.; et al. Advanced nanomaterials for hypoxia tumor therapy: Challenges and solutions. *Nanoscale* **2020**, *12*, 21497–21518. [CrossRef] [PubMed]
126. Meng, X.; Zhang, J.; Sun, Z.; Zhou, L.; Deng, G.; Li, S.; Li, W.; Gong, P.; Cai, L. Hypoxia-triggered single molecule probe for high-contrast NIR II/PA tumor imaging and robust photothermal therapy. *Theranostics* **2018**, *8*, 6025–6034. [CrossRef] [PubMed]
127. Qiu, G.Z.; Jin, M.Z.; Dai, J.X.; Sun, W.; Feng, J.H.; Jin, W.L. Reprogramming of the tumor in the hypoxic niche: The emerging concept and associated therapeutic strategies. *Trends Pharmacol. Sci.* **2017**, *38*, 669–686. [CrossRef] [PubMed]
128. Wallace, D.C. Mitochondria and cancer. *Nat. Rev. Cancer* **2012**, *12*, 685–698. [CrossRef]
129. Yang, Z.; He, Y.; Lee, J.H.; Park, N.; Suh, M.; Chae, W.S.; Cao, J.; Peng, X.; Jung, H.; Kang, C.; et al. A self-calibrating bipartite viscosity sensor for mitochondria. *J. Am. Chem. Soc.* **2013**, *135*, 9181–9185. [CrossRef]
130. Liu, F.; Yuan, Z.; Sui, X.; Wang, C.; Xu, M.; Li, W.; Chen, Y. Viscosity sensitive near-infrared fluorescent probes based on functionalized single-walled carbon nanotubes. *Chem. Commun.* **2020**, *56*, 8301–8304. [CrossRef]
131. Dou, K.; Huang, W.; Xiang, Y.; Li, S.; Liu, Z. Design of activatable NIR-II molecular probe for in vivo elucidation of disease-related viscosity variations. *Anal. Chem.* **2020**, *92*, 4177–4181. [CrossRef]
132. Kwon, H.; Kim, M.; Meany, B.; Piao, Y.; Powell, L.R.; Wang, Y. Optical probing of local pH and temperature in complex fluids with covalently functionalized, semiconducting carbon nanotubes. *J. Phys. Chem. C* **2015**, *119*, 3733–3739. [CrossRef]
133. Lau, J.T.; Lo, P.C.; Jiang, X.J.; Wang, Q.; Ng, D.K. A dual activatable photosensitizer toward targeted photodynamic therapy. *J. Med. Chem.* **2014**, *57*, 4088–4097. [CrossRef]
134. Zheng, Z.; Chen, Q.; Dai, R.; Jia, Z.; Yang, C.; Peng, X.; Zhang, R. A continuous stimuli-responsive system for NIR-II fluorescence/photoacoustic imaging guided photothermal/gas synergistic therapy. *Nanoscale* **2020**, *12*, 11562–11572. [CrossRef]
135. Xu, Q.; Lee, K.A.; Lee, S.; Lee, K.M.; Lee, W.J.; Yoon, J. A highly specific fluorescent probe for hypochlorous acid and its application in imaging microbe-induced HOCl production. *J. Am. Chem. Soc.* **2013**, *135*, 9944–9949. [CrossRef]
136. He, Y.; Wang, S.; Yu, P.; Yan, K.; Ming, J.; Yao, C.; He, Z.; El-Toni, A.M.; Khan, A.; Zhu, X.; et al. NIR-II cell endocytosis-activated fluorescent probes for in vivo high-contrast bioimaging diagnostics. *Chem. Sci.* **2021**, *12*, 10474–10482. [CrossRef]
137. Teng, L.; Song, G.; Liu, Y.; Han, X.; Li, Z.; Wang, Y.; Huan, S.; Zhang, X.B.; Tan, W. Nitric oxide-activated "dual-key-one-lock" nanoprobe for in vivo molecular imaging and high-specificity cancer therapy. *J. Am. Chem. Soc.* **2019**, *141*, 13572–13581. [CrossRef]
138. Teng, K.X.; Niu, L.Y.; Kang, Y.F.; Yang, Q.Z. Rational design of a "dual lock-and-key" supramolecular photosensitizer based on aromatic nucleophilic substitution for specific and enhanced photodynamic therapy. *Chem. Sci.* **2020**, *11*, 9703–9711. [CrossRef]
139. Tang, Y.; Li, Y.; Hu, X.; Zhao, H.; Ji, Y.; Chen, L.; Hu, W.; Zhang, W.; Li, X.; Lu, X.; et al. "Dual lock-and-key"-controlled nanoprobes for ultrahigh specific fluorescence imaging in the second near-infrared window. *Adv. Mater.* **2018**, *30*, 1801140. [CrossRef]

Review

Applications of Aptamer-Bound Nanomaterials in Cancer Therapy

Liangxi Zhu [1], Jingzhou Zhao [1], Zhukang Guo [1], Yuan Liu [1], Hui Chen [2], Zhu Chen [2] and Nongyue He [1,2,*]

[1] State Key Laboratory of Bioelectronics, School of Biological Science and Medical Engineering, Southeast University, Nanjing 210096, China; zhuliangxi@seu.edu.cn (L.Z.); zhaojingzhou@seu.edu.cn (J.Z.); guo724kk@foxmail.com (Z.G.); yuanliu11@seu.edu.cn (Y.L.)

[2] Hunan Key Laboratory of Biomedical Nanomaterials and Devices, Hunan University of Technology, Zhuzhou 412007, China; huier_88@vip.163.com (H.C.); chenzhu@hut.edu.cn (Z.C.)

* Correspondence: nyhe@seu.edu.cn

Abstract: Cancer is still a major disease that threatens human life. Although traditional cancer treatment methods are widely used, they still have many disadvantages. Aptamers, owing to their small size, low toxicity, good specificity, and excellent biocompatibility, have been widely applied in biomedical areas. Therefore, the combination of nanomaterials with aptamers offers a new method for cancer treatment. First, we briefly introduce the situation of cancer treatment and aptamers. Then, we discuss the application of aptamers in breast cancer treatment, lung cancer treatment, and other cancer treatment methods. Finally, perspectives on challenges and future applications of aptamers in cancer therapy are discussed.

Keywords: aptamers; nanomaterials; cancer; treatment; targeting

1. Introduction

Cancer is one of the key threats to human health. Cancer is a disease caused by abnormal cell growth. Cancer cells may spread to different tissues and organs [1]. According to statistics from the World Health Organization, tens of millions of people worldwide are diagnosed with cancer each year. Most cancers are caused by the external environment, such as smoking, radiation, and environmental pollution. Until now, the main way to treat cancer in developed countries has been chemotherapy, radiotherapy, and surgical treatment [2,3]. However, these cancer treatments have a series of problems, such as easy recurrence, poor treatment effects, and large side effects. Hence, people are looking forward to developing new strategies to treat cancers.

In recent years, nanotechnology has developed rapidly, and more and more nanomaterials with excellent physical and chemical properties have been discovered [4] in the fields of electronics, magnetics, and optics. Many nanomaterials have been used by researchers in the field of biomedicine [5–13]. For example, nanomaterials with photothermal conversion properties are used for tumor treatment [14–16] and mesoporous materials are used for drug delivery [17–19], while magnetic nanomaterials have been found to have excellent superparamagnetic properties, and carbon nanomaterials have been found to feature broad absorbance regions [20–23]. The focus on nanomaterial applications has been moving from the cellular level toward the tissue level, including cellular imaging, drug delivery, and cancer diagnosis/therapy. Via the enhanced permeability and retention (EPR) effect, most nanomaterials accumulate in a targeted region. People have also proposed various strategies to improve the flow of nanomaterials in blood and improve their stability and biocompatibility [24,25]. However, the EPR effect is not always effective, and the targeting ability of nanomaterials is not good. These problems are accompanied by the issue of harmful toxicity to nontarget tissues or organs. Thus, this problem hinders the development of nanomaterials in bioapplications. Therefore, active targeting strategies have been considered as more powerful tools for cancer treatments, and much attention has been focused on selecting targeted moieties to endow antitumor agents with specific recognition of tumor cells via the affinity between ligands and receptors. There are many receptors overexpressed on the membrane of tumor cells. Based on this, various specific ligands have been discovered, including antibodies, transferrin, folic acid, and peptides. Apart from these ligands, aptamers have become one kind of the most attractive biomolecules.

It is well known that aptamers have a highly selective recognition ability that helps them recognize targets [26–28]. Aptamers are single-stranded DNAs or RNAs evolved through systematic evolution of ligands by exponential enrichment (SELEX) technology [29,30] and they are widely employed, as shown in Figure 1. Aptamers can fold into various secondary structures, further forming three-dimensional structures, which contribute to the interaction between aptamers and their targets through various forces, such as hydrophobic interaction and electrostatic attraction [31,32]. Moreover, compared with traditional targeting ligands, aptamers have their advantages, such as good reproducibility, convenient modification, small size, low toxicity, good stability, and low molecular weight [33–35]. In addition, aptamers have higher rates of tumor penetration, retention, and homogenous distribution, while the attachment process of aptamers to the surface of nanomaterials is more amenable and reproducible. In principle, an aptamer sequence can be deleted, added, or united flexibly to endow aptamers with tunable recognition ability to better identify targets. Stimulus–response strategies including light, pH, ligand binding, and other cues which have been developed into the design of aptamer-based nanomaterials [27–29]. In short, these merits make aptamers ideal candidates for disease diagnosis and therapeutics owing to their ability to deliver therapeutic cargoes into cancer cells, diseased tissue, and organs. Hence, the introduction of aptamers into the construction of nanomaterial systems will promote their bioapplications.

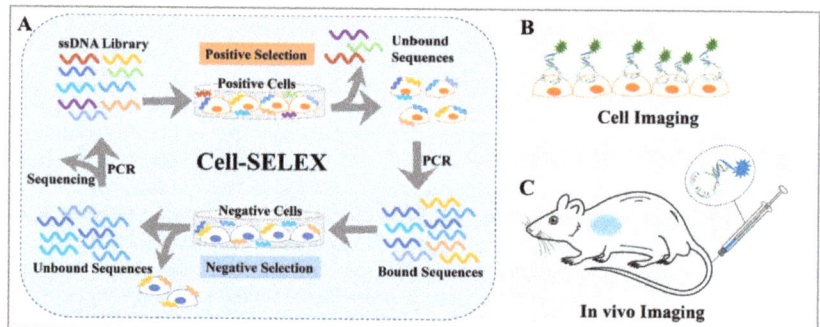

Figure 1. (**A**) Basic principle of the cell-SELEX method and applications for (**B**) cell imaging and (**C**) in vivo imaging. Reprinted with permission from [29]. Copyright © 2020 American Chemical Society.

In this review, we summarize several current applications of aptamer-bound nanomaterials in cancer therapy (Scheme 1). Today, there are various methods to treat cancer. Here, however, we mainly focus on photothermal therapy (PTT), photodynamic therapy (PDT), and improved drug delivery systems (DDSs). We chose several common cancers, including breast cancer, lung cancer, liver cancer, cervical cancer, gastric cancer, colorectal cancer, and prostate cancer. In the first section, we introduce the combination of aptamers with nanomaterials in different cancer therapies. Next, we list several commonly used nanomaterials in cancer therapies (Table 1). Finally, we discuss perspectives on challenges and future applications of aptamer-bound nanomaterials in cancer therapy.

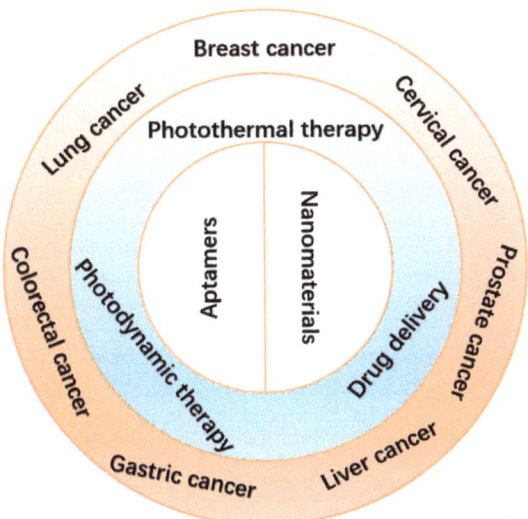

Scheme 1. Schematic illustration of the applications of aptamer-bound nanomaterials in cancer therapy.

2. Therapeutic Method

2.1. Photothermal Thearpy

PTT has gained much attention in the field of cancer therapy since Goldman used laser irradiation to remove tumors in 1966 [36–43]. Owing to its minimal toxicity, noninvasiveness, convenient operation, and low recurrence, PTT showed great promise for cancer therapy. After photosensitizing agents accumulate in the tumor site, a near-infrared (NIR) laser is used. Because of the high photothermal conversion efficiency of photosensitizing agents, the absorbed optical energy is converted into thermal energy, leading to either partial or complete ablation of the target tissues. For effective ablation of the tumors, PTT often requires the tumor center to reach high temperatures (≥ 50 °C). Because of the deep-tissue penetration of NIR light, various organic and inorganic phototherapeutic agents have been developed. For example, Guo et al. prepared B quantum dots (BQDs) with good biocompatibility. BQDs can effectively produce photothermal effects under NIR light irradiation. Both in vitro and in vivo experimental studies showed that BQDs-PEG can significantly kill cancer cells and inhibit tumor growth through photothermal effects [36]. Leng et al. synthesized core-shell and dumbbell-like gold nanorods-Cu_7S_4 heterostructures. Both gold nanorods-Cu_7S_4 heterostructures exhibited significantly enhanced photothermal conversion efficiency (η = 56% and 62%) and good photothermal stability. The in vitro photothermal ablation of cancer cells featured low cytotoxicity and effective photothermal treatments [37]. Wang et al. developed a new dual-targeted small-molecule organic photothermal agent for enhanced photothermal therapy, which can target both biotin and mitochondria. The in vivo photothermal therapy experiments indicated that the dual-targeted photothermal agent performed much better in tumor inhibition [39].

2.2. Photodynamic Therapy

During the past few decades, PDT has been an attractive therapeutic method for tumor therapy [44–48] because it is noninvasive, spatially selective, and has negligible toxicity. After photosensitizers accumulate in the tumor site, reactive oxygen species (ROS) are produced by NIR laser irradiation. Toxic ROS damages tumor cells via the oxidation of protein, DNA, or RNA. Different photosensitizers have been prepared to overcome both accumulation issues and the hypoxic environment in the tumor. For example, Zhou et al. developed an activatable ROS generation system by modulating a biochemical reaction between linoleic acid hydroperoxide and catalytic iron (II) ions. The engineered nanoparticles were capable of inducing apoptotic cancer death both in vitro and in vivo through ROS generation [46]. Yang et al. prepared a Pt nanoenzyme with functionalized nanoplatform black phosphorus/Pt-Ce6/PEG nanosheets for synergistic photothermal and enhanced photodynamic therapy, in which the Pt nanoenzyme would decompose H_2O_2 into oxygen to enhance the photodynamic effect [48].

2.3. Drug Delivery Systems

A DDS can be defined as a method or process using the principles of chemistry, engineering, and biology for administering pharmaceutical compounds. Due to their special properties, DDSs have shown tremendous promise to improve the diagnostic and therapeutic effects of drugs, especially in enhancing the pharmaceutical effects of drugs and reducing the side effects of therapeutics in the treatment of various disease conditions. Various types of DDSs have been extensively investigated as potential drug carriers for the treatment of many diseases [49–60]. For example, Wang et al. successfully constructed a simple and novel gas generator with a high drug loading level that markedly facilitated doxorubicin release and O_2 diffusion for amplifying PDT/PTT/chemotherapy combination therapy [49]. Lei et al. constructed a multifunctional mesoporous silica nanoparticle-based drug delivery platform to load indocyanine green and doxorubicin for NIR-triggered drug release and chemo/photothermal therapy [51].

Although the applications of nanomaterials in PTT, PDT, and DDS have been great successes, improving the targeting and specificity of these therapeutic methods is still a big challenge.

3. Aptamer-Bound Nanomaterials Used in Different Cancer Therapies

3.1. Breast Cancer Therapy

The incidence of breast cancer in female cancers worldwide is 24.2%, of which 52.9% occur in developing countries [61,62]. Developing an optimized treatment method is therefore of great importance. Conventional treatments include surgery, chemotherapy, and radiotherapy. Chemotherapy and radiotherapy treatments have serious side effects. For example, during chemotherapy, the patients' white blood cells may be reduced, and there may be problems with their blood clotting function. It will also affect the normal function of the liver. Radiotherapy is a local treatment method that uses radiation to treat tumors. This process also risks damaging normal tissues and insufficiently killing off cancer cells. Moreover, radiation therapy may also cause wound complications [63–65]. Therefore, the novel treatment involving binding of nanomaterials with aptamers will contribute to the development of breast cancer therapy [66,67].

3.1.1. Photothermal Therapy

Molybdenum disulfide (MoS_2) is a typical transition metal disulfide, which is a class of two-dimensional nanomaterials, that has many biomedical applications, owing to its simple preparation, good stability, large surface area, excellent water dispersibility, and biocompatibility [68–70]. MoS_2 has especially high NIR absorbance. Because of this, it has high photothermal conversion efficiency. Based on this, many researchers have applied it in photothermal therapy [71,72]. However, its ability to recognize specific tumor cells needs to be improved. For example, Pang et al. developed a new method [73] which used MoS_2 as the photothermal agent. Firstly, by mixing MoS_2 and bovine serum albumin (BSA), researchers obtained MoS_2-BSA nanosheets. Next, EDC and NHS were used to activate the free carboxyl group on the MoS_2-BSA surface, and subsequently aptamers were modified to obtain composite MoS_2-BSA-Apt nanosheets, which were stable and biocompatible. Owing to the good affinity between aptamers and receptors on the cell surface [74,75], the composition would distinguish MCF-7 human breast cancer cells from other cells and then enter the cells through endocytosis. Under the irradiation from an 808 nm laser, the heat generated kills cancer cells, which is on the basis of MoS_2 nanosheets. After MoS_2-BSA-Apt was cultured with MCF-7 human breast cancer cells and MCF-10A human breast cancer cells, the fluorescence-inverted microscope results showed that MCF-7 human breast cancer cells had uniformly distributed green fluorescence signals, while MCF-10A human breast cancer cells had almost no fluorescence signal. The results showed that under the same laser irradiation time, MoS_2-BSA-Apt exhibited a better cell-killing effect than MoS_2-BSA, indicating that MoS_2-BSA-Apt could target MCF-7 human breast cancer cells. In addition, MoS_2 nanosheets have magnetism, fluorescence, and other properties. These properties can be used to develop a combined accurate diagnosis and treatment platform that integrates photothermal therapy with imaging diagnosis.

Graphene oxides (GOs) and gold nanoparticles (AuNPs) have been widely employed in cancer therapies because of their excellent photothermal conversion efficiencies and biocompatibilities [76,77]. GOs are excellent nanomaterials as drug carriers due to their high loading capacity. Therefore, creating a nanomatrix by combining GOs with AuNPs in a single system may enhance the photothermal effects on tumors. Considering the high loading capacities of GOs, Yang et al. anchored AuNPs on GOs to enhance the photothermal effect [78]. As shown in Figure 2, to improve the targeting capability of this nanomatrix, thiolated MUC1 aptamers were immobilized on the surface of AuNPs via strong Au−S bond, which can specifically recognize breast cancer cells. Next, the Apt–AuNPs were absorbed onto GOs. Then, the generated heat kills cancer cells with irradiation of the laser. To evaluate the targeting of Apt-AuNPs-GOs, researchers used MCF-7 cells (MUC1-positive cell lines) and EA.hy926 cells (MUC1-negative cell lines). RB molecules were loaded onto Apt-AuNPs-GOs to form fluorescent RB-Apt-AuNPs-GOs. The results indicated that EA.hy926 cells showed very weak fluorescence while MCF-7 cells showed strong red fluorescence. Owing to the excellent loading capacities of GOs, this strategy could be extended to the construction of heat shock protein inhibitor-loaded Apt-AuNPs-GOs to strengthen the effect of photothermal therapy.

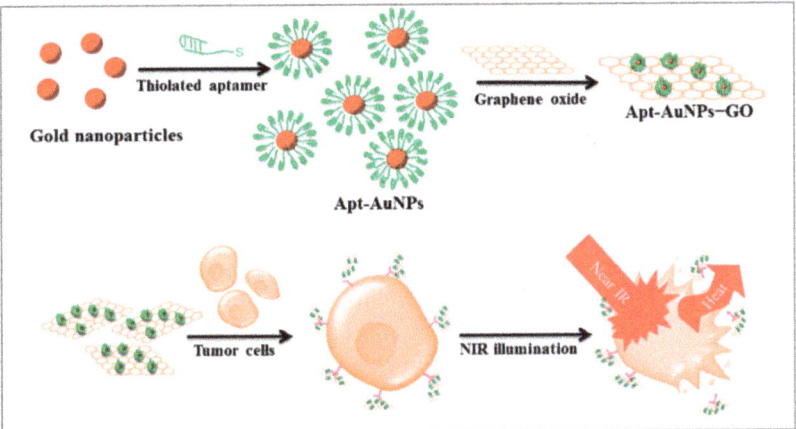

Figure 2. Schematic representation of the preparation of Apt-AuNP-GO and its application with NIR laser irradiation for photothermal therapy of cancer cells. Reprinted with permission from [78]. Copyright © 2015 American Chemical Society.

As shown in Figure 3, Wu et al. also proposed a method with metal nanomaterials [79]. Ag-Au nanostructures have high photothermal conversion efficiencies and are applied in PTT. By modifying the S2.2 aptamer on the surface of Ag-Au nanostructures, the Apt-Ag-Au nanostructures could interact with breast cancer cells on whose membrane MUC1 proteins are overexpressed and realized photothermal therapy. Besides, Ag-Au nanostructures are attractive surface-enhanced Raman scattering (SERS) substrates because of the synergism of these metals, the tunability of the plasmon resonance, and the superior SERS activity. The synthesized Apt-Ag-Au nanostructures will contribute to developing a protocol to specifically recognize and sensitively detect the cancer cells and facilitate the synergistic treatment of diagnosis and photothermal therapy.

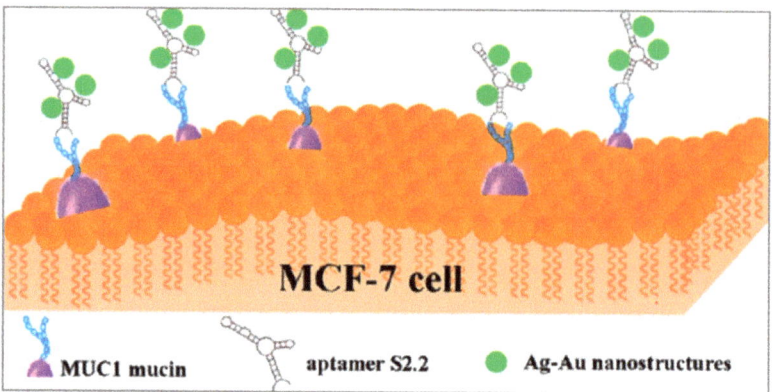

Figure 3. Schematic representation of the preparation of Apt-Ag-Au nanostructure and its application. Reprinted with permission from [79]. Copyright © 2012 American Chemical Society.

3.1.2. Photodynamic Therapy

Manganese dioxide (MnO_2) has been applied more and more in biomedical areas due to its excellent loading capacities and convenient surface functionalization [80,81]. Owing to its quenching property, MnO_2 has been used as a carrier of photosensitizer to construct novel activatable PDT systems. MnO_2 is a unique type of tumor microenvironment-responsive nanomaterial that can react with GSH, and thus overcome the problems of PDT treatment [82–84]. Liu et al. proposed a new strategy [85] that used photosensitizer HMME with mesoporous MnO_2 ($mMnO_2$) functioning as the carrier of HMME. The photosensitizers were in the quenching state when loaded on the surface of $mMnO_2$ nanoparticles and sealed by the aptamers on the particle surface. The aptamers were able to selectively recognize the specific membrane protein MUC1 on the tumor cell, and when this happened the photosensitizers were released. When it interacted with normal cells lacking MUC1, the HMME were not released and the PDT did not work. On the contrary, in the presence of MUC1-overexpressed breast cancer cells, the aptamer bound with MUC1 protein, and HMME was released [86,87]. Then ROS was produced under laser irradiation, which killed cancer cells. To examine the tumor-targeting release of HMME, researchers used MCF-7 cells and Hs578bst cells. The confocal laser scanning microscopy (CLSM) results showed that the fluorescence of HMME in MCF-7 cells was very bright, while it was very low in Hs578bst cells. Furthermore, after irradiation of the laser, the ROS level in MCF-7 (56.4%) was much higher than that in Hs578Bst (1.24%), which confirmed the HMME imaging results. Compared with the conventional PDT method, this constructed system provides a simple but effective approach for the selective killing of tumor cells, with infinitesimal toxicity to normal cells, and paves a new way for utilizing PDT in precise cancer treatment.

Upconversion nanoparticles (UCNPs) are often used as photosensitizer energy donors and delivery vectors in PDT therapy. Moreover, UCNPs functionalized with recognition moieties can also be conferred with the cell-targeting ability for the specific delivery of photosensitizer to enhance the efficiency of PDT. Jin et al. developed a novel method [88]. As shown in Figure 4, a long, single-stranded DNA (ssDNA) with an AS1411 aptamer and a DNAzyme was prepared using rolling circle amplification (RCA). UCNPs functioned as the carrier on which to load the ssDNA. The multivalence of the ssDNA endowed the upconversion nanoplatform with high recognition and drug loading capacity and DNAzyme inhibited the expression of survivin by gene interfering tools. In this nanosystem, AS1411 aptamer was not only used to load the photosensitizer $TMPyP_4$, but it also functioned as the targeting agent to recognize the nucleolin that was overexpressed on breast cancer cells. PDT was triggered by NIR irradiation and generated ROS to kill the cancer cells. To evaluate the targeting of this nanosystem, the uptake of UCNP-ApDz-TMPyP$_4$ in MCF-7 cells (the target cancer cell) and BRL 3A cells (the control cell) was detected by flow cytometry and CLSM. The CLSM results showed that the fluorescence intensity in the MCF-7 cells was significantly stronger than that of BRL 3A cells, which were consistent with the flow cytometry studies. To evaluate the cytotoxicity of MCF-7 cells, MTT assays, LIVE/DEAD viability/cytotoxicity assay were performed. The results showed that UCNP@PVP did not show any cytotoxicity, while in the UCNP-ApDz-TMPyP$_4$ groups the cell survival rate was 36.3%. Emerging evidence has indicated that PDT is always adversely attenuated by the development of cancer cells resistance. However, this multifunctional upconversion nanoplatform, collaborating with PDT and DNAzyme-based gene therapy when used on tumor tissues, exhibits excellent antitumor response in vivo and in vitro and might act as an admirable alternative strategy for treating cancer.

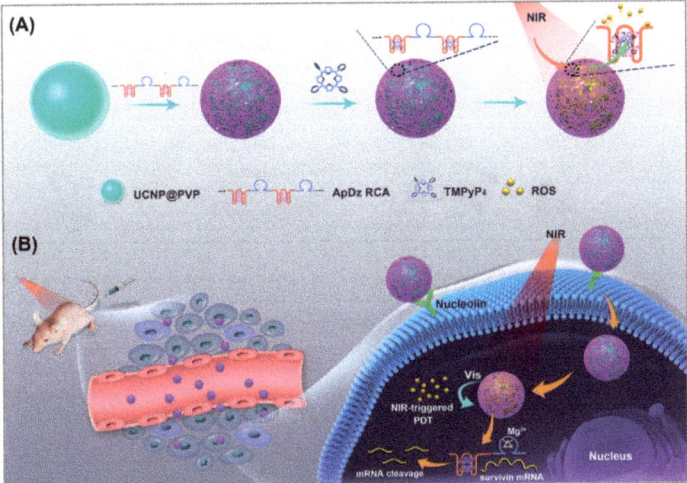

Figure 4. Illustration of: (A) the synthesis of the multifunctional DNA polymer-assisted upconversion therapeutic nanoplatform; and (B) the targeted photodynamic nanoplatform for highly efficient photodynamic therapy. Reprinted with permission from [88]. Copyright © 2020 American Chemical Society.

3.1.3. Drug Delivery System

Similar to the application of mMnO$_2$ by Liu [85], Si et al. proposed another strategy [89]. With the help of mesoporous silica nanoparticles (MSNs), they loaded DNA sensor-capped doxorubicin (DOX). DNA sensors on the targeted nanoparticles could trigger DOX release through a conformational switch induced by MUC-1 protein. They modified the aptamer on the surface of MSNs, which can recognize MUC-1 protein [90]. When the composition was endocytosed into MCF-7 cells, in which MCU-1 was overexpressed, DOX would be released and would kill cancer cells. This caused a significant difference in cell viability between breast cancer MCF-7 and normal breast Hs578bst cells (24.8% and 86.0%). The selectivity and efficiency of treatment were improved greatly. Although the MUC-1 adaptor has been used for drug delivery, the adaptor only served as a targeted ligand and could not accomplish the controlled release of drugs. In this nanosystem, DOX release could be specifically "turned on" in tumor cells according to the MUC1-induced conformational change. This nanosystem provides a new idea for the drug delivery system.

Gene therapy is a promising therapeutic strategy to combat many serious gene-related diseases. Liu et al. developed a novel drug delivery system to combine gene therapy with chemotherapy [91]. During long-term chemotherapy, tumors show drug resistance.

Researchers have found that the p53 gene can enhance the sensitivities of drug-resistant tumors to chemotherapeutics. It is well known that DNA nanostructure can be designed to assemble a variety of functional components. As shown in Figure 5, a biocompatible triangle DNA origami was chosen to efficiently load DOX (TOD) and p53 genes (TODP), and then the MUC1 aptamer was modified on the surface of the DNA nanostructure to improve targeted delivery and controlled release. To evaluate the targeting of TODP, delivery vectors with aptamers and without aptamers were used. The biodistribution of these delivery vectors was studied in an animal imaging system utilizing the fluorescent signal of Cy5.5-labeled DNA origami. The results showed that those with aptamers located mainly in the tumor tissue had substantially higher intensity than those without aptamers. Furthermore, with the application of DNA nanostructure, additional functional groups such as RNA-based drugs, gene editing systems, and imaging diagnosis components may also be introduced into this codelivery system for synergistic theranostics. We think this is a promising platform for the development of a new generation of therapeutics for the treatment of cancers.

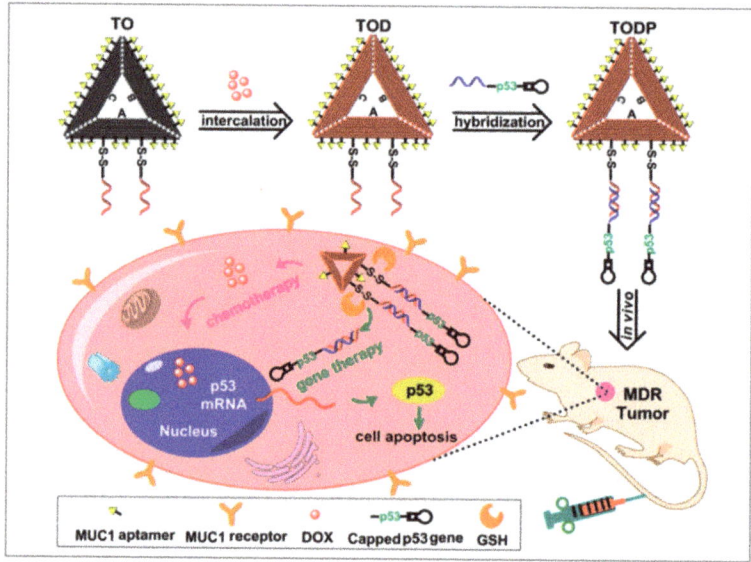

Figure 5. Schematic illustration showing the DNA nanostructure-based combination of gene therapy and chemotherapy. Reprinted with permission from [91]. Copyright © 2018 American Chemical Society.

3.1.4. Photothermal Therapy/Photodynamic Therapy/Chemotherapy

Among various combined treatments, combined chemotherapy/phototherapy is a practicable and promising strategy for cancer treatments due to its favorable synergistic effects and clinical realizability. Xu et al. developed a strategy to combine PTT, PDT, and chemotherapy [92]. As shown in Figure 6, they assembled DOX, indocyanine green (ICG), and bovine serum albumin (BSA) molecules to form nanosized DOX/ICG/BSA nanoparticles. To improve the targeting of the nanoparticles, AS1411 aptamers and a cell-penetrating peptide (KALA) were modified on the surface of the DOX/ICG/BSA nanoparticles through electrostatic interaction. Finally, under the irradiation of the laser, phototherapy was applied and DOX was released to realize chemotherapy. Researchers chose nucleolin overexpressed MCF-7 cells to study the targeting of DOX/ICG/BSA/KALA/Apt. After 4 h of incubation, compared with DOX/ICG/BSA, DOX/ICG/BSA/KALA/Apt showed higher intracellular concentrations of both DOX and ICG. However, in no nucleolin overexpressed 293T cells, the cellular uptakes of DOX/ICG/BSA and DOX/ICG/BSA/KALA/Apt were nearly the same. Besides, the cytotoxicity of DOX/ICG/BSA/KALA/Apt was stronger than DOX/ICG/BSA in MCF-7 cells. Studies showed improved antitumor efficiency of DOX/ICG/BSA/KALA/Apt nanoparticles and demonstrated that the functional theranostic system had great promise in tumor treatment. Although integrated therapeutic systems have developed a lot, the toxicity, biodegradability, and cumbersome assembly processes are still big problems. In future study, researchers should focus on the facile and biocompatible assembly of multifunctional nanosystems.

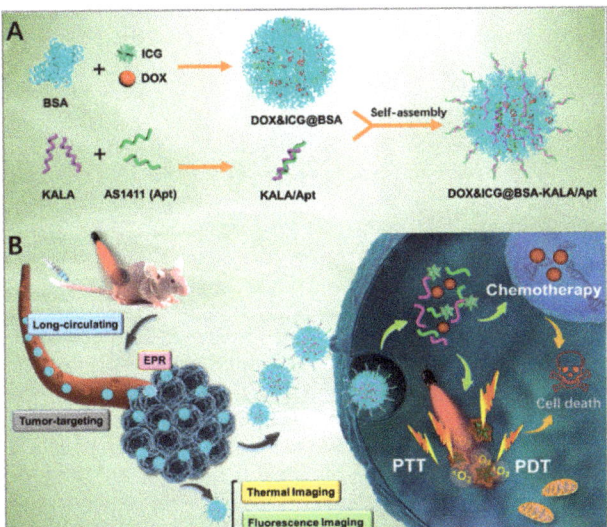

Figure 6. Schematic illustration of: (**A**) the preparation of DOX/ICG/BSA/KALA/Apt nanoparticles; and (**B**) the theranostic process based on DOX/ICG/BSA/KALA/Apt nanoparticles. Reprinted with permission from [92]. Copyright © 2019 American Chemical Society.

3.2. Lung Cancer Therapy

Lung cancer is one of the malignant tumors with the fastest increase in morbidity and mortality, and therefore lung cancer is the greatest threat to human health and life [3]. Many countries have reported that the incidence and mortality of lung cancer have increased significantly. Among all malignant tumors, both the incidence and mortality of lung cancer are highest in men [93,94]. Therefore, the development of an efficient treatment strategy is urgently needed. Chemotherapy is the dominant treatment method among conventional treatments. Though chemotherapy can work in the initial stage, cancer cells will develop drug resistances, while its side effects are unavoidable, causing damage to normal cells [95,96]. Hence, developing an efficient, specific, and less toxic cancer cell treatment platform is the problem that needs to be addressed.

3.2.1. Drug Delivery System

Researchers have made great efforts to develop a range of drug delivery systems [97–99]. It is well known that magnetic particles have many biomedical applications [100–104], such as detection [105,106], drug delivery [107], etc. Magnetic materials, especially Fe_3O_4, have been considered as effective drug delivery systems due to their hollow structures. Moreover, Fe_3O_4 is also used as a PTT agent. Zhao et al. used Fe_3O_4 nanoparticles (Fe_3O_4 NPs) as the carrier to load DOX to realize combined chemotherapy/PTT. A carbon layer was applied to cover the magnetite to improve its stability and photothermal conversion efficiency [108]. To improve the targeting of Fe_3O_4/carbon/DOX nanoparticles (Fe_3O_4/C/DOX NPs), the aptamers (sgc8) were modified on the surface of Fe_3O_4/C/DOX NPs. Furthermore, magnetite is sensitive to acidic conditions. Therefore, under laser irradiation and with a pH of about 5.5, the released DOX and generated heat will kill cancer cells. The CLSM results showed that the fluorescence of A549 cells treated with Apt/Fe_3O_4/C/DOX NPs was significantly enhanced compared with that of the Fe_3O_4/C/DOX NPs alone, which suggests that the sgc8 aptamer facilitated the uptake of NPs. The results in vitro and in vivo demonstrated that the targeted chemo–photothermal combination therapy led to the complete eradication of tumors. As Fe_3O_4 is frequently used as a contrast agent for T_2-weighted magnetic resonance (MR) imaging for diagnostic and therapeutic applications, it is promising to combine imaging diagnosis and treatment in this nanosystem. Furthermore, considering the large effective surface area of Fe_3O_4, more functional groups can be used to develop multidrug loading.

3.2.2. Photodynamic Therapy

PDT is also widely applied in lung cancer therapy. Nano metal-organic frameworks (NMOFs) have received attention for their wide applicability in drug loading and delivery. Owing to their structural advantages, NMOFs are capable of controlling drug interactions with biological systems, and the drug loading efficiency is affected by the morphology and size of NMOFs. The coordination bands of NMOFs are also weak, which contribute to their biodegradability and expand their clinical applications. Zhang et al. developed a novel method to realize the combination of chemotherapy with PDT [109]. Figure 7 shows a representation of how they first synthesized NMOFs. To achieve an excellent drug loading efficiency of NMOFs, they used NMOFs to load DOX, which is prepared for chemotherapy. Furthermore, they also modified the aptamers of A549 cells on the surface of NMOFs, which can distinguish between A549 cells and normal cells. After the irradiation of the laser, DOX and ROS will kill cancer cells when released. Alongside simultaneous treatment with chemotherapy and PDT, the viability was 30% for A549 cells, whereas it was 45% for MCF-7 cells, indicating PDT and chemotherapy's targeting effects for A549 cells of this nanosystem. Moreover, the CLSM results also showed that the fluorescence of DOX was obvious for the A549 cells while weak for the MCF-7 cells. The facile aptamer functionalization of NMOFs offers an opportunity to develop target-directed therapeutic nanosystems. Notably, the aptamer was modified with fluorescein at the terminus to trace A549 cells by targeting nanosystems towards the cells. Based on this, researchers are encouraged to facilitate the integration of diagnosis and treatment nanosystems.

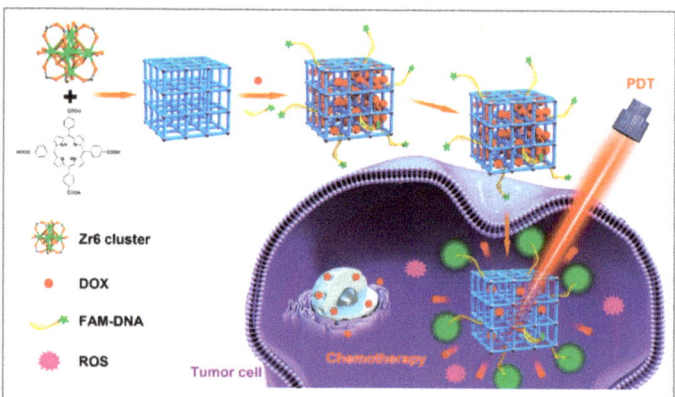

Figure 7. Schematic representation of DNA-functionalized NMOFs for targeted drug delivery and PDT. Reprinted with permission from [109]. Copyright © 2019 American Chemical Society.

Owing to their ultra-high surface areas, graphene quantum dots (GQDS) have emerged as effective cargo nanovectors for drug loading. Moreover, GQDS are widely used in phototherapy owing to the high NIR absorption. Meanwhile, GQDs are relatively common as bioimaging and fluorescent labels because of their biocompatibility. Based on the enormous unique physicochemical properties of GQDS, Cao et al. developed another method [110]. As shown in Figure 8, they used PEGylated GQDS to load porphyrin-derivative photosensitizers, and then AS1411 aptamers were modified on the surface of GQDs so that the nanosystem could selectively target A549 cells. In this nanosystem, GQDs functioned as the drug carrier and photothermal agent. Under the irradiation of the laser, PTT and PDT occurred. The toxic ROS and the heat generated killed the cancer cells. Researchers used HDF cells and A549 cells to verify the targeting of this nanosystem. After irradiation of the laser, almost all the A549 cells were apoptotic, while HDF cells showed no apoptosis. Furthermore, this nanosystem could realize intracellular miRNA biomarker detection and fluorescence-guided PTT/PDT synergetic therapy, which holds great potential in developing combined diagnostics with therapeutics. Notably, chemotherapy can be integrated into this nanosystem with the high drug loading efficiency of GQDs to enhance the treatment effect.

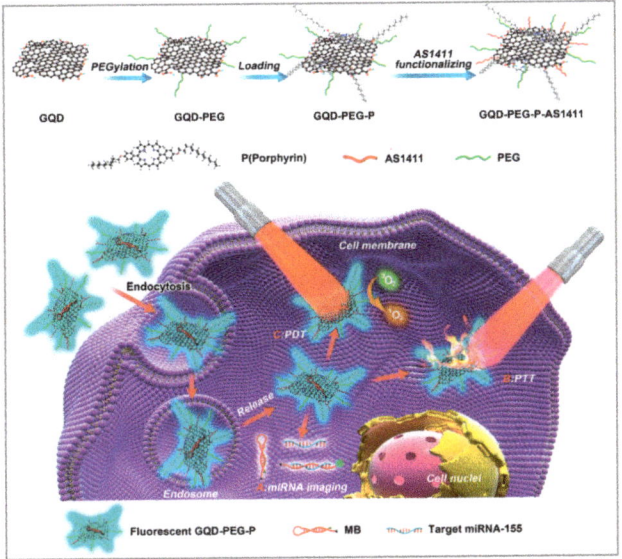

Figure 8. Synthesis of GQD-PEG-P for the combined photothermal/photodynamic therapy. Reprinted with permission from [110]. Copyright © 2016 American Chemical Society.

3.3. Liver Cancer Therapy

Liver cancer is one of the most common malignant tumors worldwide. It ranks fifth in incidence rate and second in fatality rate in China. About half of the newly diagnosed cases in the world come from China. According to relevant predictions of the World Health Organization, the number of deaths due to liver cancer in 2030 will reach one million. Globally, liver cancer has the third highest fatality rate, second only to lung cancer and gastric cancer [3]. Liver cancer is difficult to detect at the early stage and develops rapidly, however, various devices developed recently may be modified and applied to detect it [111–113]. Conventional treatments of liver cancer include surgical resection, liver transplantation, radiotherapy, and chemotherapy, but these treatments suffer from side effects of varying degrees.

Photodynamic Therapy

In recent decades, black phosphorus quantum dots (BPQDs) have been widely applied in biomedical areas, owing to their excellent photocatalysis activities in PDT and broad photo-absorption in PTT. Lan et al. proposed a PDT method to treat liver cancer [114]. As shown in Figure 9, they firstly synthesized a BPQDs-hybridized mesoporous silica framework (BMSF) and Pt nanoparticles (PtNPs). After TLS11a aptamer was decorated on the surface of BMSF-Pt, the Apt-BMSF-Pt could actively target liver cancer cells. Following irradiation, the Apt-BMSF-Pt would generate ROS to kill cancer cells. After HepG2 cells were incubated with Apt-BMSF-Pt and BMSF-Pt, several apoptosis cells could be observed in BMSF-Pt-treated HepG2 cells in the presence of laser irradiation. In contrast, a large number of dead cells were observed in Apt-BMSF-Pt-treated HepG2 cells with laser irradiation, attributing to the efficient uptake by aptamer-mediated endocytosis. The in vivo studies also indicated that Apt-BMSF-Pt had a better tumor growth inhibition effect than BMSF-Pt. Significantly, the BMSF-Pt could self-supply oxygen under H_2O_2 conditions to enhance PDT. Because the tumor microenvironment is hypoxic, this nanoplatform can be developed into a targeting nanocatalyst for the self-regulation of precise cancer phototherapy.

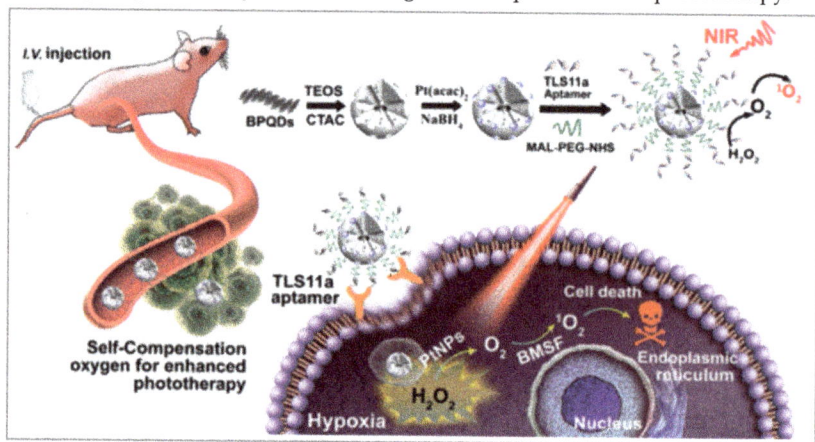

Figure 9. Schematic illustration of the self-regulation of precise cancer phototherapy. Reprinted with permission from [114]. Copyright © 2019 American Chemical Society.

3.4. Cervical Cancer Therapy

Cervical cancer is one of the most common female malignancies. According to statistics, there are about 570,000 new cervical cancer patients worldwide each year, and about 310,000 people die from the disease. Cancer-related mortality is high among all gynecological malignancies. Cervical cancer is ranked first among all female malignant tumors, and fourth among all female malignant tumor-related death rates [3]. Early cervical cancer is generally treated with surgery, but the overall prognosis of patients with recurrence and advanced cervical cancer is still not optimistic. Therefore, developing an efficient treatment method is a big challenge.

3.4.1. Drug Delivery System

Compared with conventional passive targeting in the tumor vasculature, an active targeting strategy via the specific binding of ligands to tumor markers greatly improves intracellular accumulation. Zhang et al. proposed a nanosystem to combine tumor imaging and drug delivery [115]. Persistent luminescence nanoparticles (PLNPs) have aroused widespread interest in fluorescence bioimaging because of their unique optical properties. They used PLNPs as the core, which contributed to tumor imaging. Polyamide-amine (PAMAM) was then modified on the surface of PLNPs to provide many groups for further AS1411 aptamer functionalization, while DOX was loaded onto the nanoplatform via an acid-sensitive hydrazone bond (PLNPs-PAMAM-AS1411/DOX). To examine the targeted cellular uptake of PLNPs-PAMAM-AS1411/DOX, HeLa cells (overexpressed nucleolin) and 3T3 cells (low nucleolin expression) were selected. CLSM imaging and flow cytometry analysis showed a much stronger luminescence signal compared to 3T3 cells after incubation with PLNPs-PAMAM-AS1411/DOX. The in vivo studies also indicated that

PLNPs-PAMAM-AS1411/DOX had a better tumor growth inhibition effect than PLNPs-PAMAM-DOX. The intracellular-controlled release of DOX via a pH-sensitive hydrazone killed tumor cells effectively. Owing to the AS1411 aptamer, the nanoplatform had excellent specificity for tumor cells, whose membrane nucleolin were overexpressed [116,117]. This nanoplatform has the potential for bioimaging and the treatment of tumors with high specificity and efficiency.

3.4.2. Photodynamic Therapy

Apart from an improved drug delivery system, PDT was also applied in cervical cancer therapy. Cheng et al. reported a novel strategy to develop a kind of hydrophilic photosensitizer with good tumor specificity and high ROS generation efficiency under the stimulation of NIR light [118]. In this work, they mixed DNA G-quadruplexes with a hydrophilic porphyrin (TMPipEOPP)$^{4+}\cdot 4I^{-}$. MnO_2 was used to consume GSH in order to reduce ROS consumption and generate O_2 to enhance PDT. As shown in Figure 10, when the G-quadruplex/porphyrin complexes were further assembled with AS1411 aptamer and MnO_2 nanosheets, CLSM results showed that HeLa cells incubated with MnO_2/AS1411/TMPipEOPP showed brighter red fluorescence than TMPipEOPP alone, demonstrating that MnO_2/AS1411/TMPipEOPP could be efficiently internalized into cancer cells when combined with AS1411 aptamers. The in vivo experiments showed better tumor growth inhibition in MnO_2/AS1411/TMPipEOPP + NIR groups when compared with TMPipEOPP + NIR groups. As an essential part of PDT, this work proposed a facile way to improve the penetration depth and PDT efficacy of photosensitizers. Based on this, researchers can try to prepare more kinds of photosensitizers to develop PDT.

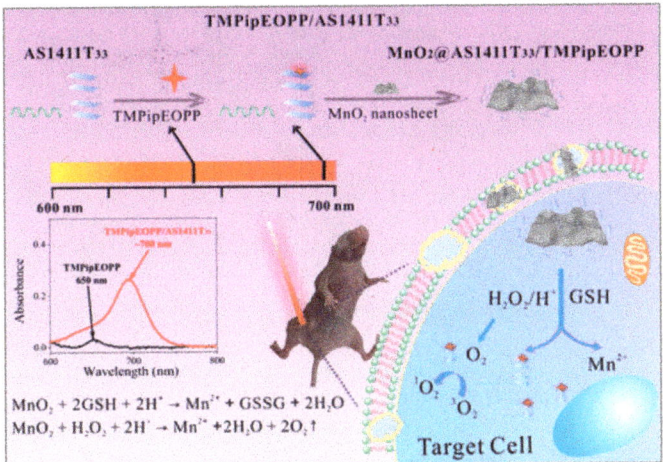

Figure 10. Preparation of G-quadruplex/porphyrin composite photosensitizer and its application in PDT treatment of solid tumors after assembly with MnO2 nanosheet. Reprinted with permission from [118]. Copyright © 2019 American Chemical Society.

Presently, due to their attractive biological optical properties, porphyrin compounds have been widely used as photosensitizers in PDT. To improve their limited tumor accumulation, many delivery systems have been developed, but biocompatibility is still a big problem. As shown in Figure 11, Chu et al. proposed a strategy [119]. They first synthesized Fmoc-H/Zn^{2+} nanoparticles (FZ-NPs) to load porphyrin/Gquadruplex composite photosensitizers (FZO-NPs), and then the AS1411 aptamer was modified on the surface of FZO-NPs (FZOA-NPs). With the NIR light irradiation, ROS would be produced to kill cancer cells. The nanoassembly which they prepared showed a great potential in tumor treatment, and this has been demonstrated in both in vitro and in vivo experiments. Notably, the preparation process only involved amino acid, nucleic acid, Zn^{2+}, and biocompatible porphyrin, obviously reducing the potential for toxic effects to occur during the synthesis process, which was beneficial to the further application. Overall, this study provided a simple and green way to prepare highly biocompatible and highly efficient NIR nanocomposite photosensitizers for clinical tumor PDT treatment.

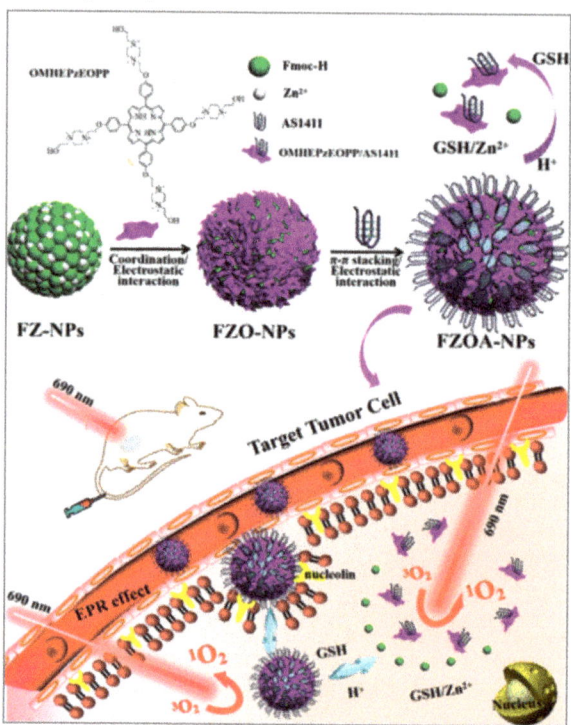

Figure 11. Preparation of FZOA-NPs and their application in tumor PDT treatment. Reprinted with permission from [119]. Copyright © 2010 American Chemical Society.

3.5. Gastric Cancer Therapy

Gastric cancer is one of the most common malignant tumors, and is a leading cause of cancer-related death worldwide [3,120]. Although researchers have made great effort to develop cancer therapy methods, the treatment efficiency is still not satisfactory. Currently, the dominant method is to use chemical drugs, but the specificity and reverse side effects are problems that remain to be solved [121,122].

Gold nanoparticles (AuNPs) are used more and more widely for their highly efficient internalization and good biocompatibility [123–126]. Moreover, AuNPs are easily modifiable with targeting ligands, and have their own high photothermal conversion efficiency. Zhang et al. developed a new method to treat gastric cancer [127]. Researchers modified the AS1411 aptamer on the surface of AuNPs. AS1411 aptamer can strongly bind to nucleolin, which is a new molecular target for gastric cancer treatment [128–130]. The nanosystem, therefore, showed tumor-specific targeting. Furthermore, with the modification of hairpin DNA, DOX was loaded by intercalation. Under the irradiation of the laser, the generated heat and drug release exhibited excellent cancer therapy effects. AGS cells (high nucleolin expressing) and L929 cells (low nucleolin expressing) were used to verify the targeting of this nanosystem. CLSM results showed that the red fluorescence was almost parallel in both AGS and L929 cells incubated with free DOX. When cells were treated with AS1411-based nanoparticles, stronger red fluorescent intensity was observed in the AGS cells, but less in the L929 cells. This multifunctional nanosystem may be promising for gastric cancer cells treated in the future.

3.6. Colorectal Cancer Therapy

Colorectal cancer is the third most frequently diagnosed cancer and its fatality rate is very high, especially in many developing countries [3]. In recent years, researchers found that the incidence age of colorectal cancer is getting younger and younger. Moreover physical inactivity and smoking are the most common factors in the occurrence of colorectal cancer [131]. Therefore, it is urgent to look for an effective treatment method.

With the application of AuNPs, Go et al. developed one strategy [132] where cellular prion protein PrP^C, which is demonstrated to be overexpressed in colorectal cancer, functioned as the site of action [133–135]. They modified PrP^C aptamer on the surface of gold nanoparticles, followed by hybridization of its complementary DNA for DOX loading. The flow cytometry results revealed that the uptake was significantly increased in PrP^C-positive cells compared to PrP^C-negative cells. The aptamer will bind to PrP^C specifically and thus improve the drug delivery efficiency. Under laser irradiation, the heat generated and the DOX released will synergistically kill cancer cells. This nanosystem can serve as an effective therapeutic agent for colorectal cancer treatment.

3.7. Prostate Cancer Therapy

The incidence of prostate cancer ranks fourth among malignant tumors and second among male malignancies, second only to lung cancer. The mortality rate of prostate cancer is in the top ten among malignant tumors and the fifth among male malignant tumors [3]. The pathogenesis of prostate cancer is not yet clear, and the incidence is increasing year by year, and the growth rate ranks second among all tumors.

Prostate-specific membrane antigen, a protein that overexpresses in prostate cancers, is an ideal target for the treatment of prostate cancer. Tang et al. proposed a new method for treatment [136]. As shown in Figure 12, they firstly developed doxorubicin-polylactide nanoconjugates (DOX-PLA NCs) for drug delivery, which would reduce the dose-limiting toxicities associated with free DOX. A10 aptamer, which can target prostate cancer, was immobilized on the surface of DOX-PLA NCs to improve the targeting. Both in vitro and in vivo results showed that controlled drug release would kill cancer cells. The improved efficacy and reduced systemic toxicity of the targeting A10-DOX-PLA NCs demonstrated their strong potential for future clinical translation.

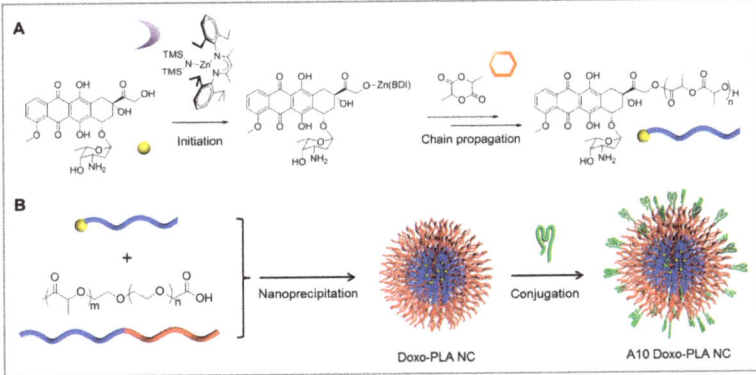

Figure 12. Preparation of A10 aptamer-functionalized doxorubicin-polylactide nanoconjugates (A10 Doxo-PLA NCs): (**A**) synthesis of Doxo-PLA polymer conjugate; (**B**) schematic illustration of formulating A10 Doxo-PLA NCs. Reprinted with permission from [136]. Copyright © 2015 American Chemical Society.

4. Commonly Used Nanomaterials Modified with Aptamers in Cancer Therapy

4.1. Gold Nanorods

Owing to their high photothermal conversion efficiency, gold nanorods (AuNRs) have been studied extensively [137–139]. Compared with nanospheres or nanoshells, AuNRs sustain strong absorption of NIR light, which is less easily absorbed by normal tissues and, therefore, provides deep-tissue penetration with high spatial precision, thus causing minimal damage to healthy tissue. As good candidates for PTT, the precise targeting of AuNRs is, indeed, of crucial importance to their effectiveness. The introduction of aptamers will greatly increase the photothermal killing effect.

Cheng et al. discovered an aptamer (KW16-13) that can specifically recognize MCF10CA1h breast cancer cells [140]. To replace the hexadecyltrimethylammonium bromide (CTAB) with PEG, mPEG-SH solution was added to the prepared AuNRs. Next, a thiolated aptamer was modified on the surface of PEG-AuNR, so that the Apt-AuNR could target the tumor cell. Then, with the irradiation of the laser, the heat generated could kill the cancer cells. Compared with AuNRs, Apt-AuNRs exhibited a better killing effect. AuNRs conjugated to KW16-13 aptamers were readily internalized by the MCF10CA1h tumor cells with minimal uptake by MCF10A normal cells. Upon NIR light irradiation, tumor cell death of >96% could be effected, compared to <1% in the normal cells. The high tumor-cell specificity exhibited under the direction of the KW16-13 aptamer made them exciting candidates for development as novel anti-tumor therapeutics.

Zheng et al. developed a new strategy with the help of exosomes to transport gold nanorods, which possess photothermal conversion properties [141]. Exosomes are nano-sized membrane vesicles endogenously secreted by cells, and they can mediate intercellular communication [142,143]. Moreover, they have excellent biocompatibility, blood circulation, and nearly have non-immunogenicity, which contributes to their quantitative use in biomedical applications [144,145]. Firstly, the exosomes were incubated with DSPE-PEG-SH and double-face-adhesive TLS11a aptamers. Thus, the exosome membranes were modified with both sulfhydryl groups and targeting aptamers. Finally, the AuNRs could be easily combined with the exosomes through the formation of Au-S bonds (Apt-Exos-AuNRs). Through the tight linkage of exosomes, AuNRs, and aptamers, it is possible to achieve both specific targeting and selective photothermal destruction of cancer cells. The fluorescence imaging revealed that the HepG2 cells were found to exhibit broader red fluorescence than the L02 cells after incubating CM-Dil-labeled Apt-Exos-AuNRs with cells, which verified the targeting capabilities of this nanosystem. Moreover, the introduction of exosomes could not only promote the internalization of AuNRs into cells, but the exosome itself could be used as a potential drug delivery vehicle for diverse clinical applications in the future.

4.2. PLGA Nanoparticles

PLGA nanoparticles are widely applied in biomedical areas for their good biocompatibility, and their degradation products are lactic acid and glycolic acid, which are relatively nontoxic. Based on this, the PLGA nano matrix is frequently used as a nano-drug carrier to release the drug in the tumor site in response to various stimuli. Although this strategy has achieved a great success, targeting is still a big problem. By modifying aptamers, the delivery efficiency has been improved a lot.

Duan et al. developed a nanoparticle-based drug delivery system to treat triple-negative breast cancer (TNBC) [146]. Researchers found that heparinase (HPA) was highly expressed in TNBC. Duan et al. used PEG-functionalized PLGA nanoparticles to load paclitaxel, a drug to treat breast cancer, and then HPA aptamers (S1.5) were modified on the surface of nanoparticles. The aptamers specifically targeted HPA, with paclitaxel to be released to kill cancer cells. In future research, HPA could potentially be one of the recognized molecular targets for TNBC therapy, and this targeted drug delivery aimed towards HPA may serve as a potential method to improve the efficiency of TNBC treatments.

Wang et al. put forward a new strategy [147], where PLGA nanoparticles were used to function as a carrier to load DOX. Poly (N-vinylpyrrolidone) (PVP) was then used to stabilize PLGA nanoparticles. To improve the specificity of DOX-PLGA-PVP NPs, aptamer AS1411 was modified on the surface of DOX-PLGA-PVP NPs, to enable it to target lung cancer cells [148,149]. When the aptamer reacted with a receptor on the cancer surface, APT-DOX-PLGA-PVP NPs could enter the cell by endocytosis. Afterwards, when pH was regulated, DOX was released to kill cancer cells. The results of flow cytometer-based analysis indicated the higher cellular uptake of the DOX released from the APT-DOX-PLGA-PVP NPs, which was higher than the DOX alone treated cells. The in vivo results also showed that APT-DOX-PLGA-PVP NPs had a better tumor growth inhibition effect than free DOX alone, attributable to the AS1411 aptamers. Overall, this novel drug delivery system can enhance cancer therapy by targeting the nucleolin receptor endocytosis.

Zhang et al. developed another method based on PLGA nanoparticles [150]. Homoharringtonine (HHT), an effective anticancer agent, was loaded by PLGA nanoparticles. Next, epidermal growth factor receptor (EGFR) aptamer was modified on the surface of PLGA nanoparticles, which enabled it to specifically recognize EGFR in lung cancer cells. As the level of GSH increased, the PLGA nanomedicine was triggered and HHT was released to kill cancer cells. Compared with a free anticancer drug, experiments indicated that the PLGA nanomedicine has better therapeutic efficacy. Therefore, owing to the high loading efficiency of PLGA nanoparticles, this nanoplatform has the possibility to load a combination of drugs and to serve as a powerful tool for therapies for various types of malignant tumors.

We list several commonly used nanomaterials in cancer therapies (Table 1).

Table 1. Summary of Aptamer-Bound Nanomaterials in Cancer Therapy.

Aptamer	Nanomaterial	Application	References
MS03 aptamer	Molybdenum disulfide	Breast cancer therapy	[73]
KW16-13 aptamer	Gold nanorods	Breast cancer therapy	[140]
MUC1 aptamer	Gold nanoparticles/Graphene oxides	Breast cancer therapy	[78]
S2.2 aptamer	Ag-Au nanostructure	Breast cancer therapy	[79]
MUC1 aptamer	Mesoporous MnO_2	Breast cancer therapy	[85]
AS1411 aptamer	Upconversion nanoparticles	Breast cancer therapy	[88]
MUC1 aptamer	Mesoporous silica nanoparticles	Breast cancer therapy	[89]
S1.5 aptamer	PLGA nanoparticles	Breast cancer therapy	[146]
MUC1 aptamer	DNA nanostructure	Breast cancer therapy	[91]
AS1411 aptamer	DOX/ICG/BSA nanoparticles	Breast cancer therapy	[92]
Sgc8 aptamer	Fe_3O_4/Carbon nanoparticles	Lung cancer therapy	[108]
AS1411 aptamer	PLGA nanoparticles	Lung cancer therapy	[147]
EGFR aptamer	PLGA nanoparticles	Lung cancer therapy	[150]
Aptamer of A549 cell	Nano metal-organic frameworks	Lung cancer therapy	[109]
AS1411 aptamer	Graphene quantum dots	Lung cancer therapy	[110]
TLS11a aptamer	Gold nanorods	Liver cancer therapy	[141]
TLS11a aptamer	Black quantum dots/Mesoporous silica framework/Pt nanoparticles	Liver cancer therapy	[114]
AS1411 aptamer	Persistent luminescence nanoparticle	Cervical cancer therapy	[115]
AS1411 aptamer	Manganese dioxide nanosheets	Cervical cancer therapy	[118]
AS1411 aptamer	Fmoc-H/Zn^{2+}/OMHEPzEOPP nanoparticles	Cervical cancer therapy	[119]
AS1411 aptamer	Gold nanoparticles	Gastric cancer therapy	[127]
PrP^C aptamer	Gold nanoparticles	Colorectal cancer therapy	[132]
A10 aptamer	Polylactide nanoconjugates	Prostate cancer therapy	[136]

5. Conclusions

In the past few decades, researchers have made great efforts to look for new methods for cancer treatment. Extensive research on the combination of aptamers with nanomaterials has worked a lot in cancer treatment. This review has summarized several applications of aptamers in cancer treatment. The greatest contribution of aptamers in cancer treatment nanoplatforms includes their convenient modification on the surface of nanomaterials and their specific recognition of and binding to targets. Firstly, easy construction of nanoplatforms will have wide and beneficial applications in the future. Secondly, owing to the excellent specificity of aptamers, cancer treatment based on aptamer-modified nanomaterials will do less harm to normal cells and improve treatment efficiency. Finally, considering the many unique properties of nanomaterials, they can function as the medium for photothermal therapy and photodynamic therapy, and as a carrier to load drugs. As more and more properties of nanomaterials are studied, the combination of aptamers with nanomaterials will be applied in more areas that are not covered.

Although the combination of aptamers with nanomaterials has developed a lot in cancer treatment, most research still stopped at the animal level. As is known, the treatment methods must be tested in the human body within ethical approval for the clinical application of these aptamers. Therefore, researchers must further testify to their effectiveness. Furthermore, the current research scope is focused on cancer treatment, and therefore researchers should study other fields to expand the applications of aptamer-embedded nanomaterials, such as tumor imaging, cancer diagnosis, and so on. In the current research, researchers take advantage of different unique properties of nanomaterials, but there are still many properties that we do not know about, and researchers can try to develop this to find further applications for aptamer-embedded nanomaterials. Moreover, for safety, biocompatibility is one of the most important issues. Compared with conventional antibodies, the established technology of aptamers is improving and their manufacturing cost is high. Therefore, researchers should further invest in aptamer technology in the future, as more and more aptamers are selected, as we expect they will be. As the aptamer technology is developing, more aptamers will be used in biomedical applications, which will improve the accuracy of diagnosis and effectiveness of treatment in cancer.

Despite all of the challenges, the applications of aptamers in cancer therapy are moving in the direction of future treatment development. With the development of different subjects, aptamer-embedded nanomaterials will have further improvement.

Author Contributions: Investigation and resources, L.Z.; writing-original draft preparation, L.Z.; writing-review and editing, L.Z., J.Z., Z.G. and Y.L.; visualization, L.Z., Z.C. and H.C.; supervision, N.H.; project administration, N.H.; funding acquisition, N.H. All authors have read and agreed to the published version of the manuscript.

Funding: This work was supported by the National Key Research and Development Program of China (No. 2017YFA0205301), the National Natural Science Foundation of China (Nos. 62071119 and 61901168), and Jiangsu Provincial Key Research and Development Program (No. BA2020016).

Institutional Review Board Statement: Not applicable.

Informed Consent Statement: Not applicable.

Data Availability Statement: Data is contained within the article.

Conflicts of Interest: The authors declare no conflict of interest.

References

1. He, J.; Fu, L.-H.; Qi, C.; Lin, J.; Huang, P. Metal peroxides for cancer treatment. *Bioact. Mater.* **2021**, *6*, 2698–2710. [CrossRef] [PubMed]
2. Wang, C.; Liu, M.; Wang, Z.; Li, S.; Deng, Y.; He, N. Point-of-care diagnostics for infectious diseases: From methods to devices. *Nano Today* **2021**, *37*, 101092. [CrossRef]
3. Bray, F.; Ferlay, J.; Soerjomataram, I.; Siegel, R.L.; Torre, L.A.; Jemal, A. Global cancer statistics 2018: GLOBOCAN estimates of incidence and mortality worldwide for 36 cancers in 185 countries. *CA Cancer J. Clin.* **2020**, *70*, 313. [CrossRef] [PubMed]
4. Li, T.; Yang, J.; Ali, Z.; Wang, Z.; Mou, X.; He, N.; Wang, Z. Synthesis of aptamer-functionalized Ag nanoclusters for MCF-7 breast cancer cells imaging. *Sci. China Chem.* **2016**, *60*, 370–376. [CrossRef]
5. Barani, M.; Mukhtar, M.; Rahdar, A.; Sargazi, S.; Pandey, S.; Kang, M. Recent Advances in Nanotechnology-Based Diagnosis and Treatments of Human Osteosarcoma. *Biosensors* **2021**, *11*, 55. [CrossRef]
6. Hong, G.; Diao, S.; Antaris, A.L.; Dai, H. Carbon Nanomaterials for Biological Imaging and Nanomedicinal Therapy. *Chem. Rev.* **2015**, *115*, 10816–10906. [CrossRef]

7. Berlina, A.N.; Zherdev, A.V.; Pridvorova, S.M.; Gaur, M.; Dzantiev, B.B. Rapid Visual Detection of Lead and Mercury via Enhanced Crosslinking Aggregation of Aptamer-Labeled Gold Nanoparticles. *J. Nanosci. Nanotechnol.* **2019**, *19*, 5489–5495. [CrossRef]
8. Cheon, H.J.; Lee, S.M.; Kim, S.-R.; Shin, H.Y.; Seo, Y.H.; Cho, Y.K.; Lee, S.P.; Kim, M.I. Colorimetric Detection of MPT64 Antibody Based on an Aptamer Adsorbed Magnetic Nanoparticles for Diagnosis of Tuberculosis. *J. Nanosci. Nanotechnol.* **2019**, *19*, 622–626. [CrossRef] [PubMed]
9. Khan, R.A.; Barani, M.; Rahdar, A.; Sargazi, S.; Cucchiarini, M.; Pandey, S.; Kang, M. Multi-Functionalized Nanomaterials and Nanoparticles for Diagnosis and Treatment of Retinoblastoma. *Biosensors* **2021**, *11*, 97.
10. Jiang, Q.; Shi, Y.; Zhang, Q.; Li, N.; Zhan, P.; Song, L.; Dai, L.; Tian, J.; Du, Y.; Cheng, Z.; et al. A Self-Assembled DNA Origami-Gold Nanorod Complex for Cancer Theranostics. *Small* **2015**, *11*, 5134–5141. [CrossRef] [PubMed]
11. Chen, W.; Kang, Y.; Qin, L.; Jiang, J.; Zhao, Y.; Zhao, Y.; Yang, Z. Aptasensor for the Detection of Ochratoxin A Using Graphene Oxide and Deoxyribonuclease I-Aided Signal Amplification. *J. Nanosci. Nanotechnol.* **2021**, *21*, 4573–4578. [CrossRef] [PubMed]
12. Gu, M.; Liu, J.; Li, D.; Wang, M.; Chi, K.; Zhang, X.; Deng, Y.; Ma, Y.; Hu, R.; Yang, Y. Development of Ochratoxin Aptasensor Based on DNA Metal Nanoclusters. *Nanosci. Nanotechnol. Lett.* **2019**, *11*, 1139–1144. [CrossRef]
13. Liu, D.-L.; Li, Y.; Sun, R.; Xu, J.-Y.; Chen, Y.; Sun, C.-Y. Colorimetric Detection of Organophosphorus Pesticides Based on the Broad-Spectrum Aptamer. *J. Nanosci. Nanotechnol.* **2020**, *20*, 2114–2121. [CrossRef] [PubMed]
14. Xia, Y.; Wu, X.; Zhao, J.; Zhao, J.; Li, Z.; Ren, W.; Tian, Y.; Li, A.; Shen, Z.; Wu, A. Three dimensional plasmonic assemblies of AuNPs with an overall size of sub-200 nm for chemo-photothermal synergistic therapy of breast cancer. *Nanoscale* **2016**, *8*, 18682–18692. [CrossRef] [PubMed]
15. Chen, C.H.; Wu, Y.-J.; Chen, J.-J. Gold Nanotheranostics: Photothermal Therapy and Imaging of Mucin 7 Conjugated Antibody Nanoparticles for Urothelial Cancer. *BioMed Res. Int.* **2015**, *2015*, 813632. [CrossRef]
16. Jia, Q.; Zhao, Z.; Liang, K.; Nan, F.; Li, Y.; Wang, J.; Ge, J.; Wang, P. Recent advances and prospects of carbon dots in cancer nanotheranostics. *Mater. Chem. Front.* **2019**, *4*, 449–471. [CrossRef]
17. Lu, Q.; Lu, T.; Xu, M.; Yang, L.; Song, Y.; Li, N. SO_2 prodrug doped nanorattles with extra-high drug payload for "collusion inside and outside" photothermal/pH triggered-gas therapy. *Biomaterials* **2020**, *257*, 120236. [CrossRef]
18. Lantero, E.; Belavilas-Trovas, A.; Biosca, A.; Recolons, P.; Moles, E.; Sulleiro, E.; Zarzuela, F.; Ávalos-Padilla, Y.; Ramírez, M.; Fernàndez-Busquets, X. Development of DNA Aptamers Against Plasmodium falciparum Blood Stages Using Cell-Systematic Evolution of Ligands by EXponential Enrichment. *J. Biomed. Nanotechnol.* **2020**, *16*, 315–334. [CrossRef]
19. Ma, X.; Zhao, Y.; Ng, K.W.; Zhao, Y. Integrated Hollow Mesoporous Silica Nanoparticles for Target Drug/siRNA Co-Delivery. *Chem. Eur. J.* **2013**, *19*, 15593–15603. [CrossRef]
20. Yu, S.H.; Kim, T.H. T-T Mismatch-Based Electrochemical Aptasensor for Ultratrace Level Detection of Hg^{2+} Using Electrochemically Reduced Graphene Oxide-Modified Electrode. *J. Biomed. Nanotechnol.* **2019**, *15*, 1824–1831. [CrossRef] [PubMed]
21. Duan, Q.; Yang, M.; Zhang, B.; Li, Y.; Zhang, Y.; Li, X.; Wang, J.; Zhang, W.; Sang, S. Gold nanoclusters modified mesoporous silica coated gold nanorods: Enhanced photothermal properties and fluorescence imaging. *J. Photochem. Photobiol. B Biol.* **2021**, *215*, 112111. [CrossRef] [PubMed]
22. Zhao, J.; Wang, A.; Si, T.; Hong, J.-D.; Li, J. Gold nanorods based multicompartment mesoporous silica composites as bioagents for highly efficient photothermal therapy. *J. Colloid Interface Sci.* **2019**, *549*, 9–15. [CrossRef] [PubMed]
23. Liu, J.; Detrembleur, C.; De Pauw-Gillet, M.-C.; Mornet, S.; Jérôme, C.; Duguet, E. Gold Nanorods Coated with Mesoporous Silica Shell as Drug Delivery System for Remote Near Infrared Light-Activated Release and Potential Phototherapy. *Small* **2015**, *11*, 2323–2332. [CrossRef]
24. Sivaram, A.J.; Wardiana, A.; Howard, C.; Mahler, S.M.; Thurecht, K.J. Recent Advances in the Generation of Antibody–Nanomaterial Conjugates. *Adv. Healthc. Mater.* **2017**, *7*, 1700607. [CrossRef]
25. Li, J.; Zheng, C.; Cansiz, S.; Wu, C.; Xu, J.; Cui, C.; Liu, Y.; Hou, W.; Wang, Y.; Zhang, L.; et al. Self-assembly of DNA Nanohydrogels with Controllable Size and Stimuli-Responsive Property for Targeted Gene Regulation Therapy. *J. Am. Chem. Soc.* **2015**, *137*, 1412–1415. [CrossRef]
26. Liu, M.; Xi, L.; Tan, T.; Jin, L.; Wang, Z.; He, N. A novel aptamer-based histochemistry assay for specific diagnosis of clinical breast cancer tissues. *Chin. Chem. Lett.* **2020**, *32*, 1726–1730. [CrossRef]
27. Liu, Y.; Yang, G.; Li, T.; Deng, Y.; Chen, Z.; He, N. Selection of a DNA aptamer for the development of fluorescent aptasensor for carbaryl detection. *Chin. Chem. Lett.* **2021**, *32*, 1957–1962. [CrossRef]
28. Huang, R.; He, L.; Li, S.; Liu, H.; Jin, L.; Chen, Z.; Zhao, Y.; Li, Z.; Deng, Y.; He, N. A simple fluorescence aptasensor for gastric cancer exosome detection based on branched rolling circle amplification. *Nanoscale* **2019**, *12*, 2445–2451. [CrossRef]
29. Lin, N.; Wu, L.; Xu, X.; Wu, Q.; Wang, Y.; Shen, H.; Song, Y.; Wang, H.; Zhu, Z.; Kang, D.; et al. Aptamer Generated by Cell-SELEX for Specific Targeting of Human Glioma Cells. *ACS Appl. Mater. Interfaces* **2020**, *13*, 9306–9315. [CrossRef] [PubMed]
30. He, L.; Huang, R.; Xiao, P.; Liu, Y.; Jin, L.; Liu, H.; Li, S.; Deng, Y.; Chen, Z.; Li, Z.; et al. Current signal amplification strategies in aptamer-based electrochemical biosensor: A review. *Chin. Chem. Lett.* **2021**, *32*, 1593–1602. [CrossRef]
31. Kim, Y.; Yang, J.; Hur, H.; Oh, S.; Lee, H. Highly Sensitive Colorimetric Assay of Cortisol Using Cortisol Antibody and Aptamer Sandwich Assay. *Biosensors* **2021**, *11*, 163. [CrossRef]
32. Malam, Y.; Loizidou, M.; Seifalian, A. Liposomes and nanoparticles: Nanosized vehicles for drug delivery in cancer. *Trends Pharmacol. Sci.* **2009**, *30*, 592–599. [CrossRef]

33. Hianik, T. Advances in Electrochemical and Acoustic Aptamer-Based Biosensors and Immunosensors in Diagnostics of Leukemia. *Biosensors* **2021**, *11*, 177. [CrossRef] [PubMed]
34. Liu, M.; Zhang, B.; Li, Z.; Wang, Z.; Li, S.; Liu, H.; Deng, Y.; He, N. Precise discrimination of Luminal A breast cancer subtype using an aptamer in vitro and in vivo. *Nanoscale* **2020**, *12*, 19689–19701. [CrossRef] [PubMed]
35. Liu, M.; Khan, A.; Wang, Z.; Liu, Y.; Yang, G.; Deng, Y.; He, N. Aptasensors for pesticide detection. *Biosens. Bioelectron.* **2019**, *130*, 174–184. [CrossRef]
36. Guo, T.; Tang, Q.; Guo, Y.; Qiu, H.; Dai, J.; Xing, C.; Zhuang, S.; Huang, G. Boron Quantum Dots for Photoacoustic Imaging-Guided Photothermal Therapy. *ACS Appl. Mater. Interfaces* **2020**, *13*, 306–311. [CrossRef] [PubMed]
37. Leng, C.; Zhang, X.; Xu, F.; Yuan, Y.; Pei, H.; Sun, Z.; Li, L.; Bao, Z. Engineering Gold Nanorod-Copper Sulfide Heterostructures with Enhanced Photothermal Conversion Efficiency and Photostability. *Small* **2018**, *14*, e1703077. [CrossRef] [PubMed]
38. Ha, M.; Nam, S.H.; Sim, K.; Chong, S.-E.; Kim, J.; Kim, Y.; Lee, Y.; Nam, J.-M. Highly Efficient Photothermal Therapy with Cell-Penetrating Peptide-Modified Bumpy Au Triangular Nanoprisms using Low Laser Power and Low Probe Dose. *Nano Lett.* **2020**, *21*, 731–739. [CrossRef] [PubMed]
39. Wang, H.; Chang, J.; Shi, M.; Pan, W.; Li, N.; Tang, B. A Dual-Targeted Organic Photothermal Agent for Enhanced Photothermal Therapy. *Angew. Chem. Int. Ed.* **2018**, *58*, 1057–1061. [CrossRef] [PubMed]
40. Hu, K.; Xie, L.; Zhang, Y.; Hanyu, M.; Yang, Z.; Nagatsu, K.; Suzuki, H.; Ouyang, J.; Ji, X.; Wei, J.; et al. Marriage of black phosphorus and Cu^{2+} as effective photothermal agents for PET-guided combination cancer therapy. *Nat. Commun.* **2020**, *11*, 1–15. [CrossRef]
41. Kennedy, L.C.; Bickford, L.R.; Lewinski, N.A.; Coughlin, A.J.; Hu, Y.; Day, E.S.; West, J.L.; Drezek, R.A. A New Era for Cancer Treatment: Gold-Nanoparticle-Mediated Thermal Therapies. *Small* **2011**, *7*, 169–183. [CrossRef]
42. Chen, J.; Wang, D.; Xi, J.; Au, L.; Siekkinen, A.; Warsen, A.; Li, Z.-Y.; Zhang, H.; Xia, Y.; Li, X. Immuno Gold Nanocages with Tailored Optical Properties for Targeted Photothermal Destruction of Cancer Cells. *Nano Lett.* **2007**, *7*, 1318–1322. [CrossRef] [PubMed]
43. Piao, J.-G.; Wang, L.; Gao, F.; You, Y.-Z.; Xiong, Y.; Yang, L. Erythrocyte Membrane Is an Alternative Coating to Polyethylene Glycol for Prolonging the Circulation Lifetime of Gold Nanocages for Photothermal Therapy. *ACS Nano* **2014**, *8*, 10414–10425. [CrossRef] [PubMed]
44. Agostinis, P.; Berg, K.; Cengel, K.A.; Foster, T.H.; Girotti, A.W.; Gollnick, S.O.; Hahn, S.M.; Hamblin, M.R.; Juzeniene, A.; Kessel, D.; et al. Photodynamic Therapy of Cancer. *CA Cancer J. Clin.* **2011**, *61*, 250–281. [CrossRef]
45. Li, X.; Kolemen, S.; Yoon, J.; Akkaya, E. Activatable Photosensitizers: Agents for Selective Photodynamic Therapy. *Adv. Funct. Mater.* **2016**, *27*, 1604053. [CrossRef]
46. Zhou, Z.; Song, J.; Tian, R.; Yang, Z.; Yu, G.; Lin, L.; Zhang, G.; Fan, W.; Zhang, F.; Niu, G.; et al. Activatable Singlet Oxygen Generation from Lipid Hydroperoxide Nanoparticles for Cancer Therapy. *Angew. Chem.* **2017**, *129*, 6592–6596. [CrossRef]
47. He, T.; Jiang, C.; He, J.; Zhang, Y.; He, G.; Wu, J.; Lin, J.; Zhou, X.; Huang, P. Manganese-Dioxide-Coating-Instructed Plasmonic Modulation of Gold Nanorods for Activatable Duplex-Imaging-Guided NIR-II Photothermal-Chemodynamic Therapy. *Adv. Mater.* **2021**, *33*, 2008540. [CrossRef]
48. Yang, X.; Liu, R.; Zhong, Z.; Huang, H.; Shao, J.; Xie, X.; Zhang, Y.; Wang, W.; Dong, X. Platinum nanoenzyme functionalized black phosphorus nanosheets for photothermal and enhanced-photodynamic therapy. *Chem. Eng. J.* **2020**, *409*, 127381. [CrossRef]
49. Wang, X.; Mao, Y.; Sun, C.; Zhao, Q.; Gao, Y.; Wang, S. A versatile gas-generator promoting drug release and oxygen replenishment for amplifying photodynamic-chemotherapy synergetic anti-tumor effects. *Biomaterials* **2021**, *276*, 120985. [CrossRef]
50. Kwon, O.S.; Song, H.S.; Conde, J.; Kim, H.-I.; Artzi, N.; Kim, J.-H. Dual-Color Emissive Upconversion Nanocapsules for Differential Cancer Bioimaging In Vivo. *ACS Nano* **2016**, *10*, 1512–1521. [CrossRef] [PubMed]
51. Lei, Q.; Qiu, W.-X.; Hu, J.-J.; Cao, P.-X.; Zhu, C.-H.; Cheng, H.; Zhang, X.-Z. Multifunctional Mesoporous Silica Nanoparticles with Thermal-Responsive Gatekeeper for NIR Light-Triggered Chemo/Photothermal-Therapy. *Small* **2016**, *12*, 4286–4298. [CrossRef] [PubMed]
52. Ojha, T.; Pathak, V.; Shi, Y.; Hennink, W.E.; Moonen, C.T.; Storm, G.; Kiessling, F.; Lammers, T. Pharmacological and physical vessel modulation strategies to improve EPR-mediated drug targeting to tumors. *Adv. Drug Deliv. Rev.* **2017**, *119*, 44–60. [CrossRef]
53. Ali, Z.; Wang, J.; Tang, Y.; Liu, B.; He, N.; Li, Z. Simultaneous detection of multiple viruses based on chemiluminescence and magnetic separation. *Biomater. Sci.* **2016**, *5*, 57–66. [CrossRef] [PubMed]
54. Kaur, H. Aptamer Conjugated Quantum Dots for Imaging Cellular Uptake in Cancer Cells. *J. Nanosci. Nanotechnol.* **2019**, *19*, 3798–3803. [CrossRef] [PubMed]
55. Wan, H.; Zhao, Q.; Zhao, P.; He, B.; Jiang, T.; Zhang, Q.; Wang, S. Versatile hybrid polyethyleneimine–mesoporous carbon nanoparticles for targeted delivery. *Carbon* **2014**, *79*, 123–134. [CrossRef]
56. Yu, S.; Bi, X.; Yang, L.; Wu, S.; Yu, Y.; Jiang, B.; Zhang, A.; Lan, K.; Duan, S. Co-Delivery of Paclitaxel and PLK1-Targeted siRNA Using Aptamer-Functionalized Cationic Liposome for Synergistic Anti-Breast Cancer Effects In Vivo. *J. Biomed. Nanotechnol.* **2019**, *15*, 1135–1148. [CrossRef]
57. Zhen, D.; Zhong, F.; Yang, D.; Cai, Q.; Liu, Y. Photoelectrochemical aptasensor based on a ternary $CdS/Au/TiO_2$ nanotube array for ultrasensitive detection of cytochrome c. *Mater. Express* **2019**, *9*, 319–327. [CrossRef]

58. Sun, T.; Zhang, Y.S.; Pang, B.; Hyun, D.C.; Yang, M.; Xia, Y. Engineered Nanoparticles for Drug Delivery in Cancer Therapy. *Angew. Chem. Int. Ed.* **2014**, *53*, 12320–12364. [CrossRef] [PubMed]
59. Torchilin, V.P. Multifunctional, stimuli-sensitive nanoparticulate systems for drug delivery. *Nat. Rev. Drug Discov.* **2014**, *13*, 813–827. [CrossRef] [PubMed]
60. Xi, Z.; Huang, R.; Li, Z.; He, N.; Wang, T.; Su, E.; Deng, Y. Selection of HBsAg-Specific DNA Aptamers Based on Carboxylated Magnetic Nanoparticles and Their Application in the Rapid and Simple Detection of Hepatitis B Virus Infection. *ACS Appl. Mater. Interfaces* **2015**, *7*, 11215–11223. [CrossRef]
61. Fang, X.; Cao, J.; Shen, A. Advances in anti-breast cancer drugs and the application of nano-drug delivery systems in breast cancer therapy. *J. Drug Deliv. Sci. Technol.* **2020**, *57*, 101662. [CrossRef]
62. Hu, J.J.; Liu, M.D.; Gao, F.; Chen, Y.; Peng, S.Y.; Li, Z.H.; Cheng, H.; Zhang, X.Z. Photo-controlled liquid metal nanoparticle-enzyme for starvation/photothermal therapy of tumor by win-win cooperation. *Biomaterials* **2019**, *217*, 119303. [CrossRef]
63. Cheng, Y.-J.; Zhang, A.-Q.; Hu, J.-J.; He, F.; Zeng, X.; Zhang, X.-Z. Multifunctional Peptide-Amphiphile End-Capped Mesoporous Silica Nanoparticles for Tumor Targeting Drug Delivery. *ACS Appl. Mater. Interfaces* **2017**, *9*, 2093–2103. [CrossRef] [PubMed]
64. Wan, X.; Min, Y.; Bludau, H.; Keith, A.; Sheiko, S.S.; Jordan, R.; Wang, A.Z.; Sokolsky-Papkov, M.; Kabanov, A.V. Drug Combination Synergy in Worm-like Polymeric Micelles Improves Treatment Outcome for Small Cell and Non-Small Cell Lung Cancer. *ACS Nano* **2018**, *12*, 2426–2439. [CrossRef] [PubMed]
65. Tan, Y.; Li, Y.; Qu, Y.-X.; Su, Y.; Peng, Y.; Zhao, Y.; Fu, T.; Wang, X.-Q.; Tan, W. Aptamer-Peptide Conjugates as Targeted Chemosensitizers for Breast Cancer Treatment. *ACS Appl. Mater. Interfaces* **2020**, *13*, 9436–9444. [CrossRef] [PubMed]
66. Hu, X.; Zhang, M.; Xue, Q.; Cai, T. ATP Aptamer-Modified Quantum Dots with Reduced Glutathione/Adenosine Triphosphate Dual Response Features as a Potential Probe for Intracellular Drug Delivery Monitoring of Vesicular Nanocarriers. *J. Biomed. Nanotechnol.* **2019**, *15*, 319–328. [CrossRef]
67. Liu, M.; Yu, X.; Chen, Z.; Yang, T.; Yang, D.; Liu, Q.; Du, K.; Li, B.; Wang, Z.; Li, S.; et al. Aptamer selection and applications for breast cancer diagnostics and therapy. *J. Nanobiotechnology* **2017**, *15*, 1–16. [CrossRef]
68. Choi, H.; Shin, C. Negative Capacitance Transistor with Two-Dimensional Channel Material (Molybdenum disulfide, MoS_2). *Phys. Status Solidi* **2019**, *216*, 1900177. [CrossRef]
69. Zhao, X.; Li, Z.; Zhang, J.; Gong, F.; Huang, B.; Zhang, Q.; Yan, Q.-L.; Yang, Z. Regulating safety and energy release of energetic materials by manipulation of molybdenum disulfide phase. *Chem. Eng. J.* **2021**, *411*, 128603. [CrossRef]
70. Shen, Y.; Shuhendler, A.J.; Ye, D.; Xu, J.-J.; Chen, H.-Y. Two-photon excitation nanoparticles for photodynamic therapy. *Chem. Soc. Rev.* **2016**, *45*, 6725–6741. [CrossRef]
71. Liu, T.; Wang, C.; Gu, X.; Gong, H.; Cheng, L.; Shi, X.; Feng, L.; Sun, B.; Liu, Z. Drug Delivery with PEGylated MoS_2 Nano-sheets for Combined Photothermal and Chemotherapy of Cancer. *Adv. Mater.* **2014**, *26*, 3433–3440. [CrossRef]
72. Kim, J.; Kim, H.; Kim, W.J. Single-Layered MoS_2 -PEI-PEG Nanocomposite-Mediated Gene Delivery Controlled by Photo and Redox Stimuli. *Small* **2015**, *12*, 1184–1192. [CrossRef]
73. Pang, B.; Yang, H.; Wang, L.; Chen, J.; Jin, L.; Shen, B. Aptamer modified MoS_2 nanosheets application in targeted photothermal therapy for breast cancer. *Colloids Surfaces A Physicochem. Eng. Asp.* **2020**, *608*, 125506. [CrossRef]
74. Yang, D.; Liu, M.; Xu, J.; Yang, C.; Wang, X.; Lou, Y.; He, N.; Wang, Z. Carbon nanosphere-based fluorescence aptasensor for targeted detection of breast cancer cell MCF-7. *Talanta* **2018**, *185*, 113–117. [CrossRef]
75. Liu, M.; Yang, T.; Chen, Z.; Wang, Z.; He, N. Differentiating breast cancer molecular subtypes using a DNA aptamer selected against MCF-7 cells. *Biomater. Sci.* **2018**, *6*, 3152–3159. [CrossRef] [PubMed]
76. Cole, J.; Mirin, N.A.; Knight, M.; Goodrich, G.P.; Halas, N. Photothermal Efficiencies of Nanoshells and Nanorods for Clinical Therapeutic Applications. *J. Phys. Chem. C* **2009**, *113*, 12090–12094. [CrossRef]
77. Robinson, J.T.; Tabakman, S.M.; Liang, Y.; Wang, H.; Casalongue, H.S.; Vinh, D.; Dai, H. Ultrasmall Reduced Graphene Oxide with High Near-Infrared Absorbance for Photothermal Therapy. *J. Am. Chem. Soc.* **2011**, *133*, 6825–6831. [CrossRef]
78. Yang, L.; Tseng, Y.-T.; Suo, G.; Chen, L.; Yu-Ting, T.; Chiu, W.-J.; Huang, C.-C.; Lin, C.-H. Photothermal Therapeutic Response of Cancer Cells to Aptamer–Gold Nanoparticle-Hybridized Graphene Oxide under NIR Illumination. *ACS Appl. Mater. Interfaces* **2015**, *7*, 5097–5106. [CrossRef] [PubMed]
79. Wu, P.; Gao, Y.; Zhang, H.; Cai, C. Aptamer-Guided Silver–Gold Bimetallic Nanostructures with Highly Active Surface-Enhanced Raman Scattering for Specific Detection and Near-Infrared Photothermal Therapy of Human Breast Cancer Cells. *Anal. Chem.* **2012**, *84*, 7692–7699. [CrossRef]
80. Zhou, Y.; Quan, G.; Wu, Q.; Zhang, X.; Niu, B.; Wu, B.; Huang, Y.; Pan, X.; Wu, C. Mesoporous silica nanoparticles for drug and gene delivery. *Acta Pharm. Sin. B* **2018**, *8*, 165–177. [CrossRef]
81. Siminzar, P.; Omidi, Y.; Golchin, A.; Aghanejad, A.; Barar, J. Targeted delivery of doxorubicin by magnetic mesoporous silica nanoparticles armed with mucin-1 aptamer. *J. Drug Target.* **2019**, *28*, 92–101. [CrossRef] [PubMed]
82. Zhong, L.; Gan, L.; Deng, Z.; Liu, X.; Peng, H.; Tang, H.; Liu, X.; Fang, F.; Yao, F.; Li, W.; et al. Antitumor Activity of Lipid-DNA Aptamer Modified T Lymphocytes in Carcinoma. *J. Biomed. Nanotechnol.* **2020**, *16*, 1110–1118. [CrossRef]
83. Zhao, Z.; Fan, H.; Zhou, G.; Bai, H.; Liang, H.; Wang, R.; Zhang, X.; Tan, W. Activatable Fluorescence/MRI Bimodal Platform for Tumor Cell Imaging via MnO_2 Nanosheet–Aptamer Nanoprobe. *J. Am. Chem. Soc.* **2014**, *136*, 11220–11223. [CrossRef] [PubMed]

84. Wang, Y.; Chang, K.; Yang, C.; Li, S.; Wang, L.; Xu, H.; Zhou, L.; Zhang, W.; Tang, X.; Wang, Y.; et al. Highly Sensitive Electrochemical Biosensor for Circulating Tumor Cells Detection via Dual-Aptamer Capture and Rolling Circle Amplification Strategy. *J. Biomed. Nanotechnol.* **2019**, *15*, 1568–1577. [CrossRef]
85. Liu, W.; Zhang, K.; Zhuang, L.; Liu, J.; Zeng, W.; Shi, J.; Zhang, Z. Aptamer/photosensitizer hybridized mesoporous MnO2 based tumor cell activated ROS regulator for precise photodynamic therapy of breast cancer. *Colloids Surfaces B Biointerfaces* **2019**, *184*, 110536. [CrossRef]
86. Liu, M.; Wang, Z.; Tan, T.; Chen, Z.; Mou, X.; Yu, X.; Deng, Y.; Lu, G.; He, N. An Aptamer-Based Probe for Molecular Subtyping of Breast Cancer. *Theranostics* **2018**, *8*, 5772–5783. [CrossRef] [PubMed]
87. Xi, Z.; Huang, R.; Deng, Y.; He, N. Progress in Selection and Biomedical Applications of Aptamers. *J. Biomed. Nanotechnol.* **2014**, *10*, 3043–3062. [CrossRef]
88. Jin, Y.; Wang, H.; Li, X.; Zhu, H.; Sun, D.; Sun, X.; Liu, H.; Zhang, Z.; Cao, L.; Gao, C.; et al. Multifunctional DNA Polymer-Assisted Upconversion Therapeutic Nanoplatform for Enhanced Photodynamic Therapy. *ACS Appl. Mater. Interfaces* **2020**, *12*, 26832–26841. [CrossRef]
89. Si, P.; Shi, J.; Zhang, P.; Wang, C.; Chen, H.; Mi, X.; Chu, W.; Zhai, B.; Li, W. MUC-1 recognition-based activated drug nanoplatform improves doxorubicin chemotherapy in breast cancer. *Cancer Lett.* **2019**, *472*, 165–174. [CrossRef]
90. Tang, Y.; Liu, H.; Chen, H.; Chen, Z.; Liu, Y.; Jin, L.; Deng, Y.; Li, S.; He, N. Advances in Aptamer Screening and Drug Delivery. *J. Biomed. Nanotechnol.* **2020**, *16*, 763–788. [CrossRef] [PubMed]
91. Liu, J.; Song, L.; Liu, S.; Jiang, Q.; Liu, Q.; Li, N.; Wang, Z.-G.; Ding, B. A DNA-Based Nanocarrier for Efficient Gene Delivery and Combined Cancer Therapy. *Nano Lett.* **2018**, *18*, 3328–3334. [CrossRef]
92. Xu, L.; Wang, S.-B.; Xu, C.; Han, D.; Ren, X.-H.; Zhang, X.-Z.; Cheng, S.-X. Multifunctional Albumin-Based Delivery System Generated by Programmed Assembly for Tumor-Targeted Multimodal Therapy and Imaging. *ACS Appl. Mater. Interfaces* **2019**, *11*, 38385–38394. [CrossRef]
93. Sadiq, M.; Pang, L.; Johnson, M.; Sathish, V.; Zhang, Q.; Wang, D. 2d nanomaterial ti3c2 mxene based sensor to guide lung cancer therapy and management. *Biosensors* **2021**, *11*, 40. [CrossRef]
94. Nguyen, A.T.V.; Trinh, T.T.T.; Hoang, V.T.; Dao, T.D.; Tuong, H.T.; Kim, H.S.; Park, H.; Yeo, S.-J. Peptide Aptamer of Complementarity-determining Region to Detect Avian Influenza Virus. *J. Biomed. Nanotechnol.* **2019**, *15*, 1185–1200. [CrossRef]
95. Wang, K.; Zhuang, J.; Liu, Y.; Xu, M.; Zhuang, J.; Chen, Z.; Wei, Y.; Zhang, Y. PEGylated chitosan nanoparticles with embedded bismuth sulfide for dual-wavelength fluorescent imaging and photothermal therapy. *Carbohydr. Polym.* **2018**, *184*, 445–452. [CrossRef] [PubMed]
96. Liu, Y.; Zhang, X.; Liu, Z.; Wang, L.; Luo, L.; Wang, M.; Wang, Q.; Gao, D. Gold nanoshell-based betulinic acid liposomes for synergistic chemo-photothermal therapy. *Nanomed. Nanotechnol. Biol. Med.* **2017**, *13*, 1891–1900. [CrossRef] [PubMed]
97. Lu, Z.; Zhang, Y.; Wang, Y.; Tan, G.-H.; Huang, F.-Y.; Cao, R.; He, N.; Zhang, L. A biotin-avidin-system-based virus-mimicking nanovaccine for tumor immunotherapy. *J. Control. Release* **2021**, *332*, 245–259. [CrossRef] [PubMed]
98. Wang, J.; Han, S.; Zhang, Z.; Wang, J.; Zhang, G. Preparation and performance of chemotherapy drug-loaded graphene oxide-based nanosheets that target ovarian cancer cells via folate receptor mediation. *J. Biomed. Nanotechnol.* **2021**, *17*, 960–970. [CrossRef] [PubMed]
99. Wang, M.; Kuang, R.; Huang, B.; Ji, D. Polylactic acid block copolymer grafted temozolomide targeted nano delivery in the treatment of glioma. *Mater. Express* **2021**, *11*, 627–633. [CrossRef]
100. Fang, Y.; Liu, H.; Wang, Y.; Su, X.; Jin, L.; Wu, Y.; Deng, Y.; Li, S.; Chen, Z.; Chen, H.; et al. Fast and Accurate Control Strategy for Portable Nucleic Acid Detection (PNAD) System Based on Magnetic Nanoparticles. *J. Biomed. Nanotechnol.* **2021**, *17*, 407–415. [CrossRef]
101. Dong, P.; Wang, H.; Xing, S.; Yang, X.; Wang, S.; Li, D.; Zhao, D. Fluorescent Magnetic Iron Oxide NanoparticleEncapsulated Protein Hydrogel Against Doxorubicin-Associated Cardiotoxicity and for Enhanced Cardiomyocyte Survival. *J. Biomed. Nanotechnol.* **2020**, *16*, 922–930. [CrossRef]
102. Chen, J.; Huang, F.; Gu, D.; Qu, M.; Xu, F.; Hu, Z. Phenotype and genotype heterogeneous resistance of L-forms of Mycobacterium tuberculosis by magnetic nanoparticle. *Mater. Express* **2020**, *10*, 94–101. [CrossRef]
103. Wang, H.; Luo, J.; Chen, J.; Chen, H.; Li, T.; Yang, M. Electrochemical immunosensor for a protein biomarker based on the formation of Prussian blue with magnetic nanoparticle. *Mater. Express* **2020**, *10*, 278–282. [CrossRef]
104. Zhang, L.; Shi, Y.; Chen, C.; Han, Q.; Chen, Q.; Xia, X.; Song, Y.; Zhang, J. Rapid, Visual Detection of Klebsiella pneumoniae Using Magnetic Nanoparticles and an Horseradish Peroxidase-Probe Based Immunosensor. *J. Biomed. Nanotechnol.* **2019**, *15*, 1061–1071. [CrossRef] [PubMed]
105. Zhao, H.; Lin, Q.; Huang, L.; Zhai, Y.; Liu, Y.; Deng, Y.; Su, E.; He, N. Ultrasensitive chemiluminescence immunoassay with enhanced precision for the detection of cTnI amplified by acridinium ester-loaded microspheres and internally calibrated by magnetic fluorescent nanoparticles. *Nanoscale* **2021**, *13*, 3275–3284. [CrossRef]
106. He, L.; Yang, H.; Xiao, P.; Singh, R.; He, N.; Liu, B.; Li, Z. Highly Selective, Sensitive and Rapid Detection of Escherichia coli O157:H7 Using Duplex PCR and Magnetic Nanoparticle-Based Chemiluminescence Assay. *J. Biomed. Nanotechnol.* **2017**, *13*, 1243–1252. [CrossRef]
107. Guo, L.; Chen, H.; He, N.; Deng, Y. Effects of surface modifications on the physicochemical properties of iron oxide nanoparticles and their performance as anticancer drug carriers. *Chin. Chem. Lett.* **2018**, *29*, 1829–1833. [CrossRef]

108. Zhao, C.; Song, X.; Jin, W.; Wu, F.; Zhang, Q.; Zhang, M.; Zhou, N.; Shen, J. Image-guided cancer therapy using aptamer-functionalized cross-linked magnetic-responsive Fe_3O_4@carbon nanoparticles. *Anal. Chim. Acta* **2019**, *1056*, 108–116. [CrossRef] [PubMed]
109. Zhang, Y.; Wang, Q.; Chen, G.; Shi, P. DNA-Functionalized Metal–Organic Framework: Cell Imaging, Targeting Drug Delivery and Photodynamic Therapy. *Inorg. Chem.* **2019**, *58*, 6593–6596. [CrossRef]
110. Cao, Y.; Dong, H.; Yang, Z.; Zhong, X.; Chen, Y.; Dai, W.; Zhang, X. Aptamer-Conjugated Graphene Quantum Dots/Porphyrin Derivative Theranostic Agent for Intracellular Cancer-Related MicroRNA Detection and Fluorescence-Guided Photothermal/Photodynamic Synergetic Therapy. *ACS Appl. Mater. Interfaces* **2016**, *9*, 159–166. [CrossRef]
111. Huang, L.; Su, E.; Liu, Y.; He, N.; Deng, Y.; Jin, L.; Chen, Z.; Li, S. A microfluidic device for accurate detection of hs-cTnI. *Chin. Chem. Lett.* **2021**, *32*, 1555–1558. [CrossRef]
112. Xu, X.; He, N. Application of adaptive pressure-driven microfluidic chip in thyroid function measurement. *Chin. Chem. Lett.* **2021**, *32*, 1747–1750. [CrossRef]
113. Hussain, M.; Chen, Z.; Lv, M.; Xu, J.; Dong, X.; Zhao, J.; Li, S.; Deng, Y.; He, N.; Li, Z.; et al. Rapid and label-free classification of pathogens based on light scattering, reduced power spectral features and support vector machine. *Chin. Chem. Lett.* **2020**, *31*, 3163–3167. [CrossRef]
114. Lan, S.; Lin, Z.; Zhang, D.; Zeng, Y.; Liu, X. Photocatalysis Enhancement for Programmable Killing of Hepatocellular Carcinoma through Self-Compensation Mechanisms Based on Black Phosphorus Quantum-Dot-Hybridized Nanocatalysts. *ACS Appl. Mater. Interfaces* **2019**, *11*, 9804–9813. [CrossRef]
115. Zhang, H.-J.; Zhao, X.; Chen, L.-J.; Yang, C.-X.; Yan, X.-P. Dendrimer grafted persistent luminescent nanoplatform for aptamer guided tumor imaging and acid-responsive drug delivery. *Talanta* **2020**, *219*, 121209. [CrossRef] [PubMed]
116. Guo, Z.; Liu, Y.; He, N.; Deng, Y.; Jin, L. Discussion of the protein characterization techniques used in the identification of membrane protein targets corresponding to tumor cell aptamers. *Chin. Chem. Lett.* **2020**, *32*, 40–47. [CrossRef]
117. Yang, C.; Xu, J.; Yang, D.; Wang, X.; Liu, B.; He, N.; Wang, Z. ICG@ZIF-8: One-step encapsulation of indocyanine green in ZIF-8 and use as a therapeutic nanoplatform. *Chin. Chem. Lett.* **2018**, *29*, 1421–1424. [CrossRef]
118. Cheng, M.; Cui, Y.-X.; Wang, J.; Zhang, J.; Zhu, L.-N.; Kong, D.-M. G-Quadruplex/Porphyrin Composite Photosensitizer: A Facile Way to Promote Absorption Redshift and Photodynamic Therapy Efficacy. *ACS Appl. Mater. Interfaces* **2019**, *11*, 13158–13167. [CrossRef]
119. Chu, J.-Q.; Wang, D.-X.; Zhang, L.-M.; Cheng, M.; Gao, R.-Z.; Gu, C.-G.; Lang, P.-F.; Liu, P.-Q.; Zhu, L.-N.; Kong, D.-M. Green Layer-by-Layer Assembly of Porphyrin/G-Quadruplex-Based Near-Infrared Nanocomposite Photosensitizer with High Biocompatibility and Bioavailability. *ACS Appl. Mater. Interfaces* **2020**, *12*, 7575–7585. [CrossRef]
120. Wang, C.; Meng, F.; Huang, Y.; He, N.; Chen, Z. Design and Implementation of Polymerase Chain Reaction Device for Aptamers Selection of Tumor Cells. *J. Nanosci. Nanotechnol.* **2020**, *20*, 1332–1340. [CrossRef]
121. Li, W.; Wang, S.; Zhou, L.; Cheng, Y.; Fang, J. An ssDNA aptamer selected by Cell-SELEX for the targeted imaging of poorly differentiated gastric cancer tissue. *Talanta* **2019**, *199*, 634–642. [CrossRef]
122. Zhu, G.; Chen, X. Aptamer-based targeted therapy. *Adv. Drug Deliv. Rev.* **2018**, *134*, 65–78. [CrossRef] [PubMed]
123. Mirón-Mérida, V.; González-Espinosa, Y.; Collado-González, M.; Gong, Y.; Guo, Y.; Goycoolea, F. Aptamer–Target–Gold Nanoparticle Conjugates for the Quantification of Fumonisin B1. *Biosensors* **2021**, *11*, 18. [CrossRef] [PubMed]
124. Singh, M.; Harris-Birtill, D.C.; Markar, S.R.; Hanna, G.B.; Elson, D.S. Application of gold nanoparticles for gastrointestinal cancer theranostics: A systematic review. *Nanomed. Nanotechnol. Biol. Med.* **2015**, *11*, 2083–2098. [CrossRef] [PubMed]
125. Liu, Y.; Lai, Y.; Yang, G.; Tang, C.; Deng, Y.; Li, S.; Wang, Z. Cd-Aptamer Electrochemical Biosensor Based on AuNPs/CS Modified Glass Carbon Electrode. *J. Biomed. Nanotechnol.* **2017**, *13*, 1253–1259. [CrossRef]
126. Xia, X.; Li, M.; Wang, M.; Gu, M.-Q.; Chi, K.-N.; Yang, Y.-H.; Hu, R. Development of Ochratoxin A Aptasensor Based on Au Nanoparticles@g-C3N4. *J. Biomed. Nanotechnol.* **2020**, *16*, 1296–1303. [CrossRef]
127. Zhang, Y.; Tan, J.; Zhou, L.; Shan, X.; Liu, J.; Ma, Y. Synthesis and Application of AS1411-Functionalized Gold Nanoparticles for Targeted Therapy of Gastric Cancer. *ACS Omega* **2020**, *5*, 31227–31233. [CrossRef] [PubMed]
128. Jiang, L.; Wang, H.; Chen, S. Aptamer (AS1411)-Conjugated Liposome for Enhanced Therapeutic Efficacy of miRNA-29b in Ovarian Cancer. *J. Nanosci. Nanotechnol.* **2020**, *20*, 2025–2031. [CrossRef]
129. Liu, D.-X.; Tien, T.T.T.; Bao, D.T.; Linh, N.T.P.; Park, H.; Yeo, S.-J. A Novel Peptide Aptamer to Detect Plasmodium falciparum Lactate Dehydrogenase. *J. Biomed. Nanotechnol.* **2019**, *15*, 204–211. [CrossRef]
130. Zhong, L.; Zou, H.; Huang, Y.; Gong, W.; He, J.; Tan, J.; Lai, Z.; Li, Y.; Zhou, C.; Zhang, G.; et al. Magnetic Endoglin Aptamer Nanoprobe for Targeted Diagnosis of Solid Tumor. *J. Biomed. Nanotechnol.* **2019**, *15*, 352–362. [CrossRef] [PubMed]
131. Zhang, Y.; Li, C.; Jia, R.; Gao, R.; Zhao, Y.; Ji, Q.; Cai, J.; Li, Q.; Wang, Y. PEG-poly(amino acid)s/EpCAM aptamer multifunctional nanoparticles arrest the growth and metastasis of colorectal cancer. *Biomater. Sci.* **2021**, *9*, 3705–3717. [CrossRef]
132. Go, G.; Lee, C.-S.; Yoon, Y.; Lim, J.; Kim, T.; Lee, S. PrP^C Aptamer Conjugated–Gold Nanoparticles for Targeted Delivery of Doxorubicin to Colorectal Cancer Cells. *Int. J. Mol. Sci.* **2021**, *22*, 1976. [CrossRef] [PubMed]
133. Liang, J.; Luo, G.; Ning, X.; Shi, Y.; Zhai, H.; Sun, S.; Jin, H.; Liu, Z.; Zhang, F.; Lu, Y.; et al. Differential expression of calcium-related genes in gastric cancer cells transfected with cellular prion protein. *Biochem. Cell Biol.* **2007**, *85*, 375–383. [CrossRef]

134. Li, Q.-Q.; Cao, X.-X.; Xu, J.-D.; Chen, Q.; Wang, W.-J.; Tang, F.; Chen, Z.-Q.; Liu, X.-P.; Xu, Z.-D. The role of P-glycoprotein/cellular prion protein interaction in multidrug-resistant breast cancer cells treated with paclitaxel. *Cell. Mol. Life Sci.* **2008**, *66*, 504–515. [CrossRef]
135. Go, G.; Lee, C.S. The Cellular Prion Protein: A Promising Therapeutic Target for Cancer. *Int. J. Mol. Sci.* **2020**, *21*, 9208. [CrossRef]
136. Tang, L.; Tong, R.; Coyle, V.J.; Yin, Q.; Pondenis, H.; Borst, L.B.; Cheng, J.; Fan, T.M. Targeting Tumor Vasculature with Aptamer-Functionalized Doxorubicin–Polylactide Nanoconjugates for Enhanced Cancer Therapy. *ACS Nano* **2015**, *9*, 5072–5091. [CrossRef]
137. Li, B.; Wang, Y.; He, J. Gold Nanorods-Based Smart Nanoplatforms for Synergic Thermotherapy and Chemotherapy of Tumor Metastasis. *ACS Appl. Mater. Interfaces* **2019**, *11*, 7800–7811. [CrossRef]
138. Zhang, Z.; Wang, L.; Wang, J.; Jiang, X.; Li, X.; Hu, Z.; Ji, Y.; Wu, X.; Chen, C. Mesoporous Silica-Coated Gold Nanorods as a Light-Mediated Multifunctional Theranostic Platform for Cancer Treatment. *Adv. Mater.* **2012**, *24*, 1418–1423. [CrossRef] [PubMed]
139. Choi, J.; Yang, J.; Bang, D.; Park, J.; Suh, J.-S.; Huh, Y.-M.; Haam, S. Targetable Gold Nanorods for Epithelial Cancer Therapy Guided by Near-IR Absorption Imaging. *Small* **2012**, *8*, 746–753. [CrossRef] [PubMed]
140. Chandrasekaran, R.; Lee, A.S.W.; Yap, L.W.; Jans, D.; Wagstaff, K.M.; Cheng, W. Tumor cell-specific photothermal killing by SELEX-derived DNA aptamer-targeted gold nanorods. *Nanoscale* **2016**, *8*, 187–196. [CrossRef] [PubMed]
141. Zheng, L.; Zhang, B.; Chu, H.; Cheng, P.; Li, H.; Huang, K.; He, X.; Xu, W. Assembly and in vitro assessment of a powerful combination: Aptamer-modified exosomes combined with gold nanorods for effective photothermal therapy. *Nanotechnology* **2020**, *31*, 485101. [CrossRef]
142. Huang, R.; He, L.; Xia, Y.; Xu, H.; Liu, C.; Xie, H.; Wang, S.; Peng, L.; Liu, Y.; Liu, Y.; et al. A Sensitive Aptasensor Based on a Hemin/G-Quadruplex-Assisted Signal Amplification Strategy for Electrochemical Detection of Gastric Cancer Exosomes. *Small* **2019**, *15*, e1900735. [CrossRef]
143. Yu, X.; He, L.; Pentok, M.; Yang, H.; Yang, Y.; Li, Z.; He, N.; Deng, Y.; Li, S.; Liu, T.; et al. An aptamer-based new method for competitive fluorescence detection of exosomes. *Nanoscale* **2019**, *11*, 15589–15595. [CrossRef]
144. Batrakova, E.V.; Kim, M. Using exosomes, naturally-equipped nanocarriers, for drug delivery. *J. Control. Release* **2015**, *219*, 396–405. [CrossRef]
145. Milane, L.; Singh, A.; Mattheolabakis, G.; Suresh, M.; Amiji, M.M. Exosome mediated communication within the tumor microenvironment. *J. Control. Release* **2015**, *219*, 278–294. [CrossRef] [PubMed]
146. Duan, T.; Xu, Z.; Sun, F.; Wang, Y.; Zhang, J.; Luo, C.; Wang, M. HPA aptamer functionalized paclitaxel-loaded PLGA nanoparticles for enhanced anticancer therapy through targeted effects and microenvironment modulation. *Biomed. Pharmacother.* **2019**, *117*, 109121. [CrossRef]
147. Saravanakumar, K.; Hu, X.; Shanmugam, S.; Chelliah, R.; Sekar, P.; Oh, D.-H.; Vijayakumar, S.; Kathiresan, K.; Wang, M.-H. Enhanced cancer therapy with pH-dependent and aptamer functionalized doxorubicin loaded polymeric (poly D, L-lactic-co-glycolic acid) nanoparticles. *Arch. Biochem. Biophys.* **2019**, *671*, 143–151. [CrossRef] [PubMed]
148. Guo, Z.; Wang, C.; Li, S.; Chen, Z.; Deng, Y.; He, N. Study on the Method of Isolating the Aptamer from the Surface of HepG2 Cells. *J. Nanosci. Nanotechnol.* **2020**, *20*, 3373–3377. [CrossRef] [PubMed]
149. Huang, R.; Chen, Z.; Liu, M.; Deng, Y.; Li, S.; He, N. The aptamers generated from HepG2 cells. *Sci. China Ser. B Chem.* **2017**, *60*, 786–792. [CrossRef]
150. Zhang, Z.; Cheng, W.; Pan, Y.; Jia, L. An anticancer agent-loaded PLGA nanomedicine with glutathione-response and targeted delivery for the treatment of lung cancer. *J. Mater. Chem. B* **2019**, *8*, 655–665. [CrossRef] [PubMed]

MDPI
St. Alban-Anlage 66
4052 Basel
Switzerland
Tel. +41 61 683 77 34
Fax +41 61 302 89 18
www.mdpi.com

Biosensors Editorial Office
E-mail: biosensors@mdpi.com
www.mdpi.com/journal/biosensors

www.ingramcontent.com/pod-product-compliance
Lightning Source LLC
LaVergne TN
LVHW070734100526
838202LV00013B/1232